Synthetic Peptide Vaccine Models

Synthetic Peptide Vaccine Models

Design, Synthesis, Purification, and Characterization

Edited by
Assoc. Prof. Mesut Karahan
Üsküdar University

CRC Press
Taylor & Francis Group
Boca Raton London New York

CRC Press is an imprint of the
Taylor & Francis Group, an **informa** business

First edition published 2021
by CRC Press
6000 Broken Sound Parkway NW, Suite 300, Boca Raton, FL 33487-2742

and by CRC Press
2 Park Square, Milton Park, Abingdon, Oxon, OX14 4RN

© 2021 Taylor & Francis Group, LLC

CRC Press is an imprint of Taylor & Francis Group, LLC

Library of Congress Cataloging-in-Publication Data

[Insert LoC Data here when available]

ISBN: 978-0-367-70088-1
ISBN: 978-1-003-14453-3 (ebk)

Typeset in Times
by Deanta Global Publishing Services, Chennai, India

Contents

Acknowledgements

Firstly, I would like to thank all the authors who worked with me on this book; without their efforts this project would never have come to fruition. Secondly, I would like to thank the team at the publishers, Taylor & Francis; without the initial friendly and encouraging contact from Chuck R. Crumly the project would never have been started, and without the continued chivvying along and moral support from my students Öznur Özge Özcan and Rümeysa Rabia Kocatürk the project would never have been finished. I would also like to thank Prof. Dr. Nevzat Tarhan who is the founder and rector of Üsküdar University.

Finally, I would like to thank my wife Harika Aktürk Karahan, my son Arın Karahan, and my mother Mercan Karahan and father Şemsettin Karahan for their patience, especially in the final few weeks, when I was not as available as I should have been.

In a work of this type there is never enough space for every piece of information that could be included or for every piece of work that could be referenced. Therefore, I and my fellow authors would like to apologize to all the scientists who have increased our knowledge of the new generation peptide vaccine but whose work could not be included in this book.

Assoc. Prof. Mesut Karahan

Preface

In the course of this book on the field of peptide vaccines and their roles, we have focused on nanotechnological approaches, cancer and brain disorders, their omic technology, computer aimed studies, and epigenetic roles. We start with the introduction of how vaccines are classified towards technological developments, after the foundations we have established with the history of vaccines. Subsequently, before entering the peptide vaccines in our field of study, we emphasize how amino acids required for peptide synthesis are produced, and their importance. From the occurrence of symptoms based on many clinical deficiencies of these amino acids, you can vaccinate these agents. Peptide vaccines have become a vaccine technology that stimulates cellular and humoral immunity without the risk of contamination like weakened viruses, and the friendly importance of acquired immunity. However, peptide vaccines have a disadvantage, which is that they sometimes do not show sufficient immunogenicity. In addition, peptides are very small molecules and can be destroyed by various enzymatic effects (such as proteases) without an immune response when released into the circulation. This also applies to drugs, not only vaccines. For this reason, we have also entered the methodology that explains how nanoparticles are synthesized polymerically and are used with peptides with the fundamentals of nanotechnology. On the other hand, peptides not used only as vaccines and drugs, but they also exhibit antioxidative properties. Studies conducted in the field of health regarding the usage areas of antioxidant peptides have been included and their advantages have been mentioned. In addition, antioxidant peptides can also be used in vaccine formulations. This has shown us that the versatile use of peptides, even the epitope sequences detected by computer-based programs, can be elucidated not only for immunological studies but also for oxidative stress and drug ability. Natural antioxidant epitope sequences have been discovered in nature through various studies without bioinformatics. Peptide vaccines are studied not only against viral or bacterial diseases, but also against brain and cancer diseases. Peptide studies used as vaccines and drugs in Alzheimer's, Parkinson's, and other neuropsychiatric diseases are also included in this book. Peptides also need to be immunized in a complex with adjuvants to increase their efficacy like other types of vaccines. However, adjuvants also have immunotoxic effects, and most of them cannot be said to provide 100% protection against enzymatic degradation. For all these reasons, new technologies are needed, and as we mentioned above, nanotechnology is actually a Nobel area of discovery. That is why we continue to work with peptide vaccines and nanotechnology adapted together. We never claim that peptides or many vaccine systems formulated with nanotechnological approaches can offer definitive solutions, even if we offer advantages over adjuvants. In addition, in this book, we discuss various types of nanoparticles created by nanotechnological approaches. We can call this combination between peptide epitopes and nanoparticles new generation vaccine technologies. I would like to emphasize that nanoparticles will offer great innovations in vaccine and many other fields as protective, vaccine and drug delivery systems with adjuvant properties. I believe nanotechnology will find its deepest place in our lives hundreds of years after this book is published.

However, another important issue that we want to draw attention to is that vaccines are seen by other scientific fields as drugs and take their place in multidisciplinary areas. In particular, genomic, transcriptomic, and proteomic studies should be done to understand how vaccines play a role in cellular translational and transcriptional processes. As a result of this interdisciplinary reunion "vaccinomics" research has begun and the cellular processes of vaccines have begun to be understood not only from an immunological aspect but also from other aspects. We claim that the reader of this book in the future be able to produce multidisciplinary studies that can think of not only the immunological aspects of the peptides they are working with, but also with omics, epigenetics, bioinformatics, and nanotechnological approaches. In addition, we have added the part of how to design in silico epitope and vaccine production on the basis of SARS-CoV-2 at the end of this book, as the SARS-CoV-2 pandemic has spread all over the world. If, at least, successful nanotechnological peptide vaccines are built based on the multidisciplinary approaches and principles discussed in this book, then all of the many hours that we spent in discussion, writing, correcting, and fighting with the all contributors because we were late, will have been justified.

Assoc. Prof. Mesut Karahan
İstanbul, August 2020

Editor Bio

Assoc. Prof. Dr Mesut Karahan was born in Istanbul in 1977 and is a graduate of Pertevniyal High School. He graduated from Yildiz Technical University, Faculty of Arts and Sciences, Department of Chemistry in 1999. Between 2001 and 2012, Dr Karahan worked in Yildiz Technical University. He got his Master's degree in 2002 at the same university with his study entitled "Ternary Polymer-Metal-Protein Complex Structure Models." Assoc. Prof. Dr Karahan completed his Ph.D. studies in 2009 on "Development of Functional Biopolymer Systems Containing Metal." He has been working at Üsküdar University since 2012. His working areas are chemistry, biochemistry, biodegradable biopolymers, peptides, proteins, biopolymer-peptide conjugation, artificial vaccine systems, immunology, cancer, drug production, drug delivery systems, nanomedicine, non-coding RNAs vaccine systems, diagnostic kits, biopolymer systems, peptide synthesis, bioconjugation, biopolymer-metal-peptide complexes. The research goals of the Karahan lab are to understand the relationship between the important peptide sequences and their combination with nanotechnology, and to understand the molecular mechanisms of brain and cancer diseases for creating new-generation drug and vaccine models.

Contributors

Fadime Canbolat
School of Health Services
Üsküdar University
Istanbul, Turkey

Mithat Çelebi
Faculty of Engineering
Department of Polymer Material
   Engineering
Yalova University
Yalova, Turkey

Greg Czyryca
Allosterix Pharmaceuticals LLC
Lehi, Utah

Dolunay Şakar Daşdan
Faculty of Science and Letters
Department of Chemistry
Yildiz Technical University
Istanbul, Turkey

Emine Ekici
Faculty of Health Sciences
Department of Nursing
Üsküdar University
İstanbul, Turkey

Joel Lim Whye Ern
Faculty of Pharmaceutical Sciences
Pharmaceutical Technology
UCSI University
Kuala Lumpur, Malaysia

Mustafa Kemal Gümüş
Science-Technology
Application and Research Center
Artvin Coruh University
Artvin, Turkey

Ramaz Katsarava
Institute of Chemistry and Molecular
   Engineering
Agricultural University of Georgia
Tbilisi, Georgia

Gülderen Karakuş
Faculty of Pharmacy
Department of Basic Pharmaceutical
   Sciences
Sivas Cumhuriyet University
Sivas, Turkey

Ömer Faruk Karasakal
Vocational School of Health
   Services
Institute of Science and Technology in
   Biology
Üsküdar University
İstanbul, Turkey

Rümeysa Rabia Kocatürk
Faculty of Health Sciences
Nutrition and Dietetics
Üsküdar University
Istanbul, Turkey

Muhsin Konuk
Faculty of Engineering and Natural
   Sciences
Molecular Biology and Genetics
Üsküdar University
İstanbul, Turkey

Palaniarajan Vijayaraj Kumar
Faculty of Pharmaceutical Sciences
Pharmaceutical Technology
UCSI University
Kuala Lumpur, Malaysia

Tan Shen Leng
Faculty of Pharmaceutical Sciences
Pharmaceutical Technology
UCSI University
Kuala Lumpur, Malaysia

Tee Yi Na
Faculty of Pharmaceutical Sciences
Pharmaceutical Technology
UCSI University
Kuala Lumpur, Malaysia

Sezer Okay
Department of Vaccine Technology
Vaccine Institute
Hacettepe University
Ankara, Turkey

Özgür Özay
Faculty of Science and Arts
Department of Chemistry
Çanakkale Onsekiz Mart University
Çanakkale, Turkey

Öznur Özge Özcan
Health Sciences Institute
Molecular Neuroscience
Üsküdar University
Istanbul Turkey

S. Arda Ozturkcan
School of Health Sciences
Department of Nutrition and Dietetics
Istanbul Gelisim University
Istanbul, Turkey

Nezih Pişkinpaşa
Faculty of Health Sciences
Nutrition and Dietetics
Üsküdar University
İstanbul, Turkey

Gökben Hızlı Sayar
Faculty of Medicine
Department of Psychiatry
Üsküdar University
Istanbul, Turkey

Necip Ozan Tiryakioğlu
Department of Molecular
    Biotechnology
Faculty of Science
Turkish-German University
Istanbul, Turkey

Korkut Ulucan
Faculty of Dentistry
Department of Medical Biology and
    Genetics
Marmara University
Istanbul, Turkey

Hüseyin Ünübol
Faculty of Medicine
Department of Psychiatry
Üsküdar University
Istanbul, Turkey

Fatmanur Zehra Zelka
Faculty of Health Sciences
Nutrition and Dietetics
Üsküdar University
Istanbul, Turkey

List of Figures

List of Tables

List of Tables

1 Background

Emine Ekici

CONTENTS

1.1 REVIEW OF VACCINE HISTORY

The first attempt at vaccination was carried out in the 15th century. The Turks and Chinese tried to induce immunity to smallpox. They used dried crusts from smallpox lesions and applied them to small cuts they made in the skin (Clem 2011). The first immunization attempts led to further experiments. The scientists Lady Mary Wortley Montagu in 1718 and Edward Jenner in 1798 conducted experiments to stimulate immunization. Edward Jenner (1798) did experiments to stimulate immunity for smallpox and he discovered the vaccine to the devastating disease. Vaccine and vaccinology terms came into use after the discovery of a smallpox vaccine by Edward Jenner. He called the smallpox vaccine variola vaccinae. The discovery of this vaccine led to Edward Jenner being named the Father of Vaccinology; Louis Pasteur is also referred to by the same epithet (Bailey 1996; Plotkin 2009; Riedel 2005). Vaccination studies started in the 18th century (Table 1.1) and gained momentum with Louis Pasteur. Pasteur reported in 1880 that use of an attenuated form of etiological agent from chicken cholera to create a protective inoculation against chicken cholera disease (now known as *Pasteurella multocida*) can be considerable. Later, in 1885, Pasteur discovered a vaccine against rabies. And this was the second vaccine for human use (Esparza 2012). Afterwards, vaccinology started to thrive, and many vaccines were found and still improve to this day. In this chapter we will detail the vaccine discoveries. After Pasteur's rabies vaccine development, a cholera vaccine was found by Laffinike in 1892. In 1896, the typhoid vaccine was found by Wright. The vaccine for tuberculosis was found by Calmette and Guerin in 1921 and shortly after that Ramon and Glenny found the diphtheria vaccine in 1923. In the following years, Madsen's pertussis vaccine was produced as a result of six years of work. Ramon and Zoeller produced a tetanus vaccine. In the following years, yellow fever, influenza, mumps, and oral polio vaccines were developed. In 1960, scientists Edmonston and Schwartz produced the vaccine for measles, a particularly dangerous viral disease for children. In the following years, a vaccine against rubella and mumps was found (Sünbül 2008; Qazi 2005) and Table 1.1 provides a general idea of the chronology (Plotkin 2014).

TABLE 1.1
History of Human Vaccines

Live Attenuated	Killed Whole Organisms	Purified Proteins or Polysaccharides	Genetically Engineered
18th century			
Smallpox (1798)			
19th century			
Rabies (1885)	Typhoid (1896)		
	Cholera (1896)		
	Plague (1897)		
Early 20th century, first half			
Tuberculosis (bacille Calmette–Guérin) (1927)	Pertussis (1926)	Diphtheria toxoid (1923)	
Yellow fever (1935)	Influenza (1936)	Tetanus toxoid (1926)	
	Rickettsia (1938)		
20th century, second half			
Polio (oral) (1963)	Polio (injected) (1955)	Anthrax secreted proteins (1970)	Hepatitis B surface antigen recombinant (1986)
Measles (1963)	Rabies (cell culture) (1980)	Meningococcus polysaccharide (1974)	Lyme OspA (1998)
Mumps (1967)	Japanese encephalitis (mouse brain) (1992)	Pneumococcus polysaccharide (1977)	Cholera (recombinant toxin B) (1993)
Rubella (1969)	Tick-borne encephalitis (1981)	*Haemophilus influenzae* type B polysaccharide (1985)	
Adenovirus (1980)	Hepatitis A (1996)	*H. influenzae* type B conjugate (1987)	
Typhoid (*Salmonella*TY21a) (1989)	Cholera (WC-rBS) (1991)	Typhoid (Vi) polysaccharide (1994)	
Varicella (1995)	Meningococcal conjugate (group C) (1999)	Acellular pertussis (1996)	
Rotavirus reassortants (1999)		Hepatitis B (plasma derived) (1981)	
Cholera (attenuated) (1994)			
Cold-adapted influenza (1999)			

(*Continued*)

TABLE 1.1 (CONTINUED)
History of Human Vaccines

Live Attenuated	Killed Whole Organisms	Purified Proteins or Polysaccharides	Genetically Engineered
21st Century			
Rotavirus (attenuated and new reassortants) (2006)	Japanese encephalitis (2009) (Vero cell)	Pneumococcal conjugates*(heptavalent) (2000)	Human papillomavirus recombinant (quadrivalent) (2006)
Zoster (2006)	Cholera (WC only) (2009)	Meningococcal conjugates*(quadrivalent) (2005)	Human papillomavirus recombinant (bivalent) (2009)
		Pneumococcal conjugates* (13-valent) (2010)	Meningococcal group B proteins (2013)

Source: Plotkin 2014.
*Capsular polysaccharide conjugated to carrier proteins.

1.2 IMMUNOLOGICAL BASES OF VACCINES

According to the World Health Organization (WHO), a vaccine is defined as a biological drug that improves immunity against a particular disease. A vaccine is typically similar to a disease-causing microorganism, and generally an agent made of weakened or killed forms of one of the surface proteins. The agent encourages the body's immune system to recognize, destroy, and "remember" the agent as foreign, so the immune system can easily detect and destroy it when it encounters the microorganisms (Ginglen and Doyle 2020).

Vaccines stimulate the body's own immune system to protect the person against subsequent infection or illness. Again, according to the WHO, vaccination is a proven tool to control and eradicate life-threatening infectious diseases. It is also estimated that vaccines prevent two to three million deaths and saves human lives each year. Vaccines have a prominent place in human life as one of the most cost-effective health investments with proven strategies that make them accessible to even the most difficult to reach and vulnerable populations (Ethgen et al. 2016; Orenstein and Ahmed 2017). Vaccination can be defined as the process by which a person is made immune or resistant to an infectious disease, typically by administering a vaccine (Zhang, Wang, and Wang 2015).

The easiest way to protect against antigens is vaccination. However, it was not possible to develop a vaccine against all diseases. The most important advantages are the efficiency of an ideal vaccine, ease of absorption, reliability, stability, and cheapness (Nandedkar 2009).

In other words, an ideal vaccine should be reliable and should be as effective as long as possible due to the length of stay in the body and should not cause any reactions and complications during the time it stays in the body. The production costs of vaccines should be low and the vaccines should be easy to access. Storage conditions should be created comfortably and enable storage for a long time.

The working mechanism of infectious disease vaccines is through prophylactic-controlled exposure to an infectious substance. The first exposure of a healthy person who has been vaccinated can cause a strong immune reaction and exert a strong effect in a vaccinated person. In this respect, this system has a very important place due to the effect of the vaccine on the immune system. If it is necessary to examine the immune system, it consists of two different branches: the innate and adaptive immune system (Marshall et al. 2018). After this initial exposure to an infectious substance or administration of a vaccine, activation of the innate immune system precedes the formation of adaptive immunity.

The innate immune system is known to consist of various cell types such as dendritic cells, neutrophils, monocytes, eosinophils, and macrophages, all of which function to interact with foreign molecules in a non-specific manner. In addition, innate immune cells cause infectious agents to be phagocytized, causing inflammatory cytokines to be secreted, as well as attracting and activating other immune cells through the secretion of chemical messengers such as cytokines and chemokines. These processes lead to the initiation of an active and effective immune response to fight infections (Karch and Burkhard 2016).

In general, the adaptive immune system is divided into two different categories, cellular and humoral (Skwarczynski and Toth 2016). Also, cells of this adaptive immune system respond to specific regions of infectious agents known as epitopes (peptides and peptide vaccines are most commonly used in the adaptive immune system). One or more epitopes are located in various regions of infectious agents, a larger molecule known as an antigen, and these antigens are highly related to stimulating the immune system. Humoral immune responses are very important and basically these responses are dependent on the action of glycoproteins, antibodies secreted from B cells by binding to specific epitopes by B cell-specific receptors on the cell surface, and upon the binding of the B cell receptor to a matching epitope; B cells can mature into plasma. B cells formed in this way provide an important protection by starting to secrete epitope-specific antibodies that will protect an individual from infectious diseases (Siegrist 2013). Cellular immune responses are based on the action of T immune cells. All these cells contain histocompatibility complex class I (MHC-I) molecules in their nuclear activities. When an intracellular infectious agent is formed near the cell membrane, cells may be present on surface linear epitopes of these infectious agents complexed with MHC-I because this molecule is responsible for stimulating the immune system against infectious (Moser and Leo 2010) (Figure 1.1).

1.3   THE IMPORTANCE OF VACCINATION

It is a remarkable feature of vaccines that prophylactic vaccines have a wide range of potential effects, including their use in health care, general health and well-being, cognitive development, and consequently economic productivity (Doherty et al.

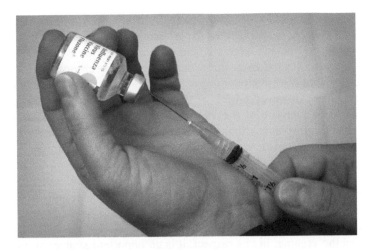

FIGURE 1.1 An influenza vaccine. (Photo by CDC on Unsplash.)

2016). In addition, the need for vaccination has been extremely important recently. The benefits of vaccination extend beyond the prevention of specific human diseases and provide care for weak individuals in society. It has become a moral imperative to ensure that beneficial vaccines are available to all segments of society because every person has the right to a healthy life (Andre et al. 2008).

On the other hand, it can be said that people avoid vaccination and sometimes tend to stay away, as a cautionary issue. This approach, especially seen in parents, depends on many factors. This refers to the delay in accepting or rejecting vaccines despite the availability of vaccination services and is an individual behavior influenced by a number of factors such as knowledge or past experience. Moreover, this refusal is also linked to historical, political, and socio-cultural contexts, but especially for newer vaccines, it often creates more hesitation (Shen and Dubey 2019). A questionnaire was used to investigate the attitude of a sample of parents towards vaccines by distributing questionnaires at six secondary schools in Messina, Italy. At the end of this survey, the results obtained can be summarized as follows. With regard to vaccines for children, the vaccines with good coverage percentages are measles-mumps-rubella and diphtheria-tetanus-whooping cough; on the other hand, very low coverage rates, especially for "new" vaccines, vary widely between HPV, meningococcal, and pneumococcal vaccines. It was observed that the vaccines were negatively correlated with both the age of the parents and their education level. A positive parental opinion was strongly influenced by the positive opinion of the doctor, while a negative parental opinion was based on direct or indirect knowledge of the persons injured by the vaccines. In addition, the data obtained showed that the parents were generally unaware or partially aware of the actual composition of the vaccines and the diseases that the vaccines prevent. Therefore, it can be said that health education and the delivery of the right information have been very important issues in improving the situation and combating common and unfounded fears about vaccines (Facciolà et al. 2019). In addition, ignorance about vaccines is not only found in parents, but there may be hesitations about vaccination including among top faculty students and doctors who personally deal with this issue. In a study on this subject, a sample consisting of an experimental group of 92 people was chosen.

Of this group, 53 people were medical students, together with 39 doctors, and information was collected from these two groups. Two groups of research have found the National Vaccination Program reliable and acknowledged the importance of vaccines. However, 64.2% of the students and 38.5% of the physicians were found to be unaware of the vaccine-preventable infectious diseases in the basic vaccination program. Although most of the interviewees did not have a personal vaccination record, it was thought that they did not have the Kerkes 2015 influenza vaccine. It was also found that in both groups there were people who refused to vaccinate for themselves or for their children. Of these, 54.7% of students (for themselves) and 43.3% of students (for their children) and 59.0% of doctors (for themselves) and 41.0% of doctors (for their children) are in this situation. On the other hand, a total of 48.7% of physicians had already helped people who refused the vaccine. In this respect, it is an important strategy to improve the knowledge of people in charge of this work, to maintain vaccine coverage, and to ethically address vaccine refusal, as it is something that can prevent vaccine rejection (Mizuta et al. 2019). It is known that vaccination has benefited many diseases and prevented the spread in millions of children, and has become the best public health intervention ever (Shen and Dubey 2019; Ginglen and Doyle 2020). In a study to investigate this problem, a systematic review was written to identify effective techniques to reduce vaccine rejection and increase the childhood vaccination rate. The results found can be summarized as follows: the articles reviewed have found that the basis of a strategy for the promotion of vaccines must first be determined, identify the perspective and needs of the target population and overcome these problems by adapting approaches to suit them to alleviate the barriers that hinder vaccine uptake. In addition, the use of technology is one of the elements that helps to maintain the effectiveness of social marketing strategies and further encourages vaccines and their benefits (Nour 2019).

Consequently, finding and producing new vaccines for diseases is an important feature for society. Every job that will benefit human health has a significant and important quality for the benefit of society, including vaccination; it is also important to improve people's knowledge of vaccines to maintain vaccine coverage and ethically address vaccine rejection; it is essential to prevent vaccine rejection and reach more people. This book has important scientific foundations for new vaccine techniques and vaccine production.

REFERENCES

Andre, F.E., R. Booy, H.L. Bock, J. Clemens, S.K. Datta, T.J. John, B.W. Lee, et al. 2008. "Vaccination greatly reduces disease, disability, death and inequity worldwide." *Bulletin of the World Health Organization* 86(2): 81–160.

Bailey, I. 1996. "Edward Jenner (1749–1823): Naturalist, scientist, country doctor, benefactor to mankind." *Journal of Medical Biography* 4: 63–70.

Clem, A.S. 2011. "Fundamentals of vaccine immunology." *Journal of Global Infectious Diseases* 3(1): 73. doi:10.4103/0974-777x.77299.

Doherty, M., P. Buchy, B. Standaert, C. Giaquinto, and D. Prado-Cohrs. 2016. "Vaccine impact: Benefits for human health." *Vaccine* 34(52): 6707–6714. https://doi.org/10.1016/j.vaccine.2016.10.025.

Esparza, J. 2012. "Review of history of vaccine development by Stanley A. Plotkin (Editor)." *Human Vaccines and Immunotherapeutics* 8(3): 289–292. doi:10.4161/hv.18745.

Ethgen, O., M. Cornier, E. Chriv, and F. Baron-Papillon. 2016. "The cost of vaccination throughout life: A western European overview." *Human Vaccines & Immunotherapeutics* 12(8): 2029–2037. doi:10.1080/21645515.2016.1154649.

Facciolà, A., G. Visalli, A. Orlando, M.P. Bertuccio, P. Spataro, R. Squeri, I. Picerno, and A. Di Pietro. 2019. "Vaccine hesitancy: An overview on parents' opinions about vaccination and possible reasons of vaccine refusal." *Journal of Public Health Research* 8(1): 1436. doi:10.4081/jphr.2019.1436.

Ginglen J.G. and M.Q. Doyle. 2020 "Immunization." In *StatPearls [Internet]*. Treasure Island: StatPearls. https://www.ncbi.nlm.nih.gov/books/NBK459331/.

Karch, C.P. and P. Burkhard. 2016. "Vaccine technologies: From whole organisms to rationally designed protein assemblies." *Biochemical Pharmacology* 120: 1–14.

Marshall, J.S., R. Warrington, W. Watson, and H.L. Kim. 2018. "An introduction to immunology and immunopathology." *Allergy, Asthma & Clinical Immunology* 14(S2). doi:10.1186/s13223-018-0278-1.

Mizuta, A.H., G.M. Succi, V. Montalli, and R. Succi. 2019. "Perceptions nn the importance of vaccination and vaccine refusal in a medical school. Percepções Acerca Da Importância Das Vacinas E Da Recusa Vacinal Numa Escola De Medicina." *Revista paulista de pediatria: orgao oficial da Sociedade de Pediatria de Sao Paulo* 37(1): 34–40. doi:10.1590/1984-0462.

Moser M. and O. Leo. 2010. "Key concepts in immunology." *Vaccine* 28(Supplement 3): 2–13.

Nandedkar, T.D. 2009. "Nanovaccines: Recent developments in vaccination." *Journal of Biosciences* 34(6): 995–1003. doi:10.1007/s12038-009-0114-3.

Nour, R. 2019. "A systematic review of methods to improve attitudes towards childhood vaccinations." *Cureus* 11(7): e5067. doi:10.7759/cureus.5067.

Orenstein, W.A., and R. Ahmed. 2017. "Simply put: Vaccination saves lives." *Proceedings of the National Academy of Sciences* 114(16): 4031–4033. doi:10.1073/pnas.1704507114.

Plotkin, S.A. 2009. "Vaccines: The fourth century." *Clinical Vaccine Immololology* 16(12): 1709–1719.

Plotkin, S. 2014. "History of vaccination." *Proceedings of the National Academy of Sciences* 111(34): 12283–12287. doi:10.1073/pnas.1400472111.

Qazi, K.R. 2005. "Heat shock proteins as vaccine adjuvants." *Wenner-Grens Institut för Experimentell Biologi Stockholm*, 71.

Riedel S. 2005. "Edward Jenner and the history of smallpox and vaccination." *Proceedings (Baylor University Medical Center)* 18(1): 21–25.

Shen, S.C. and V. Dubey. 2019. "Addressing vaccine hesitancy: Clinical guidance for primary care physicians working with parents." *Canadian Family Physician Medecin de Famille Canadien* 65(3): 175–181.

Siegrist, C.A. 2013. "Vaccine immunology." In *Vaccines* (Sixth Edition). Eds: Stanley A. Plotkin, Walter A. Orenstein, Paul A. Offit and W.B. Saunders, Elsevier, 14–32.

Skwarczynski, M. and I. Toth. 2016. "Peptide-based synthetic vaccines." *Chemical Science* 7(2): 842–854. doi:10.1039/c5sc03892h.

Sünbül, M. 2014. "Hepatitis B virus genotypes: Global distribution and clinical importance." *World journal of gastroenterology* 20(18): 5427–5434.

Zhang, L., W. Wang, and S. Wang. 2015. "Effect of vaccine administration modality on immunogenicity and efficacy." *Expert Review of Vaccines* 14(11): 1509–1523. doi:10.1586/14760584.2015.1081067.

2 Introduction to Vaccination

Nezih Pişkinpaşa and Ömer Faruk Karasakal

CONTENTS

2.1 INTRODUCTION

Vaccine scientists have taken many approaches when designing vaccines against infectious diseases. These approaches typically rely on basic knowledge about the microorganism, such as how it infects cells and how the immune system reacts to it, and practical considerations such as the regions of the world where the vaccine will be used. Empirical vaccination targets can simply be based on inactivating, attenuating, and/or targeting and disrupting the function of virulence factors such as capsule polysaccharides, toxins, and other surface proteins (Table 2.1).

2.2 CONVENTIONAL (CLASSICAL) VACCINE

Conventional (classical) vaccines are prepared by inactivation of viruses or bacteria replicated in cell cultures and by physical comparison of an adjuvant (aluminum hydroxide and/or fat adjuvant). Conventional vaccines are discussed in two groups: live attenuated vaccine and inactive vaccine (Lee et al. 2012).

TABLE 2.1
Vaccine Types

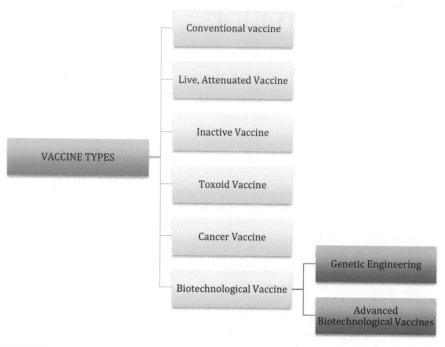

Immunity is the entire system consisting of specialized cells and tissues that protect the organism from all foreign matter and biological factors that it is exposed to from outside. The distinction between healthy and foreign substances is based on a complex system. There are basically two types of immunity in humans. They are innate immunity and acquired immunity. Natural immunity depends on inheritance, the active immunity that occurs after exposure to the antigens. Passive immunity is the transfer of protective antibodies formed by another human or animal to another human by injection. The immune system creates both passive and active immunity against antigens (Delany, Rappuoli, and De Gregorio 2013).

Vaccines create active immunity for microorganisms entering the body. Classical vaccines used today include dead or weakened microorganisms because they are used for treatment and protection. The mechanism of action of vaccines is likened to natural disease. First, the immune system is stimulated by infection. The disease agent entering the body is recognized and the necessary cellular or humoral response is created without allowing the disease to occur. Then, the necessary antibodies are produced to destroy the microbes (Delany, Rappuoli, and De Gregorio 2013).

As a result of all these processes, if the immunized individual encounters the disease-causing active living organism, the immune system will quickly produce antibodies and prevent the disease because it knows how to respond to this condition.

2.3 LIVE ATTENUATED VACCINE

Live attenuated vaccines have been based on live attenuated pathogens and contain laboratory-weakened versions of the original pathogen. The most important advantage of this type of vaccine is that both a strong cellular and an antibody response are produced with long-term protection by a single inoculation, which is often sufficient (Clem 2011). Live vaccines can be prepared with natural and artificial strains, but natural strains are safer and standard (Melnick 1978). The advantages and disadvantages of active vaccines are given in Table 2.2.

Live vaccines stimulate lymphoid and myeloid cells which are immune system cells when injected into the body. It can be said that live, attenuated vaccines for some viruses are relatively easy to create compared to others. For example, measles, chicken pox, and mumps vaccines are administered with this method. Live attenuated vaccines are more difficult for bacterial diseases. Bacteria have thousands of genes and thus are much harder to control. Also, bacteria can develop new forms of resistance to vaccines, so scientists must develop new vaccines against this resistance (Rice et al. 2017; Schroeder et al. 2017). When scientists worked on a live vaccine for a bacterium, they found that recombinant DNA technology can be used in order to extract several key genes. For example, pathogenic *Leptospira* species DNA vaccines are more efficient than live vaccines which have low efficacy, multiple side-effects, poor immunological memory, and lack of cross-protection against different serovars of *Leptospira* (Silveira et al. 2017).

Live attenuated vaccines represent one of the most important medical innovations in human history. The most successful live attenuated vaccines, such as the small-pox vaccine, the measles vaccine, and the yellow fever vaccine, have been isolated. Today, the mechanisms governing strong live attenuated vaccines, immunogenicity, and long-term induced protective immunity remain elusive and therefore hamper the

TABLE 2.2
Advantages and Disadvantages of Live Vaccines

Advantages of Live Vaccines		Disadvantages of Live Vaccines		References
Suitable for use in immune-suppressed organisms.	They can activate every part of the immune system.	Storage conditions are unstable.	They are more sensitive to environmental conditions.	Dai et al. (2019)
Have a long and effective response.	The immune response occurs faster.	It contains all the factors that cause disease except the antigenic feature. This may cause different infections.	If lyophilized vaccines are used in combination with extra solutions, they can cause different effects.	Vartak and Sucheck (2016)

rational design of innovative vaccine strategies. In recent years, scientists have conducted studies to investigate the mechanisms of these vaccines. In particular, human immune system mice (HIS mice) have been effective in studying the life cycle and immune responses against many human tropical pathogens, but are also said to be inadequate (O'Connell and Douam 2020).

In addition, live attenuated vaccines have shown success in controlling virus spread and also in other applications such as cancer immunotherapy, and in terms of how prior exposure to viruses will affect vaccine efficacy, pre-existing antibodies derived from vaccination have been shown to interfere with vaccine immunogenicity of measles, adenovirus. and influenza live attenuated vaccines. On the other hand, it has been noted that pre-existing antibodies, specifically while pre-existing antibodies are present at neutralizing stages, can also boost the efficacy of flaviviral attenuated vaccines which include those of the dengue and yellow fever viruses (Mok and Chan 2020).

Studies on these vaccines are ongoing and there are different vaccine production studies against many diseases. In one study, the cDNA copy of a ZIKV live attenuated vaccine genome was designed as a DNA plasmid to develop a DNA-initiated live attenuated vaccine against the ZIKV virus and delivered to mice. As a result, it has prevented many symptoms of this disease, so it was emphasized that the live attenuated vaccine can be an effective protection against RNA viruses because it is both easy and powerful (Zou et al. 2018).

In another study, it was found that MTBVAC, a live attenuated *Mycobacterium tuberculosis* strain against tuberculosis, can create trained immunity through induction of glycolysis and glutaminolysis and accumulate traces of histone methylation in its promoters. As a result, it provided a strong MTBVAC-induced heterologous protection against a fatal threat by *Streptococcus pneumoniae* in an experimental mouse pneumonia model (Tarancón et al. 2020). In one study, the protection of live attenuated BPZE1 vaccine against *B. pertussis* disease and the effectiveness of intra-nasal administration of this vaccine was studied. To examine the immune responses induced in humans (n = 12) receiving BPZE1, the magnitude of the antibody responses was studied and the resulting findings and the killing mechanisms induced by BPZE1 were found to have a high potential to improve vaccine efficacy and protection against *B. pertussis* transmission (Lin et al. 2020).

On the other hand, a live attenuated vaccine has been produced for African swine fever seen in pigs. In this study, Chinese ASFV HLJ/18 was used as the backbone and a number of gene deleted viruses were produced. HLJ/18-7GD with seven deleted genes provided complete protection to pigs from African swine fever (Chen et al. 2020).

2.4 INACTIVE VACCINE

Inactivated vaccines are produced by killing the disease-causing microbe with chemicals, heat, or radiation. The main reason for inactivation is the inability to attenuate the microorganism (attenuation) and the risk of returning to the mutant form. Most of the vaccines licensed to date have been developed based on "isolate, inactivate, and inject" the causative agent of disease (Pasteur's principles) (Plotkin 2009). Inactivated vaccines are usually used with combinations or used as adjuvants. The adjuvant is used to increase the effect of the drug or the active substance of the vaccine (Altınsoy 2007). There are considerable differences at titer when the adjuvant is added and when the adjuvant is not added. Bacterial cell wall components, liposomes, and synthetic polymers often show an adjuvant effect. Inactive vaccines generally do not tend to cause infection; active vaccines can rarely cause local, specific, and risk-free infections in small areas (Pelit Arayıcı 2015).

Most inactivated vaccines, however, stimulate a weaker immune system response than live vaccines, so therefore several additional doses should be taken, or booster shots have to be taken to maintain individual immunity. Moreover, inactive vaccine combination can be used for producing antibiotics against multidrug-resistant forms of bacteria. Vaccines are generally a prophylactic treatment method but can be produced for alternative antibacterial treatments (antibiotic) as well (Zhang et al. 2017). Examples for inactivated vaccines are polio, yellow fever, rubella, cholera, mumps, bacille Calmette-Guerin (BCG), and measles (Delany, Rappuoli, and Seib 2013).

In a recent vaccine study, an attempt was made to produce an inactive vaccine against botulinum disease, which is characterized by brain damage in humans as a result of toxin produced by *Clostridium botulinum* bacteria. In the study, full-length botulinum neurotoxin/C1/E1 and /F1 holoproteins against substantial monovalent challenges of the parental toxins was used. It has also been demonstrated that this may be a potential feature in the treatment of this disease (Webb et al. 2020).

2.5 TOXOID VACCINE

This vaccine type is produced with the toxoids from microorganisms. The disease-causing agents (toxoids) are removed and used to make vaccines. An example of a toxoid vaccine is diphtheria vaccine and this contains a chemically modified bacterial toxin that retains immunogenic properties. The production method of toxoid vaccine starts with the toxoid being held in formaldehyde at 37°C for 3–4 weeks, or physical methods (heat, pressure, etc.) can be used. With this, the disease factors are removed and only immunogenic properties remain, and finally toxoid vaccines are created. When the immune system encounters the toxoid vaccine the formation of antibodies neutralizes the bacterial toxin that is stimulated so immune responses are activated. However, the immune response is not very high because it does not contain any bacteria itself. (Ivory and Chadee 2014; Lollini et al. 2006). Examples of toxoid vaccines are diphtheria, tetanus, *B. pertussis* (Delany, Rappuoli, and Seib 2013; Ivory and Chadee 2014; Lollini et al. 2006).

In a study, tetanus toxoid was utilized in combination with DNA vaccines designed from tumor-based antigens. In this study, mice preconditioned with tetanus toxoid were immunized with a granulocyte-macrophage colony stimulating factor (GM-CSF) overexpressing the tumor cell-based vaccine (GVAX). In this way, it has been observed that cervical tumor growth is reduced, and the immune response is increased for human papillomavirus. As a result, it has been revealed that preconditioning with tetanus toxoid before vaccination with a tumor cell-based vaccine overexpressing GM-CSF may be a powerful strategy to target human papillomavirus-associated specific E7 cervical malignancy (Alson et al. 2020). Also, in another study, a phase 2 randomized controlled trial of meningococcal tetanus toxoid-conjugate vaccine (MenACYW-TT) was conducted in elderly individuals. The safety and immunogenicity of MenACYW-TT was evaluated by comparing it with a tetravalent meningococcal polysaccharide vaccine (MPSV4) in 301 healthy adults of 56 years and older in the USA. As a result, the MenACYW-TT conjugate vaccine was well tolerated and immunogenic in adults aged 56 and over (Kirstein et al. 2020).

2.6 CANCER VACCINES

The Food and Drug Administration (FDA) endorsement of different preventive cancer growth immunizations, for example, Gardasil (Merck), Cervarix (Glaxosmithkline), and the restorative antibody Sipulencel-T (Provenge), has had a significant effect on disease antibody advancement. These antibodies have supported the opportunity of malignant growth treatment through cell immunotherapy (Banday et al. 2014). There are two sorts of cancer growth antibodies. The first of these is prophylactic immunization. The point of this immunization is to forestall cancer growth improvement in healthy people (Zhao et al. 2014). Another immunization is helpful to malignant growth antibodies. It expects to treat a current disease by fortifying the body's characteristic invulnerable reaction (Thomas and Prendergast 2016).

Cancer growth anticipation immunizations target specialists that cause or add to the advance of disease. They are like customary antibodies. Both cancer growth anticipation immunizations and customary antibodies depend on antigens. Defensive immunizations, including antibodies that target malignancy causing infections, explicitly invigorate the creation of antibodies that act on focused microorganisms and intensify their capacity to cause disease (Thomas and Prendergast 2016).

There are two types of cancer immunity vaccines: innate and adaptive immunity. Immune surveillance involves cell types in both branches of the immune system: innate (e.g. natural killer (NK) cells) and adaptive (e.g. CD8+ cytotoxic T-lymphocytes (CTLs), CD4+ T-helper cells) (Zamarron and Chen 2011). Tumor cells also support an anti-inflammatory state by tumor growth factors. For example, tumor cells may secrete interleukin-10 (IL-10) and transforming growth factor-β (TGF-β), which have immunosuppressive effects on T-lymphocytes and macrophages, verexpression of membrane complement regulators, particularly CD46, CD55, and CD59, M1 macrophages exposed to interferon-gamma (IFN-γ), macrophages (IL-1, TNF, and IL-6) may also promote metastasis (Zielinski et al. 2013).

Rabinovich, Gabrilovich, and Sotomayor (2007) watched tumor-filtrating lymphocyte-modified cytokine profiles and recommended that a host-mounted, antitumor resistant reaction against the tumor happens, yet development and movement proceeds regardless of this reaction, which known as an autocrine signal. This signal includes self-sustained stimulation of the cell due to growth factors and for this reason sometimes a cancer vaccine may not work on these types of cell. So, another treatment method must be developed (Berzofsky et al. 2017).

The most reported side effect of cancer vaccines is inflammation at the injection site such as redness, pain, swelling, skin warming, itching, and sometimes rash. People experience flu-like symptoms, such as fever, chills, weakness, dizziness, nausea or vomiting, muscle pain, fatigue, headaches, and rarely, breathing difficulties after the cancer vaccine is administered. Blood pressure may also be affected. These side effects, which usually persist for a short time, indicate that the body is reacting to the vaccine and becoming immunized as it is exposed to a virus. In rare cases it can cause serious reactions (Pazdur and Jones 2007). Further information about cancer vaccines are given in Chapter 11.

2.7   BIOTECHNOLOGICAL VACCINE

Recently, new and advanced molecular biology and biotechnological methods have been used in obtaining vaccines with fewer side effects. Drugs created with biotechnology have some disadvantages due to factors such as the need for advanced laboratories, advanced devices, high cost, and experienced personnel. However, biotechnological vaccines only have a sequence encoding the antigenicity portion. The DNA encoding the antigen is isolated and integrated into the cells that do not have the disease-specificity and the desired protein is synthesized. Biotechnological vaccines do not form an infection and turn out to eliminate the disadvantages of classical vaccines (Mzula et al. 2019).

2.7.1   GENETIC ENGINEERING

Molecular biology, and with important advances in nanotechnology, gene therapy, gene silencing such as recombinant DNA/protein vaccine techniques in vitro and in vivo environments options, are of great interest. This biological process of transferring various nucleic acids to target the cell without being degraded through biological barriers and transfection of acid into the cytoplasm or nucleus with the help of nano biomaterials constitute a safe alternative. With the use of biocompatible and low-toxic materials as transfection agents in the transfer of nucleic acids, it is promising for the treatment of many diseases, including cancer, by overcoming biological barriers, editing the target gene of interest, and achieving high efficiency. Various nucleic acid types, properties, and uses are encapsulated into nanoparticles according to its purpose in transfection studies. Nucleic acid-based gene therapy transmits the nucleic acid sequence into the cell fragmentation of the target mRNA and is therefore responsible for diseases defined as silencing genes (Videira et al. 2014; Pamukcı, Portakal, and Eroğlu 2018).

2.7.1.1   Nucleic Acid Vaccines (DNA Vaccines)

After analyzing genes from a microbe, scientists have tried to create a DNA vaccine to use the microbe genes. These vaccines, which are still in the experimental stage today, hold great promise. DNA vaccines take immunization to a new technological level. Basically, genes encoding immunity are isolated and replicated (these genes are usually bound to a plasmid). These replicated genes are integrated directly into the recipient cell. These antigenic molecules or molecules synthesized by the gene in the plasmid are synthesized by the host cell. In this way, the synthesized gene product, the antigenic protein, stimulates the immune system and leads to effective immunity (Lee, Izzard, and Hurt 2018).

DNA vaccines consist of a small, circular DNA fragments called a DNA plasmid containing genes encoding proteins of a pathogen. When the vaccine is injected into the host, the host cells read the DNA and translate it into a pathogen from the proteins. When the body realizes that the proteins are foreign, the cells act to stimulate the immune system with helper T cells that increase the production of antibodies, as well as lethal T cells that directly kill infected cells (Xenopoulos and Pattnaik 2014).

In particular, DNA vaccines use genes for important antigens. In other words, the body's own cells become vaccine-producing factories and create the necessary antigens to stimulate the immune system. Nucleic acid vaccines are an effective method of vaccination as they are applied directly and do not enter digestive processes because the nucleic acids are integrated directly. DNA vaccines are relatively easy and inexpensive and are suitable for design and production (Ferraro et al. 2011). DNA vaccines induce antigen-specific adaptive immune responses more than conventional protein/peptide-based vaccines. We can also mention that DNA vaccines are more stable, relatively easy, inexpensive, suitable for design, cost-efficient, easy to manufacture, and safe to use (Prazeres and Monteiro 2014). DNA vaccines have many applications including against infectious diseases (Maslow 2017), cancer therapy (Fioretti, Iurescia, and Rinaldi 2014), autoimmune (Zhang and Nadakuma 2018), and allergies (Scheiblhofer, Thalhamer, and Weiss 2018). DNA vaccines are subject to more than 500 clinical trials in the US; these clinical trials are, especially, viral infections (Weniger et al. 2018) and cancer (Tiptiri-Kourpeti et al. 2016). We can mention that bacterial diseases and autoimmune responses are less atopic than other topics. DNA vaccines have had many clinical trials; therefore, we can say that DNA vaccines are a good medical approach for the future (Hobernik and Bros 2018). DNA vaccine is used to induce antigen-specific immune responses; moreover, DNA vaccines can have applications that inhibit immuno-regulatory myeloid cells and tumors (Marshall and Djamgoz 2018).

Vaccines that are applied directly to the body are called naked DNA vaccines. They are applied directly into the cells with a needle-free device using high-pressure gas to direct microscopic gold particles coating the needle or syringe. Naked unformulated plasmid DNA can contribute to low therapeutic efficiency and degradation (Lechardeur, Verkman, and Lukacs 2005) (Figure 2.1).

At present, there are no DNA vaccines for human use that are confirmed. Even so, there are some veterinary use DNA-based vaccines for West Nile virus in horses (Dauphin and Zientara 2007) and canine melanoma (Atherton et al. 2016) that were approved by the FDA and the USDA. The first clinical trial with DNA vaccines in humans was against HIV. They detected potential immunogenicity, therapeutic, and prophylactic effects. However, there were no significant immune responses (MacGregor et al. 1998). In addition, a study showed that induction of CD8+ T cell responses occurs in the proteins of primates after immunization with a mixture of different *Plasmodium falciparum* encoding plasmids (Wang 1998).

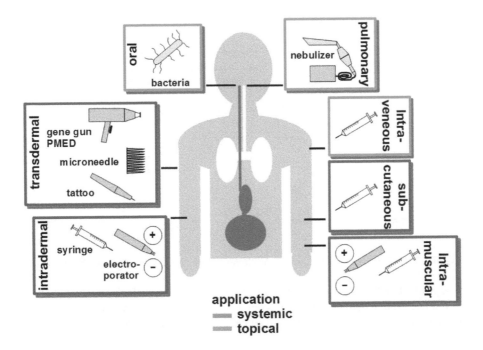

FIGURE 2.1　Routes of DNA vaccine delivery (Hobernik and Bros 2018).

Clinical trials promise that DNA vaccines can evoke efficient induction of cellular and humoral responses (Fioretti, Iurescia, and Rinaldi 2014). Then again, the reactions were not adequate to evoke noteworthy clinical advantages. Such a significant number of clinical preliminaries concentrating on DNA antibody advancement techniques achieve the greatest effectiveness at immunogenicity (Tiptiri-Kourpeti et al. 2016) (Figure 2.2).

2.7.1.2　Recombinant Vaccines

In microorganisms there are some genes that encode antigenic and other important components. These genes are integrated into the genomes of mutant microorganisms by some molecular techniques, and recombinantly these types of vaccines are obtained. These vaccines can also be incorporated into a genetic component, as well as multiple components. The method may vary according to the study. For example, recombinant mutant vaccines can be obtained by combining the gD gene of herpes simplex (HSV), the HA gene of the influenza virus, and the rabies virus glycoprotein gene of the vaccinia virus genome (Nascimento and Leite 2012).

2.7.1.3　Subunit Vaccines

Subunit vaccines contain the genes of antigens that are disinfected instead of using the entire microorganism. The presence of antigen is the most important advantage in terms of reducing side effects. The coding gene region is removed and combined with the bacterial plasmid. The product obtained by the expression of this gene is

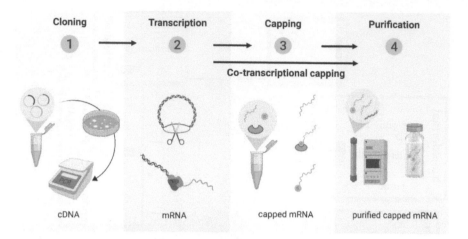

FIGURE 2.2   Here is mRNA vaccine producing is described (Versteeg et al. 2019).

used as the basis of the vaccine. Subunit vaccines include surface antigens, toxoids, and subcellular fragments. In this type of vaccine, which is produced by genetic engineering technology with low impact capacity, adjuvants are frequently included (Dai et al. 2019).

2.7.1.4   Mutant Vaccines

Microorganisms usually transform into mutant microorganisms and the mutant genomes of microorganisms can be used as mutant vaccines. Since the genes that cause the disease are removed, they do not cause infection. Hereditary pathogens are attenuated. However, there are risks of infection by returning to wild forms of mutations that may occur spontaneously (Wyand et al. 1996; Kraatz et al. 2015).

2.7.2   ADVANCED BIOTECHNOLOGICAL VACCINE

2.7.2.1   Anti-Idiotype Antibody Vaccines

Anti-idiotypic vaccines include antibodies called idiotopes with three-dimensional immunogenic regions, consisting of protein sequences that bind to cell receptors. Idiotopes are collected in idiotypes specific to the target antigen. An example of an anti-idiotype antibody is racotumomab. It is prepared in the experimental animal against a specific antigen. The resulting antibodies are administered to a different experimental animal to form anti-idiotype antibodies and may be used in immunization. Thus, a protective immunity is formed (Allen and Ansel 2013; Akbuğa 2002).

If production is given, antibodies that bind to tumor-associated antigens (TAA) are isolated and injected into mice. To the mouse immune system, TAA antibodies are antigens. They cause an immunogenic reaction that can bind to the "TAA idiotype" and produce a mouse antibody called "anti-idiotypic". The resulting mouse antibodies are inoculated into other mice. The antibodies that appear in the second group of mice have a three-dimensional binding site that mimics the original antibodies that bind to tumor-associated antigens. These antibodies are combined with an adjuvant and administered as a vaccine. Anti-idiotypic antibodies mimic the active antigen to make a copy of it. This method is used in high-risk cancer patients as well as in developing vaccines against pathogens (Büyüktanır 2010) (Figure 2.3).

2.7.3   Synthetic Peptide Vaccine

In this book synthetic peptide vaccines are explained in detail. Peptide is a word that is of Greek origin. Peptides are found naturally in the body, plants, animals, or it is synthesized in laboratories. The body can produce ribosomal and non-ribosomal peptides. Peptides can be synthesized by more than one method. Peptide synthesis depends on bonding the amino acids with peptide bonds. They are also known as amide bonds. The biological process of producing long peptides is known as protein biosynthesis. Peptides play an important role in biological and physiological processes. They can be synthesized in natural objects via genetic and bioengineering principles, or by using chemical synthesis methods. The greatest advantage of using chemical synthesis methods in vaccine studies is the elimination of the infection-induced factors caused by the antigen. This provides a more stable and specific response to the intended vaccine (Özcan et al. 2019). Peptide-based vaccination usually takes place by using an immunoadjuvant (nanoparticles or biopolymers) to induce T cell and sometimes B cell immunity. It can also be said that peptide-based

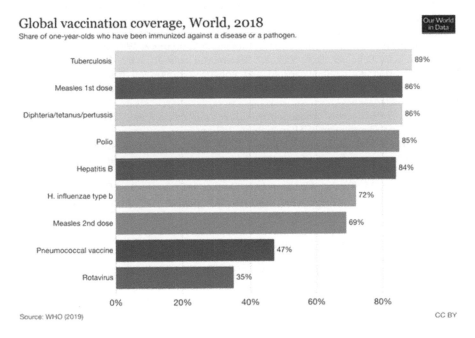

FIGURE 2.3   Total vaccinated rates at 2018 at different disease or a pathogen. (Source: Vanderslott and Dadonaite (2013) https://ourworldindata.org/grapher/global-vaccination-coverage.)

vaccines have the ability to play a role in inducing innate and adaptive immunity, but without the peptide, biopolymer, or nanoparticle system, they show weakness in immune responses. (Özcan et al. 2019). Strong targeted immune responses with a specific phenotype of many dimensions are required for a strong immunogenic trait, including T cells and B cell phebotypes, antigen-presenting cells, and phenotypes of other immune cells given in Chapter 1.

Synthesis of synthetic peptides corresponding to antigenic epitopes can be synthesized very rapidly using chemical methods, and it has been observed that such synthetic peptides mimic the structure of infectious organism epitopes and resemble their structure (Oldstone et al. 1995). Linear peptides that are components of the pathogen of interest are produced by a solid-phase peptide synthesis method with higher yield and purity (Özcan et al. 2019). Further information about peptides is given in other chapters.

As a result, there are different mechanisms for each type of vaccine described in detail above. The specificity of the antigen, the preparation process of the vaccine, and the production step depend on the vaccine type for each type of vaccine developed. It is based on how the microorganism infects the cell and how the immune system responds to it.

REFERENCES

Akbuğa, J. 2002. "Pharmaceutical biotechnology products." *Journal of In-Service Continuing Education*. 54–62.
Allen, L. and H.C. Ansel. 2013. *Ansel's pharmaceutical dosage forms and drug delivery systems* (9th Edition). United States, Lippincott Williams & Wilkins.
Alson, D., S.C. Schuyler, B.X. Yan, K. Samimuthu, and J.T. Qiu. 2020. "Combination vaccination with tetanus toxoid and enhanced tumor-cell based vaccine against cervical cancer in a mouse model." *Frontiers in Immunology* 11: 927. doi:10.3389/fimmu.2020.00927.
Altınsoy, N. 2007. *Vaccine production techniques and control.* A.Ü. Faculty of Veterinary Medicine, Department of Microbiology, Ankara.
Banday, A.H., S. Jeelaniand, and V.J. Hruby. 2014. "Cancer vaccine adjuvants – recent clinical progress and future perspectives." *Immunopharmacology and Immunotoxicology* 37(1): 1–11.
Berzofsky, J.A., M. Terabe, J.B. Trepel, I., Pastan, D.F. Stroncek, J.C. Morris, and L.V. Wood. 2017. "Cancer vaccine strategies: Translation from mice to human clinical trials." *Cancer Immunology Immunotherapy CII* 67(12): 1863–1869. doi:10.1007/s00262-017-2084-x.
Büyüktanır, Ö. 2010. "Biotechnological bacterial vaccines nowadays." *Atatürk University Journal of Veterinary Sciences* 5(2): 97–105.
Chen, W., D. Zhao, X. He, R. Liu, Z. Wang, X. Zhang, F. Li, et al. 2020. "A seven-gene-deleted African swine fever virus is safe and effective as a live attenuated vaccine in pigs." *Science China. Life Sciences* 63(5): 623–634. doi:10.1007/s11427-020-1657-9.
Clem, A.S. 2011. "Fundamentals of vaccine immunology." *Journal of Global Infectious Diseases* 3(1): 73. doi:10.4103/0974-777x.77299.
Dai, X., Y. Xiong, N. Li, and C. Jian. 2019. "Vaccine types. Vaccines - the history and future." *Intechopen.* doi:10.5772/intechopen.84626.
Dauphin, G. and S. Zientara. 2007. "West Nile virus: Recent trends in diagnosis and vaccine development." *Vaccine* 25: 5563–5576. doi:10.1016/j.vaccine.2006.12.005.

Delany, I., R. Rappuoli, and K.L. Seib. 2013. "Vaccines, reverse vaccinology, and bacterial pathogenesis." *Cold Spring Harbor Perspectives in Medicine* 3(5): a012476–a012476.

Ferraro, B., M.P. Morrow, N.A. Hutnick, T.H. Shin, C.E. Lucke, and D.B. Weiner. 2011. "Clinical applications of DNA vaccines: Current progress." *Clinical Infectious Diseases: An Official Publication of the Infectious Diseases Society of America* 53(3): 296–302. doi:10.1093/cid/cir334.

Fioretti, D., S. Iurescia, and M. Rinaldi. 2014. "Recent advances in design of immunogenic and effective naked DNA vaccines against cancer." *Recent Patents on Anti-Cancer Drug Discovery* 9: 66–82. doi:10.2174/1574891X113089990037.

Hobernik, D. and M. Bros. 2018. "DNA vaccines—how far from clinical use?" *International Journal of Molecular Sciences* 19(11): 3605. doi:10.3390/ijms19113605.

Ivory, C. and K. Chadee. 2014. "DNA vaccines: Designing strategies against parasitic infections." *Genetetic Vaccines Theraphy* 2(1): 17.

Kirstein, J., M. Pina, J. Pan, E. Jordanov, and M.S. Dhingra. 2020. "Immunogenicity and safety of a quadrivalent meningococcal tetanus toxoid-conjugate vaccine (MenACYW-TT) in adults 56 years of age and older: A phase II randomized study." *Human Vaccines & Immunotherapeutics* 16(6): 1299–1305. doi:10.1080 / 21645515.2020.1733868.

Kraatz, F., K. Wernike, S. Hechinger, P. König, H. Granzow, I. Reimann, and M. Beer. 2015. "Deletion mutants of Schmallenberg virus are avirulent and protect from virus challenge." *Journal of Virology* 89(3): 1825–1837. doi:10.1128/JVI.02729-14.

Lechardeur, D., A.S. Verkman, and G.L. Lukacs. 2005. "Intracellular routing of plasmid DNA during non-viral gene transfer." *Advance Drug Delivery Reviews* 57: 755–767. doi:10.1016/j.addr.2004.12.008.

Lee, L.Y.Y., L. Izzard, and A.C. Hurt. 2018. "A review of DNA vaccines against influenza." *Frontiers in Immunology* 9(1568): 1–8. doi:10.3389/fimmu.2018.01568.

Lee, N.H., J.A. Lee, S.Y. Park, C.S. Song, I.S. Choi, and J.B. Lee. 2012. "A review of vaccine development and research for industry animals in Korea." *Clinical and Experimental Vaccine Research* 1(1): 18. doi:10.7774/cevr.2012.1.1.18.

Lin, A., D. Apostolovic, M. Jahnmatz, F. Liang, Ols, S., Tecleab, T., C. Wu, et al. 2020. "Live attenuated pertussis vaccine BPZE1 induces a broad antibody response in humans." *The Journal of Clinical Investigation* 130(5): 2332–2346. doi:10.1172/JCI135020.

Lollini, P.L., F. Cavallo, P. Nanni, and G. Forni. 2006. "Vaccines for tumour prevention," *Nature Reviev Cancer* 6(3): 204–217.

MacGregor, R.R., J.D. Boyer, K.E. Ugen, K.E. Lacy, S.J. Gluckman M.L. Bagarazzi, M.A. Chattergoon, et al. 1998. "First human trial of a DNA-based vaccine for treatment of human immunodeficiency virus type 1 infection: Safety and host response." *Journal of Infectious Diseases* 178: 92–100. doi:10.1086/515613.

Marshall, H.T. and M.B.A. Djamgoz. 2018. "Immuno-oncology: Emerging targets and combination therapies." *Frontiers Oncology* 8: 315. doi:10.3389/fonc.2018.00315.

Maslow, J.N. 2017. "Vaccines for emerging infectious diseases: Lessons from MERS coronavirus and Zika virus." *Human Vaccines and Immunotherapeutics* 13: 2918–2930. doi:1 0.1080/21645515.2017.1358325.

Melnick, J.L. 1978. "Advantages and disadvantages of killed and live poliomyelitis vaccines." *Bulletin of World Health Organanisation* 56(1): 21.

Mok, D. and K.R. Chan. 2020. "The effects of pre-existing antibodies on live-attenuated viral vaccines." *Viruses* 12(5): 520. doi:10.3390 / v12050520.

Mzula, A., P.N. Wambura, R.H. Mdegela, and G.M. Shirima. 2019. "Current state of modern biotechnological-based aeromonas hydrophila vaccines for aquaculture: A systematic review." *BioMed Research International* 2019: 1–11. doi:10.1155/2019/3768948.

Nascimento, I.P. and L.C.C. Leite. 2012. "Recombinant vaccines and the development of new vaccine strategies." *Brazilian Journal of Medical and Biological Research* 45(12): 1102–1111. doi:10.1590/s0100-879x2012007500142.

O'Connell, A.K. and F. Douam. 2020. "Humanized mice for live-attenuated vaccine research: From unmet potential to new promises." *Vaccines* 8(1): 36. doi:10.3390 / vaccines8010036.

Oldstone, M.B.A., H. Lewicki, P. Borrow, D. Hudrisier, and J.E. Gairin. 1995. "Discriminated selection among viral peptides with the appropriate anchor residues: Implications for the size of the cytotoxic Tlymphocyte repertoire and control of viral infection." *Journal of Virolology* 69: 7423–7429.

Özcan, Ö.Ö., M. Karahan, P.V. Kumar, S.L. Tan, and Y.N. Tee. 2019. *New generation peptide-based vaccine prototype.* IntechOpen. doi:10.5772/intechopen.89115.

Pamukcı, A., H.S. Portakal, and E. Eroğlu. 2018. "New generation biomaterials used in delivery of therapeutic molecules." *Journal of Science and Technology* 11(3): 524–542 doi:10.18185/erzifbed.339405.

Pazdur, M.P. and J.L Jones. 2007. "Vaccines: An innovative approach to treating cancer." *Journal of Infusing Nursing* 30(3): 173–178.

Pelit Arayıcı, P. 2015. *Protective new generation peptide vaccine models: Development of bioconjugates of viral rabies peptides with various adjuvants.* (id) Master Thesis Yildiz Technical University.

Plotkin, S.A. 2009. "Vaccines: The fourth century." *Clinical and Vaccine Immunology* 16(12): 1709–1719.

Prazeres, D.M.F. and G.A. Monteiro. 2014. "Plasmid biopharmaceuticals." *Microbiology Spectrum* 2(6): PLAS-0022-2014. doi:10.1128/microbiolspec.PLAS-0022-2014.

Rabinovich, G.A., D. Gabrilovich, and E.M. Sotomayor. 2007. "Immunosuppressive strategies that are mediated by tumor cells." *Annual Reviews of Immunology* 25: 267–296.

Rice, P.A., W.M. Shafer, S. Ram, and A.E. Jerse. 2017. "Neisseria gonorrhoeae: Drug resistance, mouse models, and vaccine development." *Annual Review of Microbiology* 71(1): 665–686.

Scheiblhofer, S., J. Thalhamer, and R. Weiss. 2018. "DNA and mRNA vaccination against allergies." *Pediatric Allergy and Immunology* 29(7): 678–688. doi:10.1111/pai.12964.

Schroeder, M.R., S.T. Chancey, S. Thomas, W.H. Kuo, S.W. Satola, M.M. Farley, and D.S. Stephens. 2017. "A population-based assessment of the impact of 7- and 13-valent pneumococcal conjugate vaccines on macrolide-resistant invasive pneumococcal disease: Emergence and decline of streptococcus pneumoniae serotype 19A (CC320) with dual macrolide resistance mechanisms." *Clinical Infectious Diseases* 65(6): 990–998.

Silveira, M.M., T.L. Oliveira, R.A. Schuch, A.J.A. McBride, O.A. Dellagostin, and D.D. Hartwig. 2017. "DNA vaccines against leptospirosis: A literature review." *Vaccine* 35(42): 5559–5567.

Tarancón, R., J. Domínguez-Andrés, S. Uranga, A.V. Ferreira, L.A. Groh, M. Domenech, F. González-Camacho et al. 2020. "New live attenuated tuberculosis vaccine MTBVAC induces trained immunity and confers protection against experimental lethal pneumonia." *PLoS Pathogens* 16(4): e1008404. doi:10.1371 / journal.ppat.1008404.

Thomas, S. and G.C. Prendergast. 2016. "Cancer vaccines: A brief overview." *Methods in Molecular Biology* 1403: 755–761. doi:10.1007/978-1-4939-3387-7_43.

Tiptiri-Kourpeti, A., K. Spyridopoulou, A. Pappa, and K. Chlichlia. 2016. "DNA vaccines to attack cancer: Strategies for improving immunogenicity and efficacy." *Pharmacology and Therapeutics* 165: 32–49. doi:10.1016/j.pharmthera.05.004.

Vanderslott, S. and B. Dadonaite. 2013. *Vaccination.* Published online at OurWorldInData. org. https://ourworldindata.org/vaccination [Online Resource] https://ourworldindata.org/grapher/global-vaccination-coverage.

Vartak, A. and S. Sucheck. 2016. "Recent advances in subunit vaccine carriers." *Vaccines* 4(2): 12. doi:10.3390/vaccines4020012.

Videira, M., A. Arranja, D. Rafael, and R. Gaspar. 2014. "Preclinical development of siRNA therapeutics: Towards the match between fundamental science and engineered systems." *Nanomedicine: Nanotechnology, Biology and Medicine* 10(4): 689–702.

Wang, R. 1998. "Induction of antigen-specific cytotoxic T lymphocytes in humans by a malaria DNA vaccine." *Science* 282: 476–480. doi:10.1126/science.282.5388.476.

Webb, R., P.M. Wright, J.L. Brown, J.C. Skerry, R.L. Guernieri, T.J. Smith, C. Stawicki, and L.A. Smith. 2020. "Potency and stability of a trivalent, catalytically inactive vaccine against botulinum neurotoxin serotypes C, E, and F (triCEF)." *Toxicon: Official Journal of the International Society on Toxinology* 176: 67–76. doi:10.1016/j.toxicon.2020.02.001.

Weniger, B.G., I.E. Anglin, T. Tong, M. Pensiero, and J.K. Pullen. 2018. "Workshop report: Nucleic acid delivery devices for HIV vaccines: Workshop proceedings, National Institute of Allergy and Infectious Diseases Bethesda, Maryland, USA, May 21, 2015." *Vaccine* 36: 427–437. doi:10.1016/j.vaccine.2017.10.071.

Wyand, M.S., K.H. Manson, M. Garcia-Moll, D. Montefiori, and R.C. Desrosiers. 1996. "Vaccine protection by a triple deletion mutant of simian immunodeficiency virus." *Journal of Virology* 70(6): 3724–3733.

Xenopoulos, A., and P. Pattnaik. 2014. "Production and purification of plasmid DNA vaccines: Is there scope for further innovation?" *Expert Review of Vaccines* 13(12): 1537–1551. doi:10.1586/14760584.2014.968556.

Zamarron, B.F. and W. Chen. 2011. "Dual roles of immune cells and their factors in cancer development and progression." *International Journal of Biological Sciences* 7: 651–658.

Zhang, N. and K.S. Nandakuma. 2018. "Recent advances in the development of vaccines for chronic inflammatory autoimmune diseases." *Vaccine* 36: 3208–3220. doi:10.1016/j.vaccine.2018.04.062.

Zhang, W., P. Wang, B. Wang, B. Ma, and J. Wang. 2017. "A combined clostridium perfringens / trueperella pyogenes inactivated vaccine induces complete immunoprotection in a mouse model." *Biologicals* 47: 1–10.

Zhao, B., X. Li, B. Wang, B. Gao, and S. Meng. 2014. "Prophylactic cancer vaccine, from concept to reality?" *Chinese Science Bulletin* 59(10): 944–949. doi:10.1007/s11434-014-0176-y.

Zielinski, C., S. Knapp, C. Mascaux, and F. Hirsch. 2013. "Rationale for targeting the immune system through checkpoint molecule blockade in the treatment of non-small-cell lung cancer." *Annals of Oncology* 24(5): 1170–1179.

Zou, J., X. Xie, H. Luo, C. Shan, A.E. Muruato, S.C. Weaver, T. Wang, and P.Y. Shi. 2018. "A single-dose plasmid-launched live-attenuated Zika vaccine induces protective immunity." *EBioMedicine* 36: 92–102. doi:10.1016 / j.ebiom.2018.08.056.

3 Fundamentals of Modern Peptide Synthesis

Korkut Ulucan and Muhsin Konuk

CONTENTS

3.1 GENERAL INFORMATION ON AMINO ACIDS AND PEPTIDES

Amino acids are the most important building blocks that make up primary protein structures (peptides). The physicochemical properties of amino acids are largely determined by the structural bonds that make up the amino acid. Each amino acid basically consists of a central carbon atom, also known as the alpha (α) carbon bonded to a hydrogen atom, an amine group ($NH2$), and a carboxyl group ($COOH$). In addition, each amino acid contains another atom group attached to the central carbon atom known as the R group. Every alpha carbon makes a mirror arrangement of substituents with the exception of alpha carbon at amino acid glycine. In the mirror arrangement, amino acids have carbon atom (C alpha) and contain four substituents at the C atom: carboxyl group (called the C-terminus), amino group (called the N-terminus), hydrogen, and a side chain (R), and for this reason C alpha is a chiral center. Amino acids have L-enantiomer and D-enantiomer in their optic activity by rotating the polarized light plane at the same rate in opposite directions (Reece et al. 2011). There are roughly 300 types of amino acids in nature, there are 22 amino acids that constitute the monomer units of proteins and the monomers are attached via peptide bonds. They range from acidic to basic, polar to non-polar and from hydrophilic to hydrophobic. The characteristics depend on side chain R as amino acid hydrophilic chains tend to appear as water-soluble proteins as they can interact favorably in water. Amino acids that have nonpolar side chains are hydrophobic. That is, they tend to be buried inside proteins, where they are away from water and can interact favorably with each other. They have also electrically charged side chains generally seen in van der Waals bonds which have pulling and pushing power between atoms (Bischoff and Schlüter 2012) (Figure 3.1).

Amino acids in peptides and proteins are linked together with a **peptide bond**. The peptide bond is a chemical bond formed between two molecules of amino acids when the carboxyl group of one molecule reacts with the amino group of the other molecule, releasing a molecule of water. This is a dehydration synthesis reaction (also known as a *condensation reaction*), and usually occurs between amino acids. Dipeptide is the structure formed by the combination of two amino acids with amide bond. Likewise, as amino acids are added, we may have tripeptides, tetrapeptides, and other polypeptides. Gradually, when the structure moves towards quaternary structures, it is called protein (Raven et al. 2014; Reece et al. 2011).

Significant delocalization of the lone pair of electrons on the nitrogen atom gives the group a partial double bond character. The partial double bond renders the amide group planar, occurring in either the cis or trans-isomers. The geometry and dimensions of the peptide bonds are given in Figure 3.2. The peptide bond length is given in the Angstrom unit, the bond angles – in degrees. In the unfolded

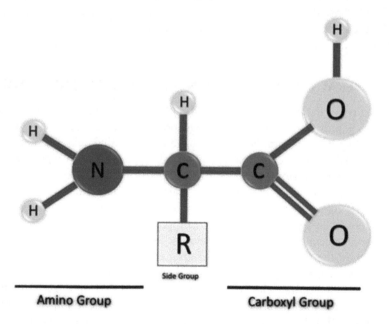

FIGURE 3.1 General structure of an amino acid. Showing the Acid Group and Amino group. "R" group represents a point where different groups can bond thus forming one of the 20 different amino acids. (Modified from: https://en.wikipedia.org/wiki/File:AminoAcidball. svg#filelinks.)

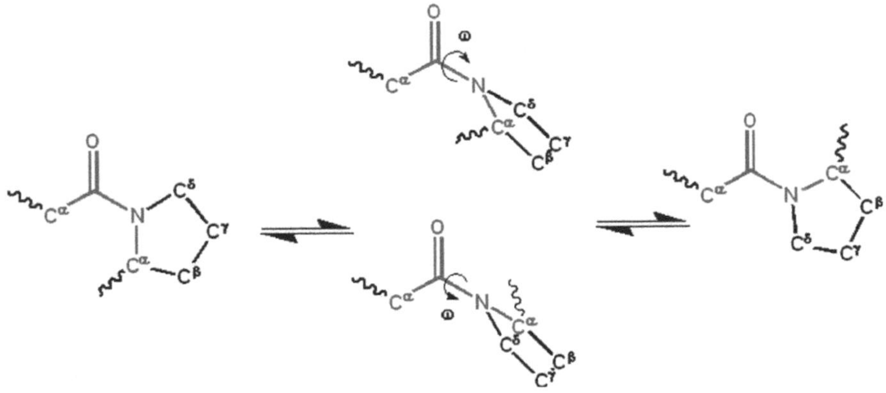

FIGURE 3.2 The geometry and dimensions of the peptide bonds. (Source: WillowW at the English language Wikipedia https://en.wikipedia.org/wiki/File:Cis_trans_isomerization_k inetics_X_Pro_peptide_bonds.png.)

time of the proteins, the peptides are isomerized, and after folding this structure appears as a single isomer (at each position of the protein). Most peptide bonds exist in trans and genus and there are transitions between them alternately. For most peptide bonds (roughly 1000:1 ratio in trans cis populations) conversion is overwhelmingly preferred. However, this symmetry between the $C\alpha$ and $C\delta$ atoms of proline (Pro) in the peptide groups can make the cis and trans isomers nearly equal in energy (they tend to have a roughly 30:1 ratio) (Raven et al. 2014; Reece et al. 2011).

3.2 FUNCTIONS OF AMINO ACIDS

Amino acids form all proteins, starting from the primary, secondary, tertiary, and quaternary, and proteins facilitate the regulation of all metabolic pathways and processes, including growth, enzymatic activity, and DNA repair of all organisms (Cesari et al. 2005; Wu 2009). As mentioned, they serve as precursors in the synthesis of proteins and formation of secondary metabolism molecules (Moran-Palacio et al. 2014) and also have many roles in the organism's health. They help to maximize the yield of food usage, control growth and immunity, and organize the protein metabolism of muscles (Yamane et al. 2007; Wu, Fang, et al. 2004; Weinert 2009). Due to those functions, amino acids are found in all parts of the body (Meletis and Barker 2005).

There are various physiological processes that amino acids participate in such as atrophic conditions, skeletal muscle function, cancer, and sarcopenia. Furthermore, they play important roles in homeostasis, cell signaling, gene expression, phosphorylation of proteins, synthesis of hormones, and also possess antioxidant abilities (antioxidant abilities are mentioned in Chapter 13) (Moran-Palacio et al. 2014; Wu 2009). The synthesis of low molecular weight nitrogenous compounds is also related to amino acids. The nitrogenous compounds have great biological importance to an organism. The existence of amino acids and of their metabolites as taurine, sero-tonin, glutathione, polyamines, and thyroid hormones is necessary for the functions of our bodies (Wu 2009).

Amino acids can be represented in two ways as essential and non-essential amino acids, but there is another group of amino acids classified in the essential group (Wu 2009. The classification of amino acids is made according to the amount of synthesis of the amino acid required for metabolism while the organism continues its vital activities (Meletis and Barker 2005). In general, amino acids that the organism can-not synthesize but are necessary for the body are called essential amino acids. For this reason, amino acids that are not produced in the organism must be supplied from the diet.

The amino acids that organisms can produce by genetic coding are called non-essential amino acids. Usually, most of the amino acids required are coded by our bodies, but when they are needed more than the usage rate, the body needs to supple-ment them externally for optimal requirements. Amino acids are responsible for the formation of proteins, muscle, connective tissue, skin, hair, nails, neurotransmitters, all enzymes, and hormones. The presence of amino acids in a certain ratio and bal-ance is a basic requirement for the proper functioning of all systems in our bodies, the repair of tissues and our mental health. Since amino acids play a role in all func-tions and structures of the body, measuring and evaluating the amount of amino acids in the blood and other tissues provides detailed information about our health. Significant changes in amino acid levels are detected in many physical and mental illnesses (Wu 2009). Amino acid analysis evaluates from blood with advanced labo-ratory methods, such as high-performance liquid chromatography (HPLC) and tan-dem mass spectrometry (MS) (Kimura et al. 2009). Everyone's amino acid profile is unique like a fingerprint and does not show major changes in a short time. Therefore, the test reflects your current amino acid profile and needs very well. Vitamins and minerals are essential for amino acids to function properly (Miyagi et al. 2011). In other words, amino acid analysis also determines which vitamins and minerals to take.

If amino acids are not synthesized enough in the body or if they are not taken from outside, the lack of amino acids and subsequently symptomatic effects are seen. In particular, these symptoms cause weight loss, insufficient growth, and develop-mental anomalies. Amino acids can be stored for use, but this will not be for long. For this reason, amino acid supplements should be added to the body in sufficient quantities, even with a daily diet (Meletis and Barker 2005). In many studies it is thought that amino acid implantation is related to the pathogenesis underlying many diseases in humans and animals, but definitive data have not been obtained, and there are treatments applied clinically with amino acid supplementation (Paul et al. 2014; Wu, Bazer, et al. 2004). The amino acids are also mostly used as thera-peutic agents. Especially in medical applications, they have a very important place in terms of being natural and supportive resources in therapeutic terms. Treatment of brain metabolism and neurotransmission imbalances are some of the most prominent areas of therapeutic applications of amino acids (these are mentioned in Chapter 10). Besides, there are many other areas where amino acids are used, such as cardiovascular and gastrointestinal (GI) health and immune function (Meletis and Barker 2005), treatment of liver diseases, fatigue, skeletal muscle damage, cancer prevention, burn, trauma and sepsis, maple urine disease, and diabetes (Tamanna and Mahmood 2014).

3.3   IMPORTANT AMINO ACIDS

Vauquelin and Robique (1806) isolated the first amino acid from *Asparagus sativus*; they named the amino acid asparagine (Vauquelin and Robique 1806). After glycine was discovered from gelatin hydrolysate (Braconnot 1820) Rose et al. described threonine from food ingredients in 1935 (Rose et al. 1935). Later, the last two of the 22 proteinogenic amino acids, selenocysteine and pyrrolysine, were discovered in 1986 (Zinoni et al. 1986) and 2002 (Srinivasan et al. 2002).

Amino acids can be classified by their structural function such as aliphatic, aromatic, neutral, acidic, basic, sulfur-containing amino acid. Apolar molecules cysteine and methionine play an especially important role to gain resistance to proteins' disulfide bond (Brosnan and Brosnan 2006). Phenylalanine is only apolar in aromatic amino acids. Amino acids such as valine and leucine are nonpolar and hydrophobic, on the other hand amino acids like serine and glutamine amino acids have water-soluble side chains and are polarized by showing hydrophilic properties. Lysine and arginine amino acids (at pH > 7) have positively charged side chains and basic amino acids; aspartate and glutamate (at pH < 7) are negatively charged and acidic. There are also amino acids with R groups, but these also change in properties depending on their structure. For example, proline has an R group that binds back to its amine (-NH2) group, giving it a ring structure. Cysteine contains a functional thiol (-SH) group and forms disulfide (covalent) bonds with other cysteine amino acids. These bonds of cysteine amino acids are in the form of bridges and give proteins secondary, tertiary, and even quaternary conformative properties, forming strong protein structures (Raven et al. 2014; Reece et al. 2011) (Figure 3.3).

3.4   IMPORTANCE TO SYNTHESIS OF AMINO ACIDS

Clinical and preclinical studies of the therapeutic and proliferative effects of amino acids are ongoing (Meletis and Barker 2005). Vaccine studies are really important. There has been a growing interest in the field of synthetic peptide vaccines over the last decade due to numerous adverse effects of other vaccine models. Peptides are becoming popular in the vaccination industry due to their applications with better potency, high specificity, low toxicity, and natural availability. Various amino acid groups are used in the synthesis of peptide vaccines (Yang and Kim 2015) and therefore for developing new techniques to synthesize new amino acids with maximum benefits. Amino acids can be synthesized in three ways. The first is extraction of protein hydrolysates, the second is chemical synthesis of amino acids, and the third is using the microorganism enzymatic and fermentation route. Through these methods. fermentation is one of the safe and developing processes for the commercial

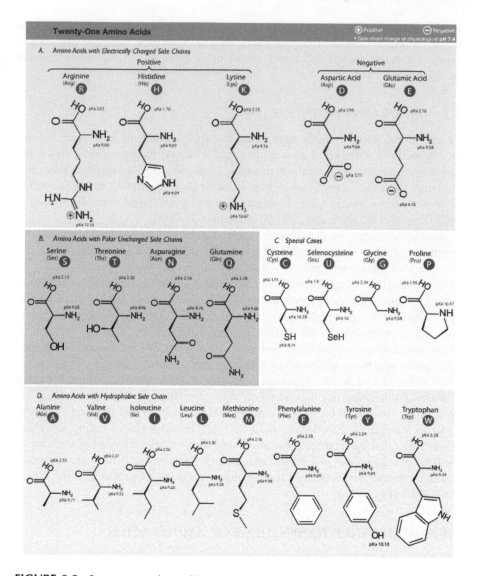

FIGURE 3.3 Important amino acids and their structures. (Image by "Dancojocari." The image is licensed under a CC BY-SA 3.0 or GFDL, from Wikimedia Commons.)

production of amino acids because of the new genetic engineering applications that are maximizing the yield and fertility of amino acids (Ikeda 2003). Essentially, amino acids exist as L-enantiomers in all organisms. On the other hand, there are important amounts of D-amino acids forms that are manufactured by bacteria with fermentation (Bardaweel 2014). The importance of D-amino acids is that the bacteria are involved in the synthesis of amino acids and amino acids are cross-linked by peptidoglycan (Vollmer, Blanot, and de Pedro 2008).

3.5 AMINO ACID SYNTHESIS

3.5.1 EXTRACTION OF PROTEIN HYDROLYSATES

This method is particularly suitable for production in the healthcare industry, however there are only a few amino acids that can be synthesized, such as L-cysteine, L-leucine, and L-tyrosine (Ikeda 2003). The amino acids are separated in the differences of physicochemical properties (such as chemical affinity and pH) (Zhang et al. 2016). Various extraction processes can be developed according to the amino acid of interest. For example, L-cysteine is commonly made from keratin found in animals and humans such as in hair, hooves, bristles, and feathers; properties are extracted using activated concentrated hydrochloric acid and charcoal (Renneberg 2008). This method has so many advantages such as the use of industrial byproducts or waste (hair, meat extracts, and plant hydrolysates) (Hauer et al. 2004). Moreover, it is a basic method, the amino acids that need it are easily available and the whole technological process is safe. However, an amino acid with low yield (Ikeda 2003) can be obtained as a result of protein degradation due to high temperature, enzyme activity, and pH conditions or due to an undesirable reason. As a result, the probability of protein values decreasing as a result of the degradation of the food from which the amino acid will be obtained is the risk of this food being preferred in this process. Also, another disadvantage is that large quantities of wastewater are generated due to the need to wash the alkaline solution from the product (Sereewatthanawut et al. 2008; Lütke-Eversloh, Santos, and Stephanopoulos 2007). In general, the main disadvantage of this method is that only natural proteins can be used as sources for amino acid production.

3.5.2 CHEMICAL SYNTHESIS

Achiral amino acids derived from glycine or racemic mixtures of D, L-methionine, or D, L-alanine have been produced by chemical synthesis. In 1850 the first chemical syntheses were reported. The name of this synthesis is the Strecker synthesis. In this method the reaction was assisted by acid catalysts, metal cyanide, and the reaction was to convert an aldehyde or ketone and amine or ammonia to α-amino acids (Strecker 1850).

Although studies have been carried out with this method, it has significant negative aspects with the presence of toxic or even lethal cyanide sources and high catalyst fees (Zuend et al. 2009). Also, the Strecker synthesis is not enantioselective and can therefore only produce a mixture of the D and L forms of amino acids (Gröger 2003). As a solution to these drawbacks, aqueous cyanide salts and chiral-thiourea catalysts were preferred (Harada 1963), and from here, methods such as asymmetric Strecker reactions were discovered and advanced (Zuend et al. 2009). In addition, such catalysts can catalyze the synthesis of a selective enantiomer by controlling the hydrocyanation process. The Strecker synthesis, a derivative of the Bucherer-Bergs method, is another important method used to obtain racemic amino acids. This method produces the hydrolysis of the racemic amino acid mixture in basic medium ketones by the reaction of aldehydes with ammonium carbonate and sodium cyanide (Hauer et al. 2004). Although it is not enantiometrically possible to obtain high purity amino acids, the advantage of being able to synthesize racemic mixtures from biochemically target compounds is important. As a result, the Bucherer-Bergs method is the most preferred method in chemical synthesis, but its negative aspects are the long reaction time and the presence of high temperatures that will cause the presence of oil deposits (Hauer et al. 2004).

3.5.3 MICROBIAL PROCESS

Another method used to synthesize amino acids is fermentation by biological catalysis, farther from the chemical described in the first method. The amino acids to be obtained are carried out by certain enzymatic reactions or the use of more than one enzyme. The preferred enzymes in this route are hydrolytic, NAD-dependent L-amino acid dehydrogenase, and ammonia lyases (Pollegioni and Servi 2003). The bacteria most commonly used as sources of these enzymes are *Saccharomyces cerevisiae, Pseudomonas dacunhae, Escherichia coli, Crypotococcus lurendii.*

The advantage of the enzymatic route is that D- and L-amino acids can be produced optically pure in higher concentrations and with the formation of a very low byproducts (Ikeda 2003). The reaction time is of great importance in the hydrolysis process, and when this time is recorded, a reaction of enzymes (alcalase and neutrase) occurs for about two days (Ramakrishnan et al. 2013). On the other hand, this method is expensive, and the enzymes are unstable (Ikeda 2003). The use of these enzyme-linked pathways is to derive the D- and L-amino acid isoforms from optically more pure and higher concentrations by minimizing the side derivatives (Ikeda 2003). Therefore, techniques have been developed to improve the performance of the process. Many activities have been provided to improve the process, for example, Hsiao et al. (1988) stated that *E. coli* immobilized with L-phenylalanine with polyazetidine was used to increase the productivity of 63% (w-w). Since L-amino acid production by chemical synthesis does not show a good result, it is used only in the production of some amino acids such as L-alanine and L-aspartic acid (Zhao et al. 2014).

The use of fermentation pathways is the current method for amino acid production. Microorganisms often prefer existing sugars as a substrate to transform them into a broad spectrum of amino acids under aerobic or anaerobic conditions. This process has some benefits over other methods. Initially, this method only produces amino acids in L form to avoid further purification steps. The advantage of this method over the other method is that it is preferred in the production of L isoform amino acids so that it does not require extra purification. In addition, it provides the conditions that prevent the degradation of amino acids and is an inexpensive method (Ugimoto 2010). However, the downside of aerobic fermentation is that the sterility and high energy consumption required for oxygen transfer is difficult, and larger reactors must be used compared to other amino acids. This means higher costs in processing (Ivanov et al. 2014).

If we have to choose the best amino acid production method needed to consider several criteria such as product market size and revenues, available technologies and capital investments, raw materials, costs, and environmental impact. To sum up, due to many advantages that were reported, fermentation has been the commonly used technique at the industrial scale (Ikeda 2003).

3.6 AMINO ACIDS AND SIDE CHAIN PROTECTION GROUPS

Protein synthesis is important to study the structure of natural proteins, to investigate how protein structure and function are formed by the amino acid sequence and to understand the synthesis mechanism. However, it is not possible to form this structure by simply mixing the amino acids together. Protective groups are often necessary to avoid undesirable reactions. Chemical peptide synthesis generally starts from the carboxyl groups (C-terminus) of the peptide and proceeds towards the amino groups (N-terminus). This is the opposite direction of protein biosynthesis. The resulting bond is the amide (peptide) bond. It is necessary to use protective groups to control the C-N coupling reaction. Certain peptide bonds can be formed by protecting the amine group (N-terminus) of one amino acid and the carboxyl group (C-terminus) of the another. The protection of side chain functional groups is necessary to prevent undesirable reactions and to form univocally the goal peptide bond. The peptide bonds are also called *eupeptide* bonds to distinguish them from *isopeptide* bonds formed by participating side chain functional groups as is the case of glutathione for example. Peptide synthesis is a multistage process comprising a number of chemical processes. To provide an overall high yield of the goal peptide it is drastically important to have yields close to quantitative at all stages of this complex process. It is also extremely important to avoid even a low racemization at all stages otherwise the obtained peptide will lack the desirable physiological activity (Isidro-Llobet, Álvarez, and Albericio 2009).

Today peptides are being synthesized using automated solid-phase peptide synthesis (see Chapter 4). The process is substantially promoted by microwave assistance (Figure 3.4).

3.7 PROTECTING GROUPS OF NA-AMINO AND C-TERMINAL CARBOXYL FUNCTIONS IN SOLID-PHASE PEPTIDE SYNTHESIS

N-protecting: There are several groups to protect the N-terminus of amino acids. They are called temporary protecting groups since in case of need they can easily be removed. The most popular and commonly used N-protecting groups are tert-butoxycarbonyl (Boc) and benzoyloxycarbonyl (Cbz or also labeled Z) (Lin et al. 2000; De Marco et al. 2013; Isidro-Llobet, Álvarez, and Albericio 2009).

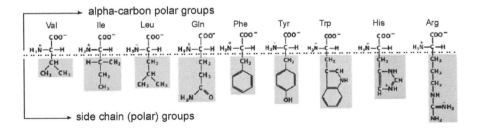

FIGURE 3.4 Example of amino acids structures. Side chain groups and alpha carbon groups (Yarus 2017).

C-protecting: These groups are responsible for protecting the C-terminus of amino acids. The classical solution method is preferred in peptide synthesis. The stability ratio in the solid phase peptide synthesis is low. The most commonly used carboxyl-protecting groups are methylethyl and benzylesters (De Marco et al. 2013; Isidro-Llobet, Álvarez, and Albericio 2009).

3.8 FUTURE PERSPECTIVE

Amino acids have very important properties in drug development as they play important roles in many physiological and pathological processes. Amino acids are important components of peptides, and any research done with amino acids actually provides a basis for peptides. Recently, many studies have been carried out with amino acids, and when looking at these studies it has been aimed at increasing the efficiency of amino acids and finding bioactive amino acids from different components, especially plants. It has also been investigated in some studies for example in in vivo, in vitro, and clinical trials that these amino acids can be used in various diseases and have beneficial properties. From this point of view, the fact that amino acids are useful in various diseases, especially peptides, provides hints that they can be useful and can guide scientists in the production of new drugs, vaccines, and carrier biopolymers. In addition, amino acids can be used to overcome the disadvantages created by the poor pharmacokinetic properties of such low molecular weight agents. For example, amino acid-based self-assembly macromolecules can act by improving the pharmacokinetic profile and accumulation of specific molecules at target sites to enhance the therapeutic effect (Vong, Trinh, and Nagasaki 2020).

In a study, amino acid composition of select cowpea varieties grown in Ethiopia was investigated and as a result, globulins and albumins were a major storage. In addition, it has been suggested that it can be used as a food additive due to its high and beneficial amino acid content (Teka et al. 2020). Another study found antioxidant properties of sulfur-containing amino acids such as methionine, cysteine, and taurine, and large-scale antioxidant properties. Based on this study, sulfur-containing amino acids have the potential to cope with many diseases, especially as it can have a beneficial pharmacological effect against cell damage caused by oxidative stress (Kim et al. 2018). Further information is given in Chapter 13 about antioxidant amino acids and peptides. Further, according to a study, it was found that amino acid properties can be affected by different applications. Amino acid loss occurs especially in the digestion phase. Therefore, important amino acid losses may occur, and new techniques are required to prevent this (Reis et al. 2020). Many studies have examined the concentration of amino acids in relation to diseases. If we examine a few of these studies: Parkinson's disease (Picca et al. 2019), kidney disease (Ikeda 2020), maple syrup urine disease (Kaur et al. 2019), acute and chronic physical disease (Katharina et al. 2019), Kawasaki disease (Shimizu et al. 2019), etc. In addition, in vitro studies are also available to determine substances that can be used for the treatment of different diseases related to the benefits of amino acids. One study investigated nanoparticles with chitosan to design a drug delivery system for controlled release of 5-amino salicylic acid (5-ASA) for inflammatory bowel disease. As a result, this amino acid has been found to be beneficial against this disease (Markam and Bajpai 2020). The distribution of positively charged amino acid residues in the antimicrobial peptide epinecidin-1 was examined and the cytotoxicity of glioblastoma and its underlying mechanisms were investigated in vitro (Su et al. 2020).

In other words, based on these studies, it has been revealed that amino acids can be a potential therapeutic agent in many diseases. The production of amino acids and the methods they are exposed to are very important. Since it is one of the important ways to find cures for diseases, it is important it is investigated thoroughly by researchers to find treatment methods.

REFERENCES

Bardaweel, S.K. 2014. "d-Amino acids: Prospects for new therapeutic agents." *Journal of Medical and Bioengineering* 3(3): 195–198.

Bischoff, R. and H. Schlüter. 2012. "Amino acids: Chemistry, functionality and selected non-enzymatic post-translational modifications." *Journal of Proteomics* 75(8): 2275–2296. doi:10.1016/j.jprot.2012.01.041.

Braconnot, H.M. 1820. "Sur la conversion des matières animales en nouvelles substances par le moyen de l'acide sulfurique." *Annales de chimie et de physique* 2(13): 113–125.

Brosnan, J.T. and M.E. Brosnan. 2006. "The sulfur-containing amino acids: An overview." *The Journal of Nutrition* 136(6 Suppl): 1636S–1640S. doi:10.1093/jn/136.6.1636S.

Cesari, M., G.P. Rossi, D. Sticchi, and A.C. Pessina. 2005. "Is homocysteine important as risk factor for coronary heart disease?" *Nutrition Metabolism Cardiovascular Disease* 15(2): 140–147.

De Marco, R., M. Spinella, A. De Lorenzo, A. Leggio, and A. Liguori. 2013. "C → N and N → C solution phase peptide synthesis using the N-acyl 4-nitrobenzenesulfonamide as protection of the carboxylic function." *Organic & Biomolecular Chemistry* 11(23): 3786. doi:10.1039/c3ob40169c.

Gröger, H. 2003. "Catalytic enantioselective Strecker reaction analogous syntheses." *Chemical Reviews* 103: 2795–2827.

Harada, K. 1963. "Asymmetric synthesis of α-Amino-acids by the strecker synthesis." *Nature* 200(4912): 1201–1201. doi:10.1038/2001201a0.

Hauer, B., M. Breuer, K. Ditrich, T. Habicher, M. Keßeler, R. Stürmer, and T. Zelinski. 2004. "Industrial methods for the production of optically active intermediates." *Angewandte Chemie - International Edition* 43: 788–824.

Hsiao, H.Y., J.F. Walter, D.M. Anderson, and B.K. Hamilton. 1988. "Enzymatic production of amino acids." *Biotechnology and Genetic Engineering Reviews* 6: 179–220.

Ikeda, M. 2003. "Amino acids production processes." In Scheper, T., Faurie, R., and Thommel, J. (Eds), *Microbial Production of L-Amino Acids*. Berlin: Springer, 1–35.

Ikeda, H. 2020. "Cross-correlation of plasma concentrations of branched-chain amino acids: A comparison between healthy participants and patients with chronic kidney disease." *Clinical Nutrition ESPEN*. 38: 201–210. doi:10.1016/j.clnesp.2020.04.014.

Isidro-Llobet, A., M. Álvarez, and F. Albericio. 2009. "Amino acid-protecting groups." *Chemical Reviews* 109(6): 2455–2504. doi:10.1021/cr800323s.

Ivanov, K., A. Stoimenova, D. Obreshkova, and L. Saso. 2014. "Biotechnology in the production of pharmaceutical industry ingredients: Amino acids." *Biotechnology and Biotechnological Equipmet* 27(2): 3620–3626.

Katharina, H., F. Dietmar, B. Michael, and S.-U. Barbara. 2019. "How acute and chronic physical disease may influence mental health - an analysis of neurotransmitter precursor amino acid levels." *Psychoneuroendocrinology* 106(2019): 95–101. doi:10.1016/j.psyneuen.2019.03.028.

Kaur, J., L. Nagy, B. Wan, H. Saleh, A. Schulze, J. Raiman, and M. Inbar-Feigenberg. 2019. "The utility of dried blood spot monitoring of branched-chain amino acids for maple syrup urine disease: A retrospective chart review study." *Clinica Chimica Acta.* 500: 195–201. doi:10.1016/j.cca.2019.10.016.

Kim, J.-H., H.-J. Jang, W.-Y. Cho, S.-J. Yeon, and C.-H. Lee. 2018. "In vitro antioxidant actions of sulfur-containing amino acids." *Arabian Journal of Chemistry.* 13(1): 1678–1684. doi:10.1016/j.arabjc.2017.12.036.

Kimura, T., Y. Noguchi, N. Shikata, and M. Takahashi. 2009. "Plasma amino acid analysis for diagnosis and amino acid-based metabolic networks." *Current Opinion in Clinical Nutrition and Metabolic Care* 12(1): 49–53. doi:10.1097/mco.0b013e3283169242.

Lin, L.S., T. Lanza, deS.E. Laszlo, Q. Truong, T. Kamenecka, and W.K. Hagmann. 2000. "Deprotection of N-tert-butoxycarbonyl (Boc) groups in the presence of tert-butyl esters." *Tetrahedron Letters* 41(36): 7013–7016. doi:10.1016/s0040-4039(00)01203-x.

Lütke-Eversloh, T., C.N.S. Santos, and G. Stephanopoulos. 2007. "Perspectives of biotechnological production of L-tyrosine and its applications. *Appl Microbiol Biotechnol* 77: 751–762. https://doi.org/10.1007/s00253-007-1243-y.

Markam, R. and A.K. Bajpai. 2020. "Functionalization of ginger derived nanoparticles with chitosan to design drug delivery system for controlled release of 5-amino salicylic acid (5-ASA) in treatment of inflammatory bowel diseases: An in vitro study." *Reactive and Functional Polymers* 149(2020): 104520. doi:10.1016/j.reactfunctpolym.2020.104520.

Meletis, C.D., and J.E. Barker. 2005. "Therapeutic uses of amino acids." *Alternative and Complementary Therapies* 11: 24–28.

Miyagi, Y., M. Higashiyama, A. Gochi, M. Akaike, T. Ishikawa, T. Miura, N. Saruki, et al. 2011. "Plasma free amino acid profiling of five types of cancer patients and its application for early detection." *PLoS ONE* 6(9): e24143. doi:10.1371/journal.pone.0024143.

Moran-Palacio, E.F., O. Tortoledo-Ortiz, G.A. Yañez-Farias, L.A. Zamora-Álvarez, N.A. Stephens-Camacho, J.G. Soñanez-Organis, L.M. Ochoa-López, and J.A. Rosas-Rodríguez. 2014. "Determination of amino acids in medicinal plants from Southern Sonora, Mexico." *Tropical Journal of Pharmaceutical Research* 13(4): 601–606.

Paul, B.D., J.I. Sbodio, R. Xu, M.S. Vandiver, J.Y. Cha, A.M. Snowman, and H.S. Solomon. 2014. "Cystathionine γ-lyase deficiency mediates neurodegeneration in Huntington's disease." *Nature* 509: 96–100.

Picca, A., R. Calvani, G. Lani, F. Marini, A. Biancolillo, J. Gervasoni, and S. Persichilli, et al. 2019. "Circulating amino acid signature in older people with Parkinson's disease: A metabolic complement to the EXosomes in PArkiNson Disease (EXPAND) study." *Experimental Gerontology* 128: 110766. doi:10.1016/j.exger.2019.110766.

Pollegioni, L. and S. Servi. 2003. *Unnatural amino acids methods and protocols.* New York, NY: Humana Press.

Ramakrishnan, V., A. E. Ghaly, M.S. Brooks, S. M. Budge. 2013. "Enzymatic extraction of amino acids from fish waste for possible use as a substrate for production of jadomycin." *Enzyme Engineering* 2(2):112.

Raven, P.H., G.B. Johnson, K.A. Mason, J.B. Losos, and S.R. Singer. 2014. "Proteins: Molecules with diverse structures and functions." In *Biology* (10th ed., AP ed.). New York: McGraw-Hill, 44–53.

Reece, J.B., L.A. Urry, M.L. Cain, S.A. Wasserman, P.V. Minorsky, and R.B. Jackson. 2011. "Proteins include a diversity of structures, resulting in a wide range of functions." In *Campbell biology* (10th ed.). San Francisco, CA: Pearson, 75–84.

Reis, G.C.L., B.M. Dala-Paula, O.L. Tavano, L.R. Guidi, H.T. Godoy, A. Beatriz, and M. Gloria. 2020. "In vitro digestion of spermidine and amino acids in fresh and processed agaricus bisporus mushroom." *Food Research International* 137: 109616. doi:10.1016/j. foodres.2020.109616.

Renneberg, R. 2008. "High grade cysteine no longer has to be extracted from hair." In Demain, A.L. (Ed.), *Biotechnology for beginners*. Amsterdam: Elsevier. 106.

Rose, W.C., R.H. McCoy, C.E. Meyer, H.E. Carter, M.Womack, and E.T. Mertz. 1935. "Isolation of the 'unknown essential' present in proteins." *The Journal of Biological Chemistry* 109: LXXVII.

Sereewatthanawut, I., S. Prapintip, K. Watchiraruji, M. Goto, M. Sasaki, and A. Shotipruk. 2008. "Extraction of protein and amino acids from deoiled rice bran by subcritical water hydrolysis." *Bioresources Technology* 99: 555–561.

Shimizu, C., J. Kim, H. Eleftherohorinou, V.J. Wright, L.T. Hoang, A.H.Tremoulet, A. Franco, et al. 2019. "HLA-C variants associated with amino acid substitutions in the peptide binding groove influence susceptibility to Kawasaki disease." *Human Immunology* 80(2019): 731–738. doi:10.1016/j.humimm.2019.04.020.

Srinivasan, G.C.M. and J.A. Krzycki. 2002. "Pyrrolysine encoded by UAG in Archaea: Charging of a UAG-decoding specialized tRNA." *Science* 296: 1459–1462.

Strecker, A. 1850. "Ueber die künstliche Bildung der Milchsäure und einen neuen, dem Glycocoll homologen Körper." *Justus Liebigs Annalen Chemie* 75: 27–45. doi:10.1002/ jlac.18500750103.

Su, B.-C., T.-H. Wu, C.-H. Hsu, and J.-Y. Chan. 2020. "Distribution of positively charged amino acid residues in antimicrobial peptide epinecidin-1 is crucial for in vitro glioblastoma cytotoxicity and its underlying mechanisms." *Chemico-Biological Interactions* 315: 108904. doi:10.1016/j.cbi.2019.108904.

Tamanna, N. and N. Mahmood. 2014. "Emerging roles of branched-chain amino acid supplementation in human diseases." *International Scholarly Research Notices* 2014: 1–8.

Teka, T.A., N. Retta, G. Bultosa, H. Admassu, and T. Astatkie. 2020. "Protein fractions, in vitro protein digestibility and amino acid composition of select cowpea varieties grown in Ethiopia." *Food Bioscience* 36: 100634. doi:10.1016/j.fbio.2020.100634.

Ugimoto, M.S. 2010. "Amino acids, production processes." In Flickinger, M.C. (Ed.), *Encyclopedia of Bioprocess Technology*. New Jersey: John Wiley & Sons. 1–11.

Vauquelin, L.N. and P.J. Robique. 1806. "The discovery of a new plant principle in *Asparagus sativus*." *Annales de Chimie* 57: 88–93.

Vollmer, W., D. Blanot, and M.A. de Pedro. 2008. "Peptidoglycan structure and architecture." *FEMS Microbiological Reviews* 32: 149–167 https://www.sciencedirect.com/science/ar ticle/pii/S0032579119409723-bbib16.

Vong, L.B., N.T. Trinh, and Y. Nagasaki. 2020. "Design of amino acid-based self-assembled nano-drugs for therapeutic applications." *Journal of Controlled Release* 326: 140–149. doi:10.1016/j.jconrel.2020.06.009.

Weinert, D.J. 2009. "Nutrition and muscle protein synthesis: A descriptive review." *The Journal of Canadian. Chiropractic Association* 53: 186–193.

Wu, G. 2009. "Amino acids: Metabolism, functions, and nutrition." *Amino Acids* 37: 1. doi:10.1007/s00726-009-0269-0.

Wu, G., F.W. Bazer, T.A. Cudd, C.J. Meininger, and T.E. Spencer. 2004. "Maternal nutrition and fetal development." *The Journal of Nutrition* 134: 2169–2172.

Wu, G., Y. Fang, S. Yang, J.R. Lupton, and N.D. Turner. 2004. "Glutathione metabolism and its implications for health." *The Journal of Nutrition* 134: 489–492.

Yamane, H., S. Tomonaga, R. Suenaga, D.M. Denbow, and M. Furuse. 2007. "Intracerebroventricular injection of glutathione and its derivative induces sedative and hypnotic effects under acute stress in neonatal chicks." *Neuroscience Letter* 418: 87–91.

Yang, H. and D.S Kim. 2015. "Peptide immunotherapy in vaccine development." *Advances in Protein Chemistry and Structural Biology* 1–14. doi:10.1016/bs.apcsb.2015.03.001.

Yarus, M. 2017. "The genetic code and RNA-amino acid affinities." *Life* 7(2): 13. doi:10.3390/life7020013.

Zhang, J., S. Zhang, X. Yang, L. Qiu, B. Gao, and J. Chen. 2016. "Reactive extraction of amino acids mixture in hydrolysate from cottonseed meal with di (2-ethylhexyl) phosphoric acid." *Journal of Chememical Technology and Biotechnology* 91: 483–489.

Zhao, G., G. Gong, P. Wang, and L. Wang. 2014. "Enzymatic synthesis of L-aspartic acid by Escherichia coli cultured with a cost-effective corn plasm medium." *Annal of Microbiology* 64: 1615–1621.

Zinoni, F., A. Birkmann, T.C. Stadtman, and A. Bock. 1986. "Nucleotide sequence and expression of the selenocysteine-containing polypeptide of formate dehydrogenase (formate-hydrogen-lyase-linked) from *Escherichia coli*." *Proceedings of the National Academy of Sciences U S A* 83: 4650–4654.

Zuend, S.J., M.P. Coughlin, M.P. Lalonde, and E.N. Jacobsen. 2009. "Scalable catalytic asymmetric Strecker syntheses of unnatural alpha-amino acids." *Nature* 461(7266): 968–970.

4 Peptide Synthesis and Characterization Stages

Mustafa Kemal Gümüş

CONTENTS

4.1 HOW PEPTIDE SYNTHESIS IS PERFORMED

Peptides are formed by binding the carboxyl group of an amino acid to the amino group of another amino acid molecule accompanied by the liberation of one molecule of water (H_2O). Peptides can be synthesized by two different methods – classical solution method and automated solid phase method (Amblard et al. 2005).

4.2 MATERIALS

1. Reaction vessel
2. Polytetrafluoroethylene (PTFE) stick (15 cm length, 0.6–0.8 cm diameter)
3. Rotor
4. Filtration flask
5. Porous frit
6. Lyophilizer
7. High pressure (performance) liquid chromatography (HPLC) equipped with reverse phase C_{18} column
8. pH indicating paper
9. Solvents [N,N-dimethylformamide (DMF), methanol (MeOH), dichloromethane (DCM)] in wash bottles
10. Diisopropylethylamide (DIPEA)
11. Piperidine solution in DMF (20:80)
12. Kaiser test solutions (ninhydrin, pyridine, phenol)
13. Fmoc-amino-acids with protected side chains
14. Trifluoroacetic acid (TFA)

15. Triisopropylsilane (TIS)
16. Tert-butyl methyl ether (MTBE)

4.3 SOLID PHASE PEPTIDE SYNTHESIS (SPPS)

Solid phase peptide synthesis from the C-terminus to N-terminus consists of five steps:

1. The first amino acid in which all functional groups except the C-terminus are protected is attached (immobilized) to the solid support via free carboxyl group (C-terminus) by interaction with the resin. The resin in most cases represents a cross-linked polystyrene that carries a chloromethyl functional group (Merrifield resin). Hence, the amino acid is attached through the link similar to a benzyl ester. After the attachment of the amino acid the resin is washed.
2. The N^α-amino group protection is removed, and the resin is washed again.
3. A solution of the protected second amino acid with activated C^α-terminus is added to the resin to form the peptide bond by interaction with the free N-terminus of the first amino acid immobilized to the solid support via C-terminus.
4. The protecting group of the last attached amino acid is removed and its N-terminus is released for subsequent coupling reaction.
5. After the peptide of the desired length is built up it is separated from the solid support (Figure 4.1).

The second, third, and fourth steps are performed repeatedly during each amino acid addition. If the fourth step is not carried out, the amine end (N-terminus) of the peptide sequence remains inactive until the protecting group is removed. In case of need, the protecting group can be removed and the N-terminus can be protected by acetylation.

There are two primary types of solid-phase peptide synthesis: (1) Fmoc (base-labile Na ensuring gathering) and (2) t-Boc (corrosive labile Na securing gathering). Every strategy incorporates various steps and amino corrosive side-chain assurance followed by division/deprotection steps. Contrasted with the t-Boc concoction technique, the Fmoc substance strategy can create peptides of a higher caliber and better yield. Debasements in t-Boc peptide science are mostly because of cleavage issues, drying out, and t-cleavage. After division from the resin, the peptide is purified and filtered by reverse phase HPLC utilizing segments, for example, C-18, C-8, and C-4. Moreover, HPLC and MS are used to guarantee the virtue of the peptide (Fields 2002; Amblard et al. 2005; Friligou et al. 2011) (Figure 4.2).

Nanodiamonds (NDs) have naturally occurring nitrogen-space (N-V) centers with a diameter of approximately 5 nm, or nitrogen impurities that can form complexes in the nuclei of ND particles such as peptides and amines. These particles can also be surface modified due to their small size and have little observed toxicity in vitro and in vivo (Knapinska et al. 2015; Chauhan, Jain, and Nagaich 2019).

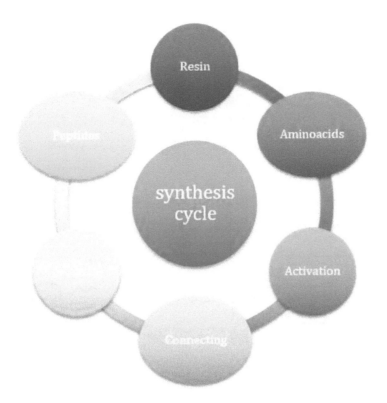

FIGURE 4.1   Schematic illustration of the cycle of peptide synthesis.

In a study with these particles, efficient solid phase conjugation of NDs to peptides and methods of characterization of ND-peptide conjugates are described. As a result, the following conclusions have been reached. The use of NDs in peptides derived from collagen has been found to support or even increase cell adhesion and viability activities of the conjugated sequence. It has been found that incorporation of NDs into peptides and proteins could potentially increase the in vivo activities of the biomolecule to which it is bound. In addition, the ND peptide was used in the continuation of the solid phase peptide synthesis without purification. The HPLC-based purification method was not found suitable for NDs. The use of convergent solid-phase synthesis of protected peptide fragments, while thus setting a limit against binding of NDs to longer peptides, allows highly pure resin-bound peptides to be obtained, with which NDs can be selectively conjugated (Knapinska et al. 2015).

In another study, a peptide epitope was synthesized for Q fever by the microwave-assisted solid-phase peptide synthesis method, purified and characterized by HPLC and mass spectrometry (LC-MS). This study was performed by the method summarized below:

For the solid-phase synthesis of a synthetic peptide (NH2 - Ser - Leu - Thr - Trp - His - Lys - His - Glu - Leu - His - Arg - Lys - COOH), which has protective properties

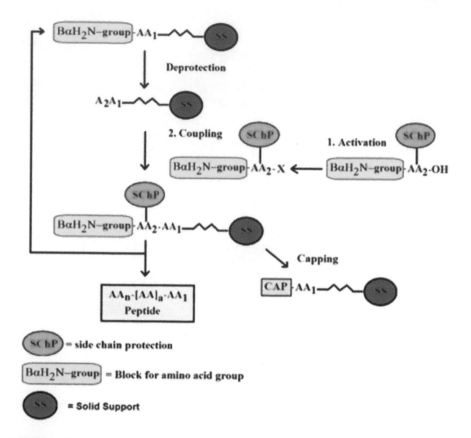

FIGURE 4.2 The stage of solid phase peptide synthesis. (Image by "Biosynjessica" is licensed under / CC BY-SA from Wikimedia Commons.)

against Q fever disease, Wang resin loaded with serine amino acid, amino acids, and binding reagents is required. Furthermore, DMF, DCM, TFA, HOBt, HCTU, DIEA, NMP, and Fmoc-protected amino acids are also necessary substances for this solid-phase synthesis. 2- (6-chloro-1H-benzotriazole-1-yl) -1,1,3,3-tetramethyluronium hexafluorophosphate/1-hydroxybenzotriazole (HCTU/HOBt) was used as activator in DMF by microwave-assisted SPPS method for antigenic peptide synthesis to be synthesized (activator) and N,N-diisopropylethylamine/N-methyl – 2-pyrrolidinone (DIEA/NMP) activator base, piperidine was used as deprotector (DEP). Wang resin loaded with serine amino acid, was inflated by waiting in DMF for three hours before synthesis started. Another important step is the separation of the peptide from the resin (breaking, cleavage). This is the process of separating the peptide chain growing on the polymeric resin from this resin. The peptide cleavage cocktail [(TFA/thioanisole/EDT/water), (87.5/5/2.5/5 mL, (v/v)] obtained from the device depending on the resin and peptides shaken slow with a shaker at room temperature with the cleavage cocktail. Also, cold ether (–20°C) was used for precipitation and separation of the peptide. The centrifuged peptide precipitation procedure was repeated, washing the final precipitate, and dried. The dried peptide was stored at −18°C. Finally, the peptide synthesized in the purification and characterization phase of the peptide was determined by HPLC (Shimadzu). By using a Shimadzu PRC-ODS C-18 column 30 × 2.1 column it was analyzed with the aid of UV detector, and purification was carried out. A LC-MS system (Shimadzu LC-MS 2010 EV) Electro Spray Ionizer (ESI) probe and Teknokroma Tracer Exel 120 ODS-A, C-18 colon (20 cm length and 0.21 cm inlet diameter) and HPLC (4.6 × 250 mm, Venusil MP A gradient of 220 nm acetonitrile/water in a C18-5 column was used for characterization of the peptide. The isoelectric point of the peptide synthesized using Innovagen's peptide was calculated: pH 10.66; Also, its net load at pH 7 was calculated as 2.3. In addition, it has been found to be well soluble in water and will be used in future vaccine prototype development studies; this is considered to be advantageous in conjugation with the polymer (Karahan 2017).

4.4   HOW TO USE MICROWAVES IN PEPTIDE SYNTHESIS

In the 1940s microwave irradiation was first used in radars during World War II, and in subsequent years was used in ovens for heating. Dr Percy Lebaron Spencer found microwaves' heating capacity by chance during laboratory tests. He noticed that the candy in his pocket warmed up as a result of exposure to microwave energy. After that invention, home-type microwave ovens were developed. Later, microwave devices for research purposes were constructed. The main reasons for the slow development of microwave technology were the lack of understanding of the principles of dielectric heating, safety, and influence on health. One of the important properties of microwaves is the fact that they do not contain enough energy to chemically change substances through ionization; thus, being an example of non-ionizing radiation (Tang 2015).

The first laboratory microwave device was used for moisture determination in 1978 and then used in chemical analysis such as luminescence powdering, sample preparation, and extraction (Hanson and Martin 1978). Afterwards, various devices were constructed for microwave laboratory applications. The microwave spectrum consists of electric and magnetic field waves. In the classification of these electromagnetic waves, certain frequency groups form specific microwave bands.

It can be said that the use of microwave technology in solid-phase peptide synthesis has been one of the most significant leaps of peptide science in the last decade. It has been determined by research on the use of microwave-assisted peptide synthesis for the easier and faster synthesis of peptides that are difficult to synthesize. Microwave rays fundamentally diminish the union season of these peptides and improve the nature of the created peptide.

In addition, routine methods have been developed and different methods have emerged to minimize the potential of side reactions, including optically inactivating cysteine and histidine residues during coupling and aspartimide formation during 9-fluorenylmethoxycarbonyl (Fmoc) deprotection. During the synthesis of peptides, there are several polar and ionic species that can be rapidly heated by microwave energy. The magnetron initiates the operation of the microwave – the magnetron begins to produce high-energy radio waves with the power that the microwave receives from the electrical source. The surfaces used in this device are made of reflective materials to target the focal point and have focusing and reflective properties. Thus, the movement of ions or polar molecules is provided by electrical and magnetic emission. A positive increase in this temperature difference can help the molecules break down due to inter-chain and chain-to-chain linkage, and helps provide easier access to the growing end of the chain, so temperature can be said to be of great importance. Therefore, because the temperature and other advantages that have been discussed the microwave irradiation can provide access to peptides previously unattainable with conventional techniques (Vanier 2013; Qvit 2014; Palasek, Cox, and Collins 2007; Petrou and Sarigiannis 2018). Moreover, the Fmoc peptide synthesis is advantageous in many ways. Fmoc building blocks are cost-effective and facilitate synthetic access to a wide range of peptide derivatives. Recently, peptide studies have increased, and the peptides found have increased thanks to this technology (Behrendt, White, and Offer 2016).

In particular, vaccination studies are carried out and microwave-assisted synthesis has recently been used in many studies, for example, in a study with self-assembling peptides. Microwave-assisted synthesis is an alternative to manual SPPS of these peptides (Brent Chesson, Alvarado, and Rudra 2018). The use of microwave-assisted solid-phase peptide synthesis vaccine candidates was found against hookworm infection (Fuaad, Skwarczynski, and Toth 2016). Also, in another study microwave-assisted solid-phase synthesis was used in the synthesis of colistin and tigecycline resistant *E. coli* bacteria and *K. pneumoniae* active antisense acpP peptide nucleic acid-peptide conjugates synthesis (Hansen et al. 2019). As a result, microwave peptide synthesis has become a preferred method (Table 4.1).

TABLE 4.1

Steps at Peptide Synthesis. Comparison Microwave Assisted and Conventional Method and Max Watt, Temperature and Time to Be Applied for Each Process

Process	Microwave Assisted			Conventional			References
	Max Watt	Heat	Time	Max Watt	Heat	Time	
Deprotection	35 W	75°C	3 min	0 W	75°C	30 min	Vanier 2013; Qvit 2014; Palasek, Cox, and Collins 2007
Coupling	20 W	75°C	5 min	0 W	75°C	2 h	Vanier 2013; Qvit 2014; Palasek, Cox, and Collins 2007
Cleavage	15 W	38°C	30 min	0 W	–	3 h	Vanier 2013; Qvit 2014; Palasek, Cox, and Collins 2007

4.5   ADVANTAGES OF USING MICROWAVES

There are known significant advantages of using microwaves in solid phase peptide synthesis, such as the removal of the Fmoc group (deprotection), coupling, and cleavage reactions take place within minutes. In normal SPPS all these reactions take a long time compared to microwave-assisted SPPS. This makes the microwave-supported solid-phase peptide synthesis more advantageous compared to the conventional method (Palasek, Cox, and Collins 2007; Sabatino and Papini 2008). Also, there are more advantages to using microwaves in solid-phase peptide synthesis, for example, the production of peptides executes with higher yields. Reactions are faster under elevated temperatures. Microwave energy is a rapid energy source for reactions. Products with high purities can be obtained. The efficient synthesis of difficult sequences can be obtained in an easy way. Racemization of histidine and cysteine can be obtained, to ensure a uniform and targeted heating of the samples. It requires less solvent and reagent. It enables the synthesis speed and purity of the peptides to be increased. It provides a low risk of ratification in slow-coupling reactions. It provides more repeatability. To obtain greener solvents, solvents should be obtained nature friendly and this provides more reliable transactions (Sabatino and Papini 2008; Petrou and Sarigiannis 2018). In a study, automatic microwave technology was used in the production of antimicrobial peptides, which are one of the unique therapeutic tools to treat various destructive diseases that affect millions of lives. The reason why this technology is preferred is that traditional peptide synthesis requires a longer time and therefore they have emphasized that it can be considered as an alternative tool that offers advantages such as automatic microwave technology, less reaction time, and higher efficiency, and also overcomes the difficulties in preparing long and difficult peptides. In overcoming these obstacles they also used a synthetic protocol that included remedial procedures (Ramesh et al. 2017).

Consequently, recently microwave synthesis has been used for its advantages on the synthesis of peptides and produced vaccine candidates. The efficiency of the production is important so in synthetic peptide production microwaves are important. In this chapter general peptide synthesis and the importance of microwave technology have also been explained.

REFERENCES

Amblard, M., J.A. Fehrentz, J. Martinez, and G. Subra. 2005. "Fundamentals of modern peptide synthesis." In Howl, J. (Ed.), *Peptide synthesis and applications. Methods in molecular biology*™, vol 298. Humana Press.

Behrendt, R., P. White, and J. Offer. 2016. "Advances in Fmoc solid-phase peptide synthesis." *Journal of Peptide Science* 22(1): 4–27. doi:10.1002/psc.2836.

Brent Chesson, C., R.E. Alvarado, and J.S. Rudra. 2018. "Microwave-assisted synthesis and immunological evaluation of self-assembling peptide vaccines." *Methods in Molecular Biology (Clifton, N.J.)* 1777: 249–259. doi:10.1007/978-1-4939-7811-3_15.

Chauhan, S., N.G. Jain, and U. Nagaich. 2019. "Nanodiamonds with powerful ability for drug delivery and biomedical applications: Recent updates on in vivo study and patents." *Journal of Pharmaceutical Analysis* 10(1): 1–12. doi:10.1016 / j.jpha.2019.09.003.

Fields, G.B. 2002. "Introduction to peptide synthesis." *Current Protocols in Protein Science,* 18.1.1–18.1.9. doi:10.1002/0471140864.ps1801s26.

Friligou, I., E. Papadimitriou, D. Gatos, J. Matsoukas, and T. Tselios. 2011. "Microwave-assisted solid-phase peptide synthesis of the 60–110 domain of human pleiotrophin on 2-chlorotrityl resin." *Amino Acids* 40(5): 1431–1440. doi:10.1007/s00726-010-0753-6.

Fuaad, A.A., M. Skwarczynski, and I. Toth. 2016. "The use of microwave-assisted solid-phase peptide synthesis and click chemistry for the synthesis of vaccine candidates against hookworm infection." *Methods in Molecular Biology (Clifton, N.J.)* 1403: 639–653. doi:10.1007/978-1-4939-3387-7_36.

Hansen, A.M., G. Bonke, W. Hogendorf, F. Björkling, J. Nielsen, K.T. Kongstad, D. Zabicka et al. 2019. "Microwave-assisted solid-phase synthesis of antisense acpP peptide nucleic acid-peptide conjugates active against colistin- and tigecycline-resistant E. coli and K. pneumoniae." *European Journal of Medicinal Chemistry* 168: 134–145. doi:10.1016/j.ejmech.2019.02.024.

Hanson, C.W. and W.J. Martin. 1978. "Microwave oven for melting laboratory media." *Journal of Clinical Microbiology* 7: 401–402.

Karahan, M. 2017. "Antigenic Peptide Synthesis and Characterization of Q Fever Disease." *Afyon Kocatepe University Journal of Science and Engineering Sciences* 17(1): 312–317. doi: 10.5578/fmbd.53815.

Knapinska, A.M., D. Tokmina-Roszyk, S. Amar, M. Tokmina-Roszyk, V.N. Mochalin, Y. Gogotsi, P. Cosme, A.C. Terentis, and G.B. Fields. 2015. "Solid-phase synthesis, characterization, and cellular activities of collagen-model nanodiamond-peptide conjugates." *Biopolymers* 104(3): 186–195. doi:10.1002 / bip.22636.

Palasek, S.A., Z.J. Cox, and J.M. Collins. 2007. "Limiting racemization and aspartimide formation in microwave-enhanced Fmoc solid phase peptide synthesis." *Journal of Peptide Science* 13(3): 143–148. doi:10.1002/psc.804.

Petrou, C. and Y. Sarigiannis. 2018. "Peptide synthesis: Methods, trends, and challenges." In Koutsopoulos, S. (Ed.), *Peptide applications in biomedicine, biotechnology and bioengineering.* Elsevier.

Qvit, N. 2014. "Microwave-assisted synthesis of cyclic phosphopeptide on solid support." *Chemical Biology & Drug Desig* 85(3): 300–305. doi:10.1111/cbdd.12388.

Ramesh, S., B.G. de la Torre, F. Albericio, H.G. Kruger, and T. Govender. 2017. "Microwave-assisted synthesis of antimicrobial peptides." *Methods in Molecular Biology (Clifton, N.J.)* 1548: 51–59. doi:10.1007/978-1-4939-6737-7_4.

Sabatino, G. and A.M. Papini. 2008. "Advances in automatic, manual and microwave-assisted solid-phase peptide synthesis." *Current Opinion in Drug Discovery & Development* 11(6): 762–770.

Tang, J. 2015. "Unlocking potentials of microwaves for food safety and quality." *Journal of Food Science* 80(8): E1776–E1793. doi:10.1111/1750-3841.12959.

Vanier, G.S. 2013. "Microwave-assisted solid-phase peptide synthesis based on the Fmoc protecting group strategy (CEM)." *Peptide Synthesis and Applications,* 1047: 235–249. doi:10.1007/978-1-62703-544-6_17.

5 Conjugation of Polymers with Biomolecules and Polymeric Vaccine Development Technologies

S. Arda Ozturkcan

CONTENTS

5.1 POLYMERIC BIOMATERIALS

Polymers are high molecular weight compounds formed by binding a large number of molecules with chemical bonds on a regular basis. "Poly" is a Greek word and means "many" and "mer" means "repetitive". Polymers are found in many areas of our lives such as clothing, contact lenses, home-building materials, and health products. Polymers are of two groups; synthetic natural polymers include nylon, polyethylene, polyester, Teflon, and epoxy while natural polymers occur in nature and can be extracted. They are often water-based. Another group of naturally occurring polymers are silk, wool, DNA, cellulose, and proteins. For this book the synthetic biopolymer class is of importance. This class includes plastics, fibers, elastomers, resins, and adhesives (Li and Fu 2017).

Biomaterials have evolved from crude wooden prostheses dating back millennia (Huebsch and Mooney 2009). Today, biomaterials are used for cell delivery, drug delivery, microcapsules, and 3D-printing. Polymeric biomaterials are generally described as very useful materials offering many advantages in biomedical, medical, and biology fields (Mann et al. 2018). Biomaterials are of two types: synthetic and natural polymers; lipids, self-assembled nanostructures, and engineered artificial cells offer unique features. Biomaterials offer benefits: control over the loading and release kinetics of multiple immune cargoes, and protection from enzymatic degradation and extreme pH. Moreover, biomaterials can be conjugated with antibodies or receptor ligands to contribute the molecular-specific target to immune cells or membrane proteins/genes. This feature can be exploited to reduce systemic and local toxicity.

Biopolymers arise from the common use of biomaterials and polymers. Polymeric structures are used in solid, liquid, and solution cases. Some polymers can also be crystallized (Song et al. 2018).

5.2 CLASSIFICATION OF BIOPOLYMERS

It is very important to classify polymers to study. Classification according to purpose are given below (Sneha et al. 2016; Younes 2017; Hassan, Bai, and Dou 2019):

1. According to molecular weights (macromolecules, oligomers).
2. Natural and synthetic polymers.
3. Inorganic or organic polymers.
4. Their forms against heat.
5. Physical and chemical structure of the chain.
6. Straight, branched, cross-linked, crystal, and amorphous polymer (Figure 5.1).
7. Polymer synthesis techniques.
8. Chain structure (copolymer, homopolymer) (Figure 5.2).

If a polymer is formed by repeating a single monomer unit, it is called a homopolymer (Figure 5.3).

If the polymer molecule is formed by two different monomers it is called a copolymer and generally used for vaccine models.

Whether homopolymer or copolymer, chains can exist in only four different forms: sequential copolymer, block copolymer, random copolymer, and graft copolymer. These forms can be used for applications that require the development of nanoparticle formulations that are suitable for *in-vivo* applications; primarily, the formulated nanoparticles should be sufficiently small, chemically and biologically inert, and stable against aggregation under physiological conditions while producing the vaccine models (Sneha et al. 2016; Younes 2017; Hassan, Bai and Dou 2019).

5.3 ADJUVANTS: IMMUNOSTIMULANT PROPERTIES OF POLYMERS

Classical vaccines are prepared by inactivation of the virus or bacteria which are grown in cell cultures and physical mixing of an adjuvant (aluminum hydroxide or fat adjuvant). Recently, $Ca_3(PO_4)_2$ salt was used in the Pasteur Institute in France, but the vaccine could not be applied because it was slightly soluble. Classical vaccines

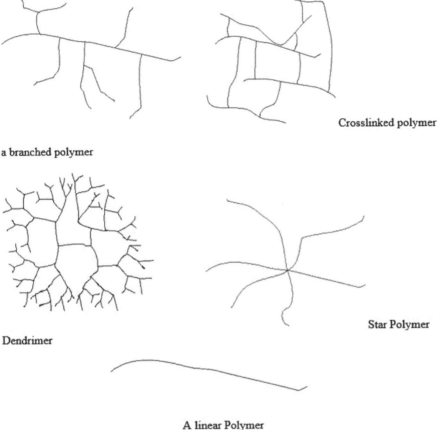

Crosslinked polymer

a branched polymer

Dendrimer

Star Polymer

A linear Polymer

FIGURE 5.1   Polymer types.

prepared with oil and aluminum adjuvants have disadvantages such as (Sivakumar et al. 2011):

1. Local reaction at injection sites.
2. Problem of allergies.
3. High toxicity.
4. T cell immune response.
5. Insufficient protection.
6. Low stability of the mixture (denaturation and temperature insensitivity).

For all these disadvantages polymers have immunostimulant properties (adjuvants). Synthetic polymers are used to increase drug activity and efficacy (Sivakumar et al. 2011). Polymeric carriers are also used for increasing drug activity and efficacy (see Chapter 6).

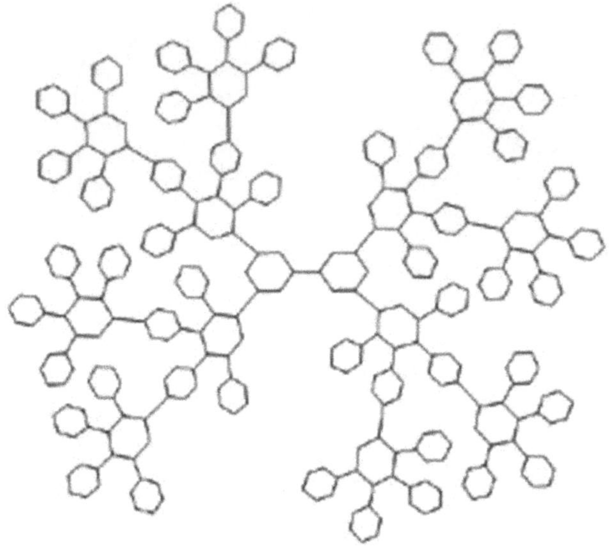

FIGURE 5.2 An example of dendrimer nanostar polymer. (Image by "Integral-formula" is licensed under a CC BY-SA, from Wikimedia, https://commons.wikimedia.org/wiki/File: Dendrimer_Nanostar.png.)

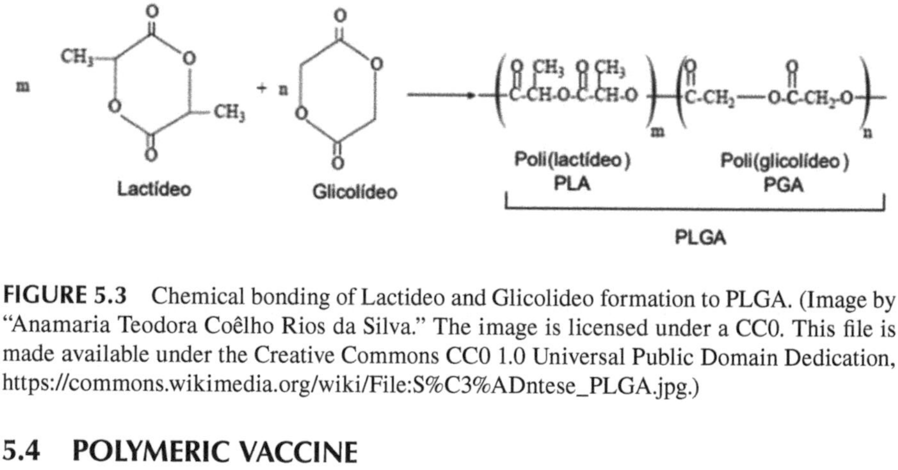

FIGURE 5.3 Chemical bonding of Lactideo and Glicolideo formation to PLGA. (Image by "Anamaria Teodora Coêlho Rios da Silva." The image is licensed under a CC0. This file is made available under the Creative Commons CC0 1.0 Universal Public Domain Dedication, https://commons.wikimedia.org/wiki/File:S%C3%ADntese_PLGA.jpg.)

5.4 POLYMERIC VACCINE

Polymers, lipids, scaffolds, microneedles, and other biomaterials are developing as technologies to improve the efficacy of vaccines against infectious disease and immunotherapies for cancer, autoimmunity, and transplantation. When the complexes or conjugates of different antigens are injected into a living organism, the number of antibodies is increased. This phenomenon has the logic of defending the body against the virus when conjugates containing the peptide antigens of the disease-causing virus are used. In short, this is called a polymeric vaccine (Sivakumar et al. 2011; Skwarczynski and Toth 2016; Bose et al. 2019).

Advantages of polymeric vaccines (Sivakumar et al. 2011; Skwarczynski and Toth 2016):

- Non toxic
- Cheap and technologically prepared by a method
- Easy production of synthetic linear epitopes (peptide and lipopolysaccharide)
- Safe (does not contain live virus)
- The knowledge of structure-function relationships

Recent studies conducted have shown an increase in importance of polymeric vaccines (Bose et al. 2019).

For example, in a systematic review, a total of 31 studies (until 2016) were accessed and reviewed to determine the potential of polymeric particles as future vaccine delivery systems/adjuvants for parenteral and non-parenteral immunization against tuberculosis. Based on the data obtained, it has been determined that poly (lactide-co-glycolide) (PLGA) and chitosan polymers are extensively used as tuberculosis vaccine shipping systems/adjuvants. The immunogenicity of tuberculosis vaccines is accelerated using herbal polymers as well as biodegradable and non-biodegradable artificial polymers. It has also been shown to improve humoral and mobile immune responses more effectively than the tuberculosis vaccine on its own (Khademi et al. 2018). One study suggested that the use of PDMAEMA: PβAE / DNA polyplexes (Pol) as the mediator of a pDNA vaccine encoding the hepatitis B surface antigen (HBsAg) to obtain an antigen-specific immune responses following DNA vaccination (not easy due to the difficulty of mediating efficient gene delivery); this Pol is designed in combination with a soluble (Glu) or β-glucan form particles (GP)s. These polyplexes were found to result in greater transfection activity than the positive control, indicating higher luciferase gene expression in the presence of GPs (COS-7 and RAW 264.7 cell lines), i.e. they showed a better effect. This has shown the potential of nanosystems to be further investigated as a platform for increasing *in-vitro* transfection capacity with GPs and as a DNA vaccination platform (Soares et al. 2019). There is also evidence that the polymers used as a result of this study can be used as components in different types of vaccines, as well as the use of additional vaccines.

5.4.1 Conjugation Methods in the Polymer Antigen Systems

The binding of the peptides to carrier proteins is required to form antibodies against peptides with poor immunogenic properties. Activation of carboxyl groups of molecules in aqueous solution with covalent bonding of amino groups to activated carboxyl groups form the basis for the conjugation reaction of a globular protein and synthetic peptides. The development of conjugation methods for the development of more effective biomedical preparations (enhancement of immunogenicity of antigens, development of effective vaccines and diagnostics, drug release, etc.) is one of the current studies. In recent years, it has been known that synthetic polymers are used for the prolongation of the drug substances as well as the prolongation of life.

These studies carried out in the international field cause the development of new kinds of biomedical preparations aimed at developing preparations (McCarthy et al. 2014).

In recent years, different techniques have been developed by our group for the synthesis of water-soluble conjugates of synthetic polymers with different antigens (organic molecules with protein, peptide and hapten) for the formation of polymer antigen conjugates' covalent bonding, electrostatic complex formation, and metal coordination via ion coordination.

Thionylchloride, carbodiimide, and irradiation in covalent bonding methods were used. Polycations (poly-4-vinylpyridine) polyanions (polyacrylic acid, poly N-isopropylacrylamide, poly-N-vinylpyrrolidone) and differently structured copolymers (copolymers of acrylic acid with N-isopropylacrylamide and N-vinylpyrrolidone) with proteins, steroid hormones (estradiol, anti-cancer betaine hapten), and the binding techniques of polypeptides of different polypeptides – hepatitis B and VP1 protein of foot-and-mouth disease virus have been developed (Anasir and Poh 2019; Mamabolo et al. 2020). Pentamers of VP1 proteins are known to be self-assembled into capsid-like particles and unable to specifically bind DNA. The surface loops of the protein interact with the sialic acid of ganglioside receptors. In another study with VP1 protein, the main structural protein of mouse polyomavirus (MPyV). In the cytoplasm, the interaction of VP1-bound microtubules and VP1, including the mitotic spindle, with the microtubules has resulted in a cell cycle block in the G2/M phase. In late phase of MPyV infection and in cells expressing VP1, microtubes have been found to be hyperacillated. Based on this, it has been shown how VP1 interacts with microtubes is that VP1 is a multifunctional protein that participates in the regulation of cell cycle progression in MPyV-infected cells (Horníková et al. 2017).

A variety of methods are attributed to the structure of the polymer and its antigen, without binding to the structure, selecting a more technological method, creating a more immunological structure and forming water-soluble conjugates (Mustafaev et al. 1998).

These methods are different and depend on the composition and components.

1. In the presence of thionylchloride, polycaboxylic anions were conjugated with steroid hormones containing hydroxyl functional groups and betulin. Immunoglobulin conjugates with immunogenic properties were synthesized in one step.
2. Polymer-peptide conjugation in micelle microreactors was performed. (Mustafaev et al. 1998).

After purifying bioconjugates for example, synthetic peptide, bovine serum albumin, and bovine serum albumin synthetic peptide physical mixtures and their characterization, binding, HPLC, GPC-VISCOTEK (gel permeation chromatography, UV, refractive index, viscosity and light scattering detectors with chromatographic system), fluorescence spectrophotometer and particle size, and zeta potential measurement are performed using the devices (see Chapter 7). Different results were obtained when the enzyme-linked immunosorbent assay test was compared to conjugation methods such as physical mixture and covalent bonding (Ciaurriz et al. 2017).

According to the result, covalent bonding gave the highest optic density (OD), under 405 nm (Das et al. 2010).

Bovine serum albumin (BSA) is the most widely used carrier protein for conjugation in antibody formation. Albumin belongs to the sub-class of proteins and about half of the total protein in the plasma is composed of albumin. This shows that albumin is the most stable and highest-resolution protein in plasma. BSA is widely used in experimental studies due to its advantages such as easy accessibility and solubility. Due to its various functional groups, it is also preferred as a carrier for various molecules, proteins, or haptens in conjugation reactions (Friedli 1996).

5.4.1.1 Physical Mixture

The importance of polyelectrolyte use in immunology:

- Analysis of functional properties of complexes that dissolve in water which is made with synthetic polyelectrolytes (PE) and biomacromolecules (protein, peptide, polysaccharide, etc.) is a remarkable immunological development.
- These types of synthetic polymers are used in important practice areas of medicine and biological-based sciences.

Polyelectrolytes have the following diverse effects: adjuvant effect – the collateral chemical materials that can go into the organism with antigens to increase immunity of the organism; prolongation – a drug's effective materials which are kept in the organism for a long time; drug release systems – slowly setting the drug free in the organism (Rudra et al. 2010; Powell, Andrianov, and Fusco 2015; Zhang, Bookstaver, and Jewell 2017; Yang et al. 2019) (Figure 5.4).

5.4.1.2 Covalent Bonding

In the covalent bonding method the peptide protein is the most preferred carbodiimide of direct cross-linking reagents used to form protein-protein conjugates. The carbodiimides, in which water-soluble and insoluble forms exist, mediate the binding

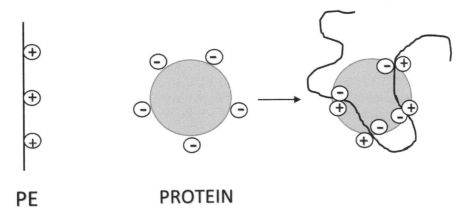

PE   PROTEIN

FIGURE 5.4 Polyelectrolytes and protein conjugation of illustration.

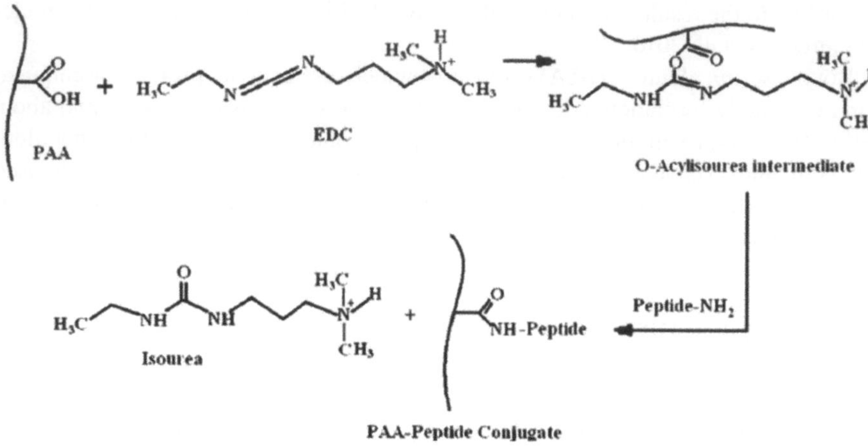

FIGURE 5.5 The activation carboxyl group of polymers, and coupling with the peptide to obtain a polymer-peptide conjugate with [1-Ethyl-3- (3-dimethylaminopropyl)] EDC method (Mansuroğlu and Mustafaeva 2012).

of the carboxylate and primary amines with the amide bond, and the phosphamide bond of the phosphate and amino groups. The water-soluble carbodiimides are more preferred in biochemical conjugates because many macromolecules of biological origin are soluble in aqueous buffer solutions. Carbodiimides which are insoluble in water are generally used in peptide synthesis and in bioconjugation reactions where the molecules must be dissolved only in organic solvents. Products and byproducts that occur in these reactions can only be dissolved in organic solvents and are not soluble in water (Hermanson 1996; McCarthy et al. 2014) (Figure 5.5).

5.4.1.2.1   1-Ethyl-3-3-(dimethylaminopropyl)carbodiimide

1-ethyl-3-(3-dimethylaminopropyl)carbodiimide hydrochloride, known as EDC or EDAC, is the most commonly used carbodiimide derivative in the conjugation of biological molecules. Its water-soluble property offers the ability to directly add the cross-linker in the reaction medium without having to dissolve it in an organic solvent. Urea derivatives, which are the byproduct of carbodiimide and the byproducts of the cross-linking reaction, are easily soluble by dialysis or gel filtration since they are water-soluble. The resulting conjugates can be purified (Sheehan, Cruickshank, and Boshart 1961; Sheehan 1965). Because EDC is not stable in the presence of water, the chemical should be stored at −20°C. When the chemical is used, it should be brought to room temperature and opened, which prevents the EDC from degrading early.

Stock solution in water can be prepared to add small amounts of EDC to the reaction medium, however this solution must be prepared very quickly and used in a short time so that the chemical does not lose its activity (Derman 2006).

During conjugation of the EDC peptides with proteins, there is also the polymerization of the molecules themselves. For example, since the peptide molecule contains both the carboxyl group and the amino group, the amino group of the peptide molecule can also attack the unstable intermediate formed by the carboxyl group activated by the carbodiimide. As a result, the peptide-protein conjugate as well as the formation of oligomers of the peptide may be involved. The optimum pH range for carbodiimides is 4.7–6. In addition, the carbodiimide reactions can be carried out with a yield of up to pH 7.5.

In conjugation reactions with EDC, water can be used as a solvent and the pH is fixed by addition of HCl to the medium. However, the use of a buffer as a solvent is much more suitable as the pH cannot be controlled during the reaction. A 0.1 M phosphate buffer is used for conjugation at neutral pHs, but amino or carboxylate-containing buffers for the reaction are not suitable (because they will interact with carbodiimide). During the activation of the carboxyl groups of proteins with EDC, some side reactions may occur. One of them is peptide polymerization. Others are protein polymerization. The reaction of EDC molecules with the exposed sulfhydryl ends in the peptide or protein structure, or the reaction of the tyrosine tip with EDC, resulting in phenolate ion formation (Derman 2006).

5.4.1.2.2 Laboratory Method in EDC

- The protein to be modified at a concentration of 10 mg/ml is dissolved in water or phosphate buffer. If more or less concentration of protein is to be studied, the concentrations of other reagents are calculated according to the ratio to be studied.
- The peptide to be conjugated is also dissolved in the same solvent. If not working with certain ratios, the small molecule is added to the reaction medium at a concentration of at least ten times the molar concentration of the protein. If possible, the required amount of peptide is added directly to the protein solution. If small amounts are to be studied, the stock solution of the peptide can be prepared, and the amount required for the reaction can be used. If the peptide solution is prepared separately, it is added to the protein solution.
- EDC is added to the medium to ten times the molar concentration of the protein in the solution. If low amounts of EDC are to be used, a stock solution of carbodiimide is prepared immediately after consumption and the required amount is added from this solution.
- Leave for two hours at room temperature.
- The resulting conjugate is purified by dialysis using 0.1 M sodium phosphate (0.15 M NaCl, pH = 7.4) using gel filtration or a suitable buffer [many conjugates].

If any turbidity is observed during conjugation, it can be removed by centrifugation at the end of the reaction. However, this turbidity during the formation of immunogenic conjugates with EDC does not cause problems in general because these immunogenic conjugates in the form of precipitates show higher immunogenicity than water-soluble ones (Hermanson 1996; Hermanson 2013). Various methods are used to purify the conjugates obtained using all these methods.

5.4.1.2.3 Water-soluble 1-Cyclobutyl-3-(2-morpholinethyl) Carbodiimide (CMC)

The water-soluble 1-cyclobutyl-3-(2-morpholinethyl) carbodiimide is used in the conjugation of carboxyl and amino group-containing molecules. The molecule is a positively charged morpholine group in its structure which provides water-soluble properties (Hermanson 1996; Hermanson 2013).

5.4.1.2.4 Dicyclohexylcarbodiimide (DCC)

Dicyclohexylcarbodiimide is a commonly used cross-linker, especially in organic synthesis applications. It has been used in peptide synthesis since 1955. DCC is soluble in organic solvents and can be used at temperatures of up to 80°C (Hermanson 1996; Hermanson 2013).

5.4.1.2.5 Diisopropylcarbodiimide (DIC)

Diisopropylcarbodiimide is another type of carbodiimide that is insoluble in water and present as liquid at room temperature (Hermanson 1996; Hermanson 2013).

5.4.1.2.6 N, N-Carbonyldiimidazole

N, N-Carbonyldiimidazole is another direct cross-linking reagent which carries two acylimidazole groups that are separated during conjugation in structure properties (Hermanson 1996; Hermanson 2013).

5.5 DEVELOPMENT OF METAL-CONTAINING FUNCTIONAL BIOPOLYMER SYSTEMS

The mechanism and functional properties of biomacromolecules (protein, polysaccharide, etc.) formed by synthetic polyelectrolytes and water-soluble and insoluble complexes were investigated (Mustafaev et al. 1998). Water-soluble polymers have recently found a wide range of applications, both in theoretical polymer chemistry and practical application. In particular, the use of water-soluble systems as physiological active substances appears to be an immunological benefit (Nadakumar and Shakya 2012). Coordination and organometallic colloids and gels have attracted unique interest from scientists for application in new technologies within the production of biopolymers. Unfortunately, in step with the regarded strategies of guidance for these biomaterials, they are nonetheless quite basic and may be stated to be the result of coincidences instead of planning. In addition to coordination bonds in manufacturing, other non-covalent interactions, which include coordination colloids and gels, hydrogen bonds, stacking, and van der Waals forces, had been discovered and are recognized to work in collaboration. Because of the collective effect of these sensitive interactions, understanding and manipulating intermolecular connections are crucial to structural design and a number of properties in the molecules' function. Therefore, fundamental studies that consist of innovation in chemistry, characterization strategies, and the improvement of effective strategies for shape-function assessment should be supported and expanded (Wang and McHale 2010).

A future of developing metal-containing polymers is envisaged, especially in the biomedical field, but some important obstacles limit the development of biopolymers as biomaterials.

One of the conditions limiting this production is the limitation of metals. This has limited the production of new metal-containing biopolymers. Of all metal elements, only about 30 metals can be used in the production of biopolymers. Most of these metals are produced from VIII, IB, and VA group elements. Secondly, the difficulties in the synthesis of metallopolymers hinder their versatility. Thirdly, the toxicity of metals has seriously limited the application of metal-containing substances in the human body (Yan et al. 2016). Radical and ionic polymerizations are regularly controlled strategies for polymers with controlled properties. However, the synthesis of metallopolymers through controlled radical or ionic polymerizations has been much less explored. On the other hand, metals in polymerization structures appreciably increase the complexity of polymerization methods (Yan et al. 2016) (Figure 5.6).

Some metals are known for their toxicity in biological systems. How to avoid toxic metals in the environment is a critical problem in making biomaterials containing metal, but besides this toxicity, there are many benefits (Yan et al. 2016).

Complex structures using transition metal for polymeric vaccines have many advantages (Sanda 2015):

1. They connect two negative molecules (polymer-protein and polymer-polymer).
2. They connect two large molecules.
3. When the polymer and antigen are conjugated with the covalent bond, there is a risk of alteration of the chemical structure of the antigenic determinant. In order to prevent this, the polyelectrolytes and antigens, which are linked together by metal, are connected to one another. Thus, there is no significant change in the antigen structure with the carrier polymer.

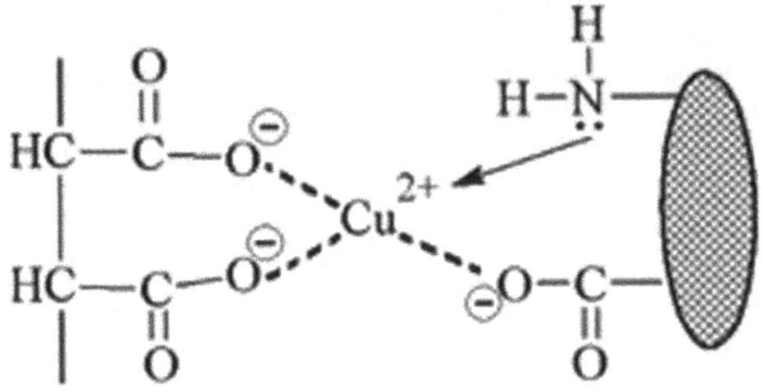

PMVEMA   BSA

FIGURE 5.6 The interaction between anionic Polyacrylic Acid (PAA) and Bovin Serum Albumin in the presence of Copper ions in aqueous solution (pH:7) (Karahan, Mustafaeva and Özeroğlu 2014).

Additionally it has been shown, that polymers containing metal are a highly promising field of study on the border of polymer technological know-how, coordination chemistry, supramolecular chemistry, and colloidal and substances technology.

For this purpose, there is a need for more information about physicochemical methods.

The fact that the variety of biopolymer production methods has recently dropped is a phenomenon that has drawn attention and restricts the production of most biopolymers.

More efficient results can be obtained if new methods are found (Wang and McHale 2010).

REFERENCES

Anasir, M.I. and C.L. Poh. 2019. "Advances in antigenic peptide-based vaccine and neutralizing antibodies against viruses causing hand, foot, and mouth disease." *International Journal of Molecular Sciences* 20(6): 1256. doi:10.3390/ijms20061256.

Bose, R.J., M. Kim, J.H. Chang, R. Paulmurugan, J.J. Moon, W.G. Koh, S.H. Lee, and H. Park. 2019. "Biodegradable polymers for modern vaccine development." *Journal of Industrial and Engineering Chemistry (Seoul, Korea)* 77: 12–24. doi:10.1016/j.jiec.2019.04.044.

Ciaurriz, P., F. Fernández, E. Tellechea, J.F. Moran, and A.C. Asensio. 2017. "Comparison of four functionalization methods of gold nanoparticles for enhancing the enzyme-linked immunosorbent assay (ELISA)." *Beilstein Journal of Nanotechnology* 8: 244–253. doi:10.3762/bjnano.8.27.

Das, R.D., S. Maji, S. Das, and C. RoyChaudhuri. 2010. "Optimization of covalent antibody immobilization on macroporous silicon solid supports." *Applied Surface Science* 256(20): 5867–5875. doi:10.1016/j.apsusc.2010.03.066.

Derman, S. 2006. *Peptide protein covalent conjugation.* (M.S.c) Yıldız Technical University Department of Bioengineering, Istanbul.

Friedli, G.L. 1996. *Interaction of deamidated soluble wheat protein (SWP) with other food proteins and metals.* Ph.D. Thesis. University of Surrey, England.

Hassan, M., J. Bai and D. Dou. 2019. "Biopolymers; definition, classification and applications." *Egyptian Journal of Chemistry* 62(9): 1725–1737. doi:10.21608/ejchem.2019.6967.1580.

Hermanson, G.T. 2013. *Bioconjugate techniques.* London: Academic Press.

Hermanson, G.T. 1996. *Bioconjugate techniques.* San Diego: Academic Press, 297–364.

Horníková, L., M. Fraiberk, P. Man, V. Janovec, and J. Forstová. 2017. "VP1, the major capsid protein of the mouse polyomavirus, binds microtubules, promotes their acetylation and blocks the host cell cycle." *The FEBS Journal* 284(2): 301–323. doi:10.1111/febs.13977.

Huebsch, N. and D.J. Mooney. 2009. "Inspiration and application in the evolution of biomaterials." *Nature* 462: 426–432.

Karahan, M., Z., Mustafaeva, and C. Özeroğlu. 2014. "The formation of polycomplexes of poly(methyl vinyl ether-co-maleic anhydride) and bovine serum albumin in the presence of copper ions." *Polish Journal of Chemical Technology* 16(3): 97–105. doi:10.2478/pjct-2014-0058.

Khademi, F., M. Derakhshan, A. Yousefi-Avarvand, and M. Tafaghodi. 2018. "Potential of polymeric particles as future vaccine delivery systems/adjuvants for parenteral and non-parenteral immunization against tuberculosis: A systematic review." *Iranian Journal of Basic Medical Sciences* 21(2): 116–123. doi:10.22038/IJBMS.2017.22059.5648.

Li, G. and Q. Fu. 2017. "Polymer materials and engineerings research at Sichuan university." *Macromolecular Rapid Communications* 38(23): 1700592.

Mamabolo, M.V., J. Theron, F. Maree, and M. Crampton. 2020. "Production of foot-and-mouth disease virus SAT2 VP1 protein." *AMB Express* 10(2): 1–9. doi:10.1186/s13568-019-0938-7.

Mann, J.L., A.C. Yu, G. Agmon, and E.A. Appel. 2018. "Supramolecular polymeric biomaterials." *Biomaterials Science* 6(1): 10–37.

Mansuroğlu, B. and Z. Mustafaeva. 2012. "Characterization of water-soluble conjugates of polyacrylic acid and antigenic peptide of FMDV by size exclusion chromatography with quadruple detection" *Materials Science and Engineering: C, Elsevier* 32(2): 112–118. doi:10.1016/j.msec.2011.10.004.

McCarthy, D.P., Z.N. Hunter, B. Chackerian, L.D. Shea, and S.D. Miller 2014. "Targeted immunomodulation using antigen-conjugated nanoparticles." *Wiley Interdisciplinary Reviews. Nanomedicine and Nanobiotechnology* 6(3): 298–315. doi:10.1002/wnan.1263.

Mustafaev, M.I., Z. Mustafaeva, E. Bermek and Y.Osada. 1998. "New amphiphilic immunogens by Cu2+ - mediated temary complexes of poly-(N-isopropylacrylamide-co-acrylic acid) and bovine serum albumine." *Journal of Bioactive and Compatible Polymers* 13: 33–49.

Nadakumar, K.S. and A. Shakya. 2012. "Polymers as immunological adjuvants: An update on recent developments." *Journal of BioScence and Biotecknology* 1(3):199–210.

Powell, B.S., A.K. Andrianov, and P.C. Fusco. 2015. "Polyionic vaccine adjuvants: Another look at aluminum salts and polyelectrolytes." *Clinical and Experimental Vaccine Research* 4(1): 23–45. doi:10.7774/cevr.2015.4.1.23

Rudra, J.S., Y.F. Tian, J.P. Jung, and J.H. Collier. 2010. "A self-assembling peptide acting as an immune adjuvant." *Proceedings of the National Academy of Sciences of the United States of America* 107(2), 622–627. doi:10.1073/pnas.0912124107

Sanda, F. 2015. "Transition metal containing polymers." In Kobayashi, S., and Müllen, K. (Eds), *Encyclopedia of polymeric nanomaterials*. Verlag Berlin Heildberg: Springer, 1–6.

Sheehan, H.L. 1965. "The frequency of post-partum hypopituitarism." *Bjog: An International Journal of Obstetrics and Gynaecology* 72(1): 103–111. doi:10.1111/j.1471-0528.1965.tb01380.x.

Sheehan, J., P. Cruickshank, and G. Boshart. 1961. "Notes- a convenient synthesis of water-soluble carbodiimides." *The Journal of Organic Chemistry* 26(7): 2525–2528. doi:10.1021/jo01351a600.

Sivakumar, S.M., M.M. Safhi, M. Kannadasan, and N. Sukumaran. 2011. "Vaccine adjuvants – current status and prospects on controlled release adjuvancity." *Saudi Pharmaceutical Journal* 19(4): 197–206. doi:10.1016/j.jsps.2011.06.003.

Skwarczynski, M. and I. Toth. 2016. "Peptide-based synthetic vaccines." *Chemical Science* 7(2): 842–854. doi:10.1039/c5sc03892h.

Sneha, M., O.S. Oluwafemi, N. Kalarikkal, S. Thomas, and S.P. Songca. 2016. *Biopolymers – Application in Nanoscience and Nanotechnology*. Ed. Farzana Khan Perveen, Intechopen. doi:10.5772/62225.

Soares, E., R. Cordeiro, H. Faneca, and O. Borges. 2019. "Polymeric nanoengineered HBsAg DNA vaccine designed in combination with β-glucan." *International Journal of Biological Macromolecules* 122: 930–939. doi:10.1016/j.ijbiomac.2018.11.024.

Song, R., M. Murphy, C. Li, K. Ting, C. Soo, and Z. Zheng. 2018. "Current development of biodegradable polymeric materials for biomedical applications." *Drug Design, Development and Therapy* 12: 3117–3145. doi:10.2147/DDDT.S165440.

Wang, X. and R. McHale. 2010. "Metal-containing polymers: Building blocks for functional (nano)materials." *Macromolecular Rapid Communications* 31(4): 331–350. doi:10.1002/marc.200900558.

Yan, Y., J. Zhang, L. Ren, and C. Tang. 2016. "Metal-containing and related polymers for biomedical applications." *Chemical Society Reviews* 45(19): 5232–5263. doi:10.1039/c6cs00026f.

Yang, J., Y. Luo, M.A. Shibu, I. Toth, and M. Skwarczyski 2019. "Cell-penetrating peptides: Efficient vectors for vaccine delivery." *Current Drug Delivery* 16: 430–443. doi:10.2174/1567201816666190123120915.

Younes, B. 2017. "Classification, characterization, and the production processes of biopolymers used in the textiles industry." *The Journal of The Textile Institute* 108(5): 674–682. doi:10.1080/00405000.2016.1180731.

Zhang, P., M.L. Bookstaver, and C.M. Jewell. 2017. "Engineering cell surfaces with polyelectrolyte materials for translational applications." *Polymers* 9(2): 40. doi:10.3390/polym9020040.

6 What Are Polymeric Carriers?

Gülderen Karakuş and Dolunay Şakar Daşdan

CONTENTS

6.1 INTRODUCTION

Since synthetic peptides with poor immunogenic properties do not have sufficient molecular sizes to form antibodies, the peptides can also bind to proteins as well as linear polymers to enhance immunogenicity in the production of antibodies against peptides.

Depending on the peptide sequence, as in the conjugation region proteins, the N-terminus of the sequence may be the C-terminus or a point within it. The conjugation point may be carboxyl (-COOH), amino (-NH$_2$), or sulfhydryl (-SH) ends. Synthetic peptides, haptens, etc., when used as a vaccine, have a significant effect. Nowadays, it is widely used to create specific antibodies. However, therapeutic agents and synthetic peptides or haptens, due to their low solubility, instability, low molecular weight, low antigenic properties, undesirable properties such as biocompatibility, and non-specificity or cytokine toxicity have greatly reduced their ability to act. However, the formation of conjugates of proteins, antigens, or actives with water-soluble polymers significantly changes the properties and immunogenicity of the therapeutic agents.

The binding of the antigenic peptides to the water-soluble polymer has multiple effects. Some of those are:

- To provide modification of peptides
- To increase the water solubility of those with hydrophobic properties
- To raise regional impact
- To increase immunogenic effects and immunoreactivity
- To be more effective in the living organism (Mustafaev and Mustafaeva 2002)

6.2   POLYMERIC CARRIERS IN LINEAR PROPERTIES

The simplest polymer types are homopolymers formed by condensation of monomers. This kind of homopolymer may be linear or a three-dimensional (spherical) structure. Generally, a linear polymer contains an alternative copolymer and block copolymer, but a random copolymer and graft copolymer are non-linear polymers. The average molecular weights should be known if the polymers are composed of linear molecules. The number of branches and the length of the branches are important if they are composed of non-linear characteristics.

Because the immunogenic properties of the synthetic peptides are poor and their molecular size is not sufficient to form antibodies, the peptides should be bound to various carriers such as those with linear properties. Polyacrylic acid (PAA), poly(N-isopropylacrylamide) (PNiPAM), poly(N-vinyl-2-pyrrolidone) (PVP), copolymers of polyvinylpyridine-polysetilpyridine, and poly(vinylpyrrolidone)-poly(acrylic acid) copolymers are linear carriers used for peptides.

6.2.1   POLYACRYLIC ACID (PAA)

Polyacrylic acid (PAA) is a potent adjuvant for primary and secondary immune response. PAA is the general name of synthetic high molecular weight homopolymers of acrylic acid. The solubility of PAA in water increases with a rise in temperature. It is an anionic polymer in a water solution at neutral pH. Many of the side chains have lost their protons and gained a negative charge. With its ability to absorb and retain water due to its hygroscopic nature, it is a polyelectrolyte which can be inflated to a very large extent above the original volume. It is a widely used model polymer for understanding the immunostimulatory properties of synthetic polyelectrolytes. PAA, which is not ionized at low pHs, can be associated with various non-ionic polymers {poly(ethylene oxide), poly(N-vinyl pyrrolidone), polyacrylamide} and some cellulose ethers, and may form hydrogenated interpolymer complexes.

6.2.2   CARBOXYMETHYLCELLULOSE (CMC)

Carboxymethylcellulose (CMC) is a *cellulose derivative* with carboxymethyl groups (-CH$_2$-COOH) widely used in the field of pharmacy, cosmetics, and food. It is an anionic polysaccharide with high water soluble cellulose-derivative. It has biocompatibility properties and has been used after drug distribution studies and surgical operations. Because of its good properties such as being non-toxic, renewable, biodegradable, and a modifiable natural polymer, CMC has been used as a polymeric carrier and is usually used in pharmaceutical applications.

6.3 POLYMERIC CARRIERS IN SPHERICAL FEATURES

Microparticle- and nanoparticle-based drug delivery systems have been a burgeoning area of interest for well over two decades. They have potential for biocompatible, biodegradable polymers, are used in the formation of vaccine vectors and have attracted attention in the field of immunology. The main reason behind the development of these particle-based vaccines is to yield safer and improved vaccines tailored to treat specific diseases through the triggering of appropriate immune effector phenotypes.

6.3.1 NANOPARTICLES

Polymeric nanoparticles (NPs) are mostly nanospheres or nanocapsule shaped, if a homopolymer it may have spherical features; monomers should have more than two functional groups.

Particles that have a size of at least 1000 nanometers (nm) and generally less than 100 nm are called nanoparticles (NPs). These particles with superior properties are used in many fields such as electricity, electronics, biotechnology, automotive, medical, etc. NPs are morphologically and physicochemically influenced by the physical and chemical properties of the source material used. The NPs used as polymeric carriers are of a solid colloidal structure. The active substance can be encapsulated, absorbed, or dissolved in the particle. As NPs, proteins are used as well as polysaccharides, polyanhydride, polycaprolactone, polyacrylic acid, and polylactic-*co*-glycolic acid (Derman, Kızılbey, and Akdeste 2013; Gürmen and Ebin 2008). Based on physical and chemical characteristics, some of the well-known classes of polymeric nanoparticles (PNPs) of therapeutic applications in treatment are given as below:

1. Pluronics®
2. PEG-PLA
3. PEG-PCL
4. PEG-Lipid
5. PEG-PLGA
6. PEG-poly(amino acids)
7. Stimuli-sensitive polymeric micelles
8. Endogenous stimuli-sensitive polymeric micelles
9. pH-sensitive polymeric micelles
10. Reduction sensitive polymeric micelles
11. Thermo-sensitive polymeric micelles
12. Exogenous stimuli-sensitive polymeric micelles
13. Light-sensitive polymeric micelles

14. Magnetic field-sensitive polymeric micelles
15. Ultra-sound sensitive polymeric micelles
16. Margination of micro-/nanoparticles: requirement for optimum drug delivery

Other hydrophilic block-forming polymers include chitosan, poly(*N*-vinyl pyrrolidone) (PVP), and poly(*N*-isopropylacrylamide) (pNIPAAm). There are various polymer blocks utilized to form micellar cores, including the class of polyethers such as poly(propylene oxide) (PPO), various polyesters such as poly(L-lactide) (PLA), poly-ξ-caprolactone (PCL), poly(lactide-*co*-glycolic acid) (PLGA), poly(β-aminoesters), polyamino acids such as poly(L-histidine) (pHis), poly(L-aspartic acid) (pAsp), and lipids such as dioleoylphosphatidylethanolamine (DOPE), distearoylphosphatidylethanolamine (DSPE). The assembly of block copolymers, in which PPO attached to PEG as A-B-A triblock co-polymers (PEO-PPO-PEO) is known as Pluronics® (Biswas et al. 2016).

Various biocompatible NP-based drug-delivery systems such as liposomes, dendrimers, micelles, silica, quantum dots, and magnetic, gold, and carbon nanotubes have already been reported for targeted treatment. Silver nanoparticles (used as an adjuvant and carrier) have significant importance to modulate ABC transporter activity for immunotherapy in multidrug resistant cancer (Kovács et al. 2016).

6.3.2 MICROPARTICLES

Microencapsulation is the coating of a solid or liquid particle or a droplet with a polymeric film material and known as the first of the microparticular systems entering our lives in recent years. Microencapsulation technology is often used in a wide range of pharmaceuticals, foodstuffs, agriculture, cosmetics, textiles, etc. In the case of polymeric vaccine techniques, the active drug substance is coated with a coating material called a wall at the so-called core. The liquid substance is reduced by more easily portable volatility, increased stability, and stability with this coating technique. In addition, the effect of time also increases. Different techniques are used for encapsulation. Depending on the physical and chemical properties of the core material, the technique to be used also varies. The coacervation method is the oldest and the most widely used method. Coacervation occurs as a result of temperature change, addition of salt, addition of another polymer or polymer-polymer interaction. Unlike microcapsules, microspheres are the carriers that enable the active substance in the drug to be delivered to the desired area in the body. Biocompatibility and nontoxicity are the most important reasons. The dimensions range in size from 1 μm to 50 μm. The polymers (natural or synthetic) generally used are chitosan, polyesters, lipids, and cellulose derivatives (Geary et al. 2015).

6.4 POLYMERIZATION METHODS

Chemical substances with small molecular mass that can form more complex and large molecules by covalent bonding are called monomers.

A large number of monomers connected to each other by covalent bonds of molecules are composed of many repeating subunits to form polymer molecule. *A polymer is* a macromolecule composed of many *monomer units* or segments. The conversion of ethylene into polyethylene, the most common plastic in the world found in items ranging from shopping bags to storage containers; polymerization is shown in (Figure 6.1).

As seen in the example of polyethylene, a large, chain-like coarse molecule has been formed by the binding of numerous small ethylene monomer ($-C_2H_4-$) repeated along the chain with chemical bonds. Therefore, polymer molecule, polymer chain, or macro molecule expressions are often used instead of each other. The formation of polymerization consists of three parts: these are the start, multiplication, and termination steps. The methods used in the synthesis of polymers are divided into two groups by taking their mechanisms into account (chain) and condensation polymerization. Radical polymerization and ionic addition polymerization due to radicalization or ionicity of the active center providing chain growth can be examined in two groups. In addition, the ionic addition polymerization takes place in different categories according to the cationic or anionic state of the active center (Nesvadba 2012) (Table 6.1).

Polymerization reactions are carried out for laboratory purposes or for industrial purposes. Polymerization of carrier and adjuvant biopolymers for synthetic peptide vaccines is provided in the laboratory by certain methods. Polymerization methods can be examined as follows:

6.4.1 Mass or Bulk Polymerization

Mass or bulk polymerization of monomers such as initiators and chain transfer agents in the form of dissolved or poorly soluble monomer phase based on the polymerization of the medium. The medium involves a monomer and an initiator as the main components, without a solvent. The polymer formed is soluble in the styrene, such as poly(methyl methacrylate) (PMMA), methyl methacrylate, in the styrene. Viscous liquid is formed when the monomer is polymerized at a rate of 10–20%. If the reaction is carried out further, it may be difficult to remove the polymer from the reaction media. The monomers which undergo condensation polymerization are usually polymerized by this method. Foreign substances are less likely to enter the polymerization medium and the polymeric product is very easy to separate (Adumitrăchioaie 2018; Nesvadba 2012).

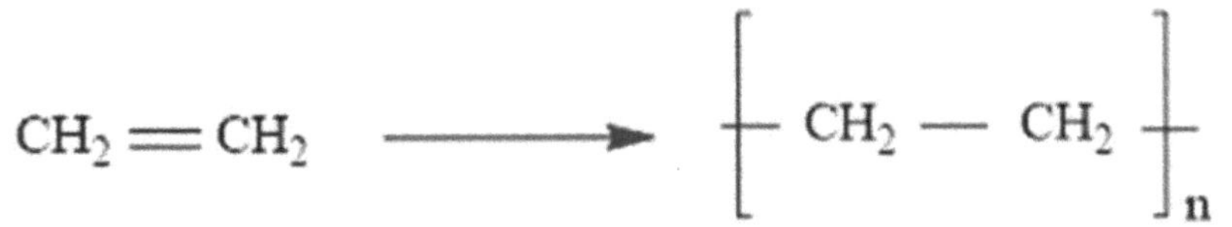

$$CH_2 = CH_2 \longrightarrow \left[CH_2 - CH_2 \right]_n$$

FIGURE 6.1 Polyethylene.

TABLE 6.1

Comparison of Polymerization Types and Information About Their Facilities

Step Growth Polymerization	Chain Growth Polymerization
Any two molecules in the environment enlarge the chain by reacting.	If monomer is added only to the active polymer chain, the chain grows.
Polymerization proceeds while the monomers remain.	The monomer is always present, and its concentration decreases over the course of the reaction.
The molar mass of the polymer increases continuously during the polymerization until full consumption of functional groups.	A high molar mass polymer is formed in the first moments of the reaction and the molar mass of the polymer does not change over the course of the reaction.
A long production polymerization time is required for high molar mass polymer.	A longer polymerization time is required for high conversions. However, the effect of this time on the molar mass of the polymer is insignificant.
Each type of molecule can react with each other in the polymerization.	Only radical species can react with each other.
One type of molecules (e.g. electrophiles) react with another type of molecules (e.g. nucleophiles).	
Polymerization has polymer chains of all sizes during polymerization.	During the polymerization, the reaction comprises monomer, high molar mass polymer, and growing active chains.
First, two molecules react with each other and the dimer is formed. The dimer reacts with the monomer and combines with the trimer or other dimer to give a tetramer and the chains grow in this way, i.e. n-mer reacts with m-mer that results in the elongation of macro-chains.	The monomers are rapidly added to the active center.
Lutz et al. (2016)	Lutz et al. (2016)
Nuyken and Pask (2013)	Nuyken and Pask (2013)
Nesvadba (2012)	Nesvadba (2012)

6.4.2 SOLUTION POLYMERIZATION

In the polymerization reaction at the beginning of solution polymerization, the reaction medium contains monomers, solvents, and initiators. In solution polymerization occurs in the non-reacting inert solvent. Because the solvent dilutes the polymerization reaction viscosity, mixing becomes easier and more efficient heat transfer can be made. If the solvent used in the polymerization dissolves both the monomer and the initiator, the polymerization starts, proceeds, and ends in a homogeneous environment. This process is called homogeneous solution polymerization. In addition, care must be taken in the choice of solvent in solution polymerizations.

By knowing the melting and boiling points of the solvent, it is important that the solvent can be removed from the polymer without any cost and without cause any threats to health. It is important. More aliphatic and aromatic hydrocarbons, esters, alcohols, and ethers are used. Water is also a solvent but most of the monomers are organic, so solution polymerization in water cannot be carried out, but polyacrylamide, poly (acrylic acid), poly (methyl methacrylate) can be synthesized in water (Nesvadba 2012).

6.4.3 SUSPENSION POLYMERIZATION

In the suspension polymerization, the monomer is formed by dispersing the monomer in a liquid which is not mixed. Suspension polymerization is one of the most widely used polymerization techniques that it has several advantages over other polymerization techniques since water is usually the continuous phase and also most commonly used liquid. The polymer is dispersed in drops of 0.01–0.5 cm diameter in the aqueous phase, that is, suspension of the monomer in water. In order to ensure that the suspension is stable and the formed polymer particles do not adhere to each other, chemicals called stabilizers are added. These suspension-forming substances wrap around the monomer. If necessary, precautions are not taken, the particles are clustered and block. As stabilizers, gelatin, kaolin, powder, bentonite, barium, calcium, and magnesium carbonates, and water-insoluble inorganic compounds such as aluminum hydroxide are used. In addition, mechanical mixing prevents the droplets from sticking together.

As initiators of polymerization, initiators are dissolved in the monomer (in the organic phase). At the end of the polymerization, the resulting powdered polymer is filtered out of the water and dried. The polymer is produced in granular form. Vinyl monomers such as styrene, methyl methacrylate, vinyl chloride, and vinyl acetate can be polymerized by these methods. This type of polymerization is called pearl or grain polymerization by looking at the final product obtained (Adumitrăchioaie 2018; Nesvadba 2012; Chaudhary and Sharma 2019).

6.4.4 EMULSION POLYMERIZATION

It is the method of polymerizing a very finely dispersed monomer in water (as the continuous phase) with the help of suitable emulsifiers to disperse various components. An emulsion polymerization system generally consists of a dispersing medium, monomer (slightly soluble in water), emulsifier, initiator and, if necessary, modifiers. In the aquatic environment, there is a surfactant and a water-soluble initiator. The monomer is dispersed into the medium by means of an emulsifying agent. The polymerization initiator is a water-soluble substance. The medium is continuously mixed, and the monomer is obtained by dispersing into very small particles.

In emulsion polymerization medium there are water, monomer, micelle builders, and initiators. The monomer is stabilized by a surfactant (such as soap) and these droplets are called micelles. One end of the micelles is hydrophobic, and the other end is of a hydrophilic character. Polymerization is carried out quickly and at very low temperatures in micelles. Measurements showed that the micelles were in the form of a rod. Each micelle consists of 50–100 emulsifying molecules. The hydrocarbon tails of these molecules, forming myelin, are oriented into the micelle and the ionic ends are facing towards the water. Micelles and water are mixed at the beginning of emulsion polymerization. A part of the micelles in the mixture is water soluble, and some of them collect together to form spherical micelles. After dissolving the micelles in the water, the monomer is added to the medium by mixing.

Some of the water-soluble monomers enter the micelles and inflate them. Others are dispersed in water as monomer drops.

The emulsifying agent attaches to the monomers in the medium and polymerization takes place. The polymerization process is terminated by wrapping around the surrounding polymers (Adumitrăchioaie 2018; Nesvadba 2012).

6.4.5 SOLID STATE POLYMERIZATION

Chemically synthesized polymer materials, including PLA, PLGA, polyurethane (PU), poly(methyl methacrylate) (PMMA), polyesters, poly(vinyl pyrrolidone) (PVP), silicone rubber, polyvinyl alcohol, etc., that are used in vaccine materials are produced through chemical methods. PLA and its copolymers are biocompatible and biodegradable and can be obtained from a wide range of raw material sources. They are renewable, non-toxic, and completely biodegradable, and recognized by the Food and Drug Administration (FDA) (Vouyiouka et al. 2005).

REFERENCES

Adumitrăchioaie, A. 2018. "Electrochemical methods based on molecularly imprinted polymers for drug detection. A review." *International Journal of Electrochemical Science* 13(2018): 2556–2576. doi:10.20964/2018.03.75.

Biswas, S., P. Kumari, P.M. Lakhani, and B. Ghosh. 2016. "Recent advances in polymeric micelles for anti-cancer drug delivery." *European Journal of Pharmaceutical Sciences* 83: 184–202.

Chaudhary, V. and S. Sharma. 2019. "Suspension polymerization technique: Parameters affecting polymer properties and application in oxidation reactions." *Journal of Polymeric Research* 26: 102. doi:10.1007/s10965-019-1767-8.

Derman, S., K. Kızılbey, and Z.M. Akdeste. 2013. "Polymeric nanoparticles." *Journal of Engineering and Natural Sciences* 31: 107–120.

Geary, S.M., Q. Hu, V.B. Joshi, N.B. Bowden, and A.K. Salem. 2015. "Diaminosulfide based polymer microparticles as cancer vaccine delivery systems." *Journal of Controlled Release* 220: 682–690.

Gürmen, S. and B. Ebin. 2008. "Nanoparticles and production methods-1." *Union of Chambers of Turkish Engineers and Architects Metallurgy and Material Engineers Chamber: Metallurgical Journal* 150: 31–38.

Kovács, D., K. Szőke, N. Igaz, G. Spengler, J. Molnár, T.Tóth, D. Madarász, et al. 2016. "Silver nanoparticles modulate ABC transporter activity and enhance chemotherapy in multidrug resistant cancer." *Nanomedicine: Nanotechnology, Biology and Medicine* 12(3): 601–610. doi:10.1016/j.nano.2015.10.015.

Lutz, J.F., J.M. Lehn, E.W. Meijer, and K. Matyjaszewski. 2016. "From precision polymers to complex materials and systems." *Nature Reviews Materials* 1(5): 1–14. doi:10.1038/natrevmats.2016.24.

Mustafaev, M.I. and Z. Mustafaeva. 2002. "Novel polypeptide-comprising biopolymer system." *Tecnology and Health Care* 10: 217–226.

Nesvadba, P. 2012. "Radical polymerization in industr." *Encyclopedia of Radicals in Chemistry, Biology and Materials* doi:10.1002/9781119953678.rad080.

Nuyken, O. and S. Pask. 2013. "Ring-opening polymerization—an introductory review." *Polymers* 5(2): 361–403. doi:10.3390/polym5020361.

Vouyiouka, S.N., E.K. Karakatsani, and C.D. Papaspyrides. 2005. "Solid state polymerization." *Progress in Polymer Science* 30(1): 10–37. doi:10.1016/j.progpolymsci.2004.11.001.

7 Conjugation and Other Methods in Polymeric Vaccines

Özgür Özay

CONTENTS

7.1 PEPTIDE-POLYMER CONJUGATION

Protein-polymer conjugates are used in many fields. These molecules are of great importance, especially for their applications in medicine, biotechnology, and nanotechnology. Covalent binding causes many advantages by binding synthetic polymers to proteins, in particular, increasing protein stability, solubility, and biocompatibility. For the development of polymer-peptide conjugates, the antigenic specific peptide sequences of the disease-causing virus must first be synthesized. This synthesis has many types of polymerization chemistry, for the new generation, polymers of various compositions, controlled molecular weight, and low polydispersity are required (Heredia and Maynard 2007). For this, solid phase peptide synthesis is important. This issue is mentioned in detail especially in Chapter 4. These polymers are carriers and can be synthesized linearly and roundly. After these processes, conjugation is carried out in methods such as physical mixture, covalent connection, electrostatic complex formation, and development of metal-containing functional biopolymer systems (Shu, Panganiban, and Xu 2013).

7.2 PURIFICATION OF CONJUGATE STRUCTURES

Purification of the conjugates is the process of removing unbound reagents and cross-linkers in the medium. The most commonly used methods are dialysis, column chromatography (see in Section 7.3.1. for HPLC method), gas pressure filtration and centrifugal filtration (Fornaguera and Solans 2018).

7.2.1 DIALYSIS

Dialysis is based on differential diffusion, which is also used in dialysis tube separation techniques, also known as Visking tubes. There is a semi-permeable membrane that facilitates the passage of small molecules in the solution. The process of cleaning and separating unreacted cross-linkers is done by dialysis and free peptides are removed from the sample of the peptide-protein conjugate through a semi-permeable membrane by selective and passive diffusion. A sample and a buffer solution (called dialysate, usually 200–500 times the volume of the sample) are placed on opposite sides of the membrane so that small contaminants are removed by changing the dialysate buffer (Walker 2009; Acar 2006; Pelit Arayıcı 2015).

7.2.2 COLUMN CHROMATOGRAPHY

Chromatography is known as an important biophysical technique that can separate, identify, and purify the blend segments for subjective and quantitative investigation. Proteins can be purified based on properties such as size and shape, total charge, hydrophobic groups present on the surface, and the ability to bind to the stationary phase. It is based on molecular properties and interaction type use mechanism, four separation technologies, ion exchange, dispersion, surface adsorption, and size exclusion. Column chromatography is one of the most used and common techniques for protein purification methods. This technique is basically used to purify biological molecules. The application of the method can be summarized as follows. The sample is separated on the column (stationary phase) and then the wash buffer is added to the column (mobile phase). It flows through the column material placed on the fiberglass support. With the help of the wash buffer, the samples are accumulated at the bottom of the column chromatography instrument, based on time and volume (Coskun 2016). Column chromatography is a powerful purification and separation process that is closely controlled to the hydrodynamic diameters of the macromolecules depending on the diameter of the pores in the filling material (see in HPLC Method) (Acar 2006; Fornaguera and Solans 2018).

7.2.3 GAS PRESSURE FILTRATION

The basic principle in gas pressure filtration depends on the molecular size and membrane potential. In order to make filtration, having the appropriate pore sizes are attached to the apparatus and the small molecules passing through the membrane are decomposed by means of pressure in the magnetic stirrer. Large molecules remain at the top of the membrane. The conjugate is thus removed and purified from undesired reagents and compounds. In principle, the porous membrane is used as in the previous process. The solution is put into the membrane tube and the centrifugation of the tube at different rotations, cycles, and times is followed by the passage of small molecules from the pores of the membrane. Thus, large molecules remain in the upper phase, small molecules accumulate in the lower phase, and the purification function is completed (Acar 2006).

7.3 OTHER METHODS THAT ARE USED FOR POLYMERIC VACCINE PRODUCTION

7.3.1 High Performance Liquid Chromatography System (HPLC)

Chromatography is the general name of the methods for separating components in complex solutions. Amino acid analysis is used for differentiation of amino acids and peptides, determination of amino acid composition of proteins, analysis of peptide sequences, and diagnosis of diseases belonging to amino acid metabolism. According to this information high performance liquid chromatography (HPLC) is a chromatographic technique that is used to separate, identify, and quantify components of a mixture, for example the separation of chemical compounds or identification of constituents of a biological sample; a typical HPLC system contains a stationary phase, a mobile phase of varying polarity, and an ultraviolet detector (Blum 2014). This technique is commonly used because it is sensitive, easily adaptable to quantitative determinations, non-volatile (suitable for decomposition of heat-degradable compounds), and has applicability with the most commonly used substances (amino acids, peptides, drugs, pesticides) (Mauldin et al. 2006; Fornaguera and Solans 2018) (Figure 7.1).

The chromatography consists of a peristaltic pump that provides transport of the solutes in the carrier phase (mobile phase) to the pump and column, and the detector suitable for the substance to be analyzed. The mobile phase generally is formed of aqueous buffer solutions or solutions formed with methanol/acetonitrile. The point of solutions is to find proper polarity by some chemicals which are HPLC-grade (HPLC-purity). If the chemicals are not HPLC-purity, analysis cannot be done to the desired efficiency. A Dezager is a pump to get rid of air bubbles. A solvent pump is generally used between 0.1–5 ml flow values that can help the mobile phase and can be moved in the HPLC columns which have a functional group with added silica or polymer-based filler, and column chromatography is a powerful purification and separation process that is closely controlled to the hydrodynamic diameters of the macromolecules depending on the diameter of the pores in the filling material. Column and additive choices and defining the appropriate gradient are the main parameters for the best resolution of peptide mixture in liquid chromatography. Frequently, C-18 columns are used in peptide separations (Acar et al. 2019; Fornaguera and Solans 2018).

In HPLC analysis, various detectors are used according to the chemical properties of the substances in the analysis such as ultraviolet (UV-Vis) detectors in Figure 7.1,

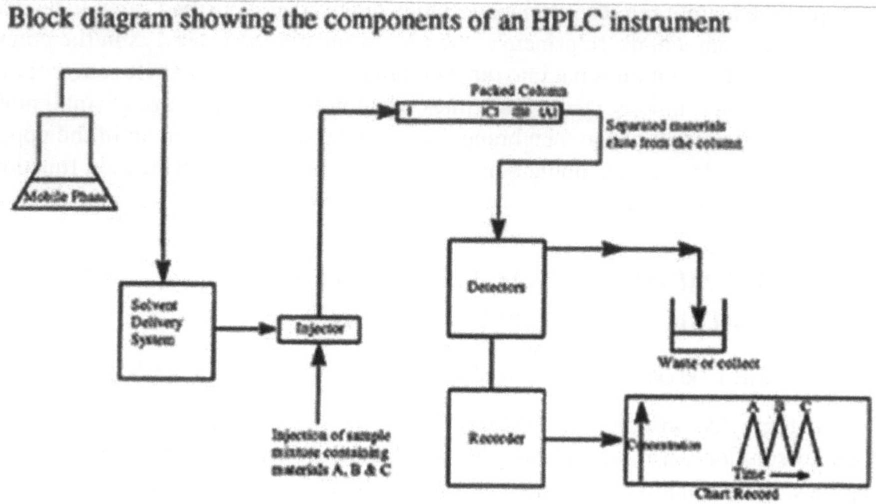

Block diagram showing the components of an HPLC instrument

FIGURE 7.1 Components of the HPLC technique used in chromatography are defined in the block diagram. (Image by "Sam.F" is licensed under a CC BY-SA, from Wikimedia Commons 1.)

diode array detector (DAD), fluorescence detector, refractive index, and conductivity detector. A UV-Vis detector has three forms but usually a multiple wavelength detector is used in our laboratory, for example antigenic peptide sequence of *Brucella abortus*, (WLAEIKQRSLMVHG) purification was performed by using an analytical reversed HPLC method and UV-PDA detector at 210 and 280 nm (Acar et al. 2019).

7.3.2 ZETA POTENTIAL MEASUREMENT

This method is commonly used for characterization of nanoparticles. The reason for negative values of the zeta potential could be ascribed to the ionized carboxyl groups appearing after a partial hydrolysis of the ester groups of the PLGA matrix. High values of the charges (positive or negative) prevent agglomeration (Suchaoin et al. 2017). In our research, a Zetasizer was used for studying the nanoparticles which were characterized by size, zeta potential, and polydispersity index (Çalman et al. 2018).

7.3.3 FOURIER TRANSFORM INFRARED SPECTROSCOPY (FTIR) MEASUREMENT

Vibrational spectroscopy, including infrared spectroscopy (IR), is the most significant tool for elucidating the structure of protein/peptide. The bands appearing in FTIR are correlated to the molecular arrangement of the peptide backbone by analyzing the most sensitive band like amide I, amide II, and amide III bonds which mainly originate from backbone vibrations such as indicated that the -C=O bond associated with this vibration. So, FTIR frequency clearly indicates changes in amide backbone conformation and the associated hydrogen bonding pattern due to formation of the

assembled structure (Banerji et al. 2017). In our laboratory, nanoparticle and peptide bonds are measured with the FTIR to provide information about the functional
groups in analyzing the compound of Zika diseases (Çalman et al. 2018)

7.3.4 FLUORESCENCE SPECTROSCOPY MEASUREMENTS

Spectroscopic methods are used in the structural characterization of organic or inorganic biomolecules. Spectroscopy is commonly defined as the study of the interaction
of electromagnetic radiation with matter (Stan Tsai 2007). Due to the high sensitivity
of fluorescence spectroscopy to other spectroscopy branches, it is more suitable to
be used for the analysis of fluorescence-specific biomolecules. It is a method that is
particularly preferred in the structural function analysis of peptides and proteins.
Fluorescence spectroscopy is a branch of spectroscopy in which the emission of an
evoked molecule as a basis is examined. It is at the top of the other spectroscopy
branches because of the degree of sensitivity. Many substances can be identified with
a sensitivity of less than one million by fluorescence spectroscopy. The selectivity of
this method is high. The operating range is in the visible area. The fluorescence event
involves two processes: absorption and emission (Lakowicz 1999).

These two processes can explain these; determination of fluorescence life (lifetime), investigation of properties of excised molecules, definition of complex formation mechanism, investigation of solvent-soluble interaction, isomerization,
diffusion, investigation of micellar properties, surface properties, polymer structure
and dynamics, enzyme structure, investigation of antigen-antibody interaction, bio-
membrane. The use of the fluorescence resonance energy transfer (FRET) method
to explain the relationship between the donor and the acceptor is an example of the
use of fluorescence spectrometers. Furthermore, this method is very important in
amino acid detection. Tryptophan (Trp, W), tyrosine (Tyr, Y), and phenylalanine
(Phe, F) are fluorescent forms of amino acids present in the structure of proteins and
peptides. Among these tryptophan is the amino acid with the highest fluorescence
(Ghisaidoobe and Chung 2014).

7.3.4.1 Absorption

When the electromagnetic wave solid passes into the liquid or gas layer, it is observed
that some wavelengths are selectively taken from the environment. In this process,
the electromagnetic wave transfers part of its energy to the atoms, ions, or molecules
that make up the sample. This phenomenon is called absorption. These particles that
absorb energy are evolved from the basic state. The transition time of absorption is
about 10^{-15} s (Skoog, Holler, and Nieman 1998; Koechner 2006).

7.3.4.2 Emission

A stimulated atom or molecule in the stimulation state can live up to approximately
10^{-8} s. Then it returns to its basic state by restoring the energy it absorbs.

When transmitting from a high energy level of a molecule to a lower energy level,
it emits more energy than a photon. This event is called emission (Koechner 2006)
(Table 7.1).

TABLE 7.1

Absorption and Fluorescence Wavelength Values of Tryptophan, Tyrosine and Phenylalanine from Aromatic Amino Acids

Aromatic Amino Acids	Wavelengths Most Absorbed by Light	Most Light Emitted by Wavelength	References
Tryptophan	220 nm	348 nm	Lesniak et al. 2013
Tyrosine	245–295 nm	300 nm	Antosiewicz and Shugar 2016
Phenylalanine	257 nm	282 nm	Lakowicz 1999; Held 2003

Fluorescence life is the time the induced level molecule passes, before it passes the basic electronic level. Most aromatic molecules have a fluorescence lifetime of 10 ns. The maximum wavelengths that aromatic amino acids absorb and fluorescence will depend on the fluorescent lifetime. In the studies, the synthesis of peptide epitopes is carried out in the structure of tryptophan and other aromatic amino acids, especially in foot and mouth disease, hepatitis B, and other diseases (Budama 2006). On synthetic polyelectrolytes and their structures (Budama et al. 2008), using different wavelengths is determined by the degree of peptide-polymer conjugation reaction using biomolecules (peptide, protein, etc.) with the same type of electric charge and, lastly, the FRET method is used to investigate the metal binding mechanism of the biopolymer (Acar et al. 2019; Karahan, Mustafaeva, and Ozer 2007). With the help of the FRET method, the intermolecular distance relationships can be examined in the distance measurement between the two places in the macromolecules (Acar et al. 2019). In addition, this method is considered to be a powerful technique to study molecular interactions in living cells with improved spatial (angstrom) and temporal (nanosecond) resolution, distance range, and sensitivity, and a wider range of biological applications (Sekar and Periasamy 2003; Carmona, Juliano, and Juliano 2009). Investigation of protein fragments in terms of these properties is thought to be of great benefit in peptide vaccine synthesis.

REFERENCES

Acar, S. 2006. *Peptide protein covalent conjugation*. Istanbul: YTU Institute of Science and Technology, Bioengineering Program.

Acar, T., P. Pelit Arayıcı, B. Ucar, M. Karahan, and Z. Mustafaeva. 2019. "Synthesis, characterization and lipophilicity study of brucella abortus' immunogenic peptide sequence that can be used in the future vaccination studies." *International Journal of Peptide Research and Therapeutics* 25: 911–918. doi:10.1007/s10989-018-9739-0.

Antosiewicz, J.M. and D.Shugar. 2016. "UV-Vis spectroscopy of tyrosine side-groups in studies of protein structure. Part 2: Selected applications." *Biophysical Reviews* 8(2): 163–177. doi:10.1007/s12551-016-0197-7.

Banerji, B., M. Chatterjee, U. Pal, and N.C. Maitim. 2017. "Formation of annular protofibrillar assembly by cysteine tripeptide: Unraveling the interactions with NMR, FTIR, and molecular dynamics." *The Journal of Physical Chemistry B* 121(26): 6367–6379.

Blum, F. 2014. "High performance liquid chromatography." *British Journal of Hospital Medicine* 75(Suppl. 2): C18–C21.

Budama, Y. 2006. "Investigation of the Interaction of Anionic Polyelectrolytes with Bovine Serum Albumin by Fluorescence Method." (M.Sc.) Yildiz Technical University Department of Bioengineering, Istanbul.

Budama, Y., E. Karabulut, Z.O. Ozdemir, and Z. Mustafaeva. 2008. *Investigation of polyelectrolyte-antigenic peptide conjugates by fluorescence spectroscopy*, 30 EPS, 31 Agust-05 Sep: Helsinki, Finland.

Carmona, A.K., M.A. Juliano, and L. Juliano. 2009. "The use of Fluorescence Resonance Energy Transfer (FRET) peptidesfor measurement of clinically important proteolytic enzymes." *Anais da Academia Brasileira de Ciências* 81(3): 381–392. doi:10.1590/S0001-37652009000300005.

Çalman, F., P. Pelit Arayıcı, H.K. Büyükbayraktar, M. Karahan, Z. Mustafaeva, and R. Katsarava. 2018. "Development of vaccine prototype against zika virus disease of peptide-loaded PLGA nanoparticles and evaluation of cytotoxicity." *International Journal of Peptide Research and Therapeutics* 25: 1057–1063.

Coskun, O. 2016. "Separation techniques: Chromatography." *Northern Clinics of Istanbul* 3(2): 156–160. doi: 10.14744 / nci.2016.32757.

Fornaguera, C. and C. Solans. 2018. "Analytical methods to characterize and purify polymeric nanoparticles." *International Journal of Polymer Science* 2018: 1–10. doi:10.1155/2018/6387826.

Ghisaidoobe, A.B. and S.J. Chung. 2014. "Intrinsic tryptophan fluorescence in the detection and analysis of proteins: A focus on förster resonance energy transfer techniques." *International Journal of Molecular Sciences* 15(12), 22518–22538. doi:10.3390/ijms151222518.

Held, P. 2003. *Quantitation of peptides and amino acids with a synergy™HT using UV fluorescence*. Winooski, VT: B.-T. Instruments.

Heredia, K.L. and H.D. Maynard. 2007. "Synthesis of protein-polymer conjugates." *Organic & Biomolecular Chemistry* 5:45–53. doi:10.1039/b612355d.

Karahan, M., Z. Mustafaeva, and H. Ozer. 2007. "Polysaccharide-protein covalent conjugates and their ternary metal complexes." *Asian Journal of Chemistry* 19(3): 1837–1845.

Koechner, W. 2006 "Energy transfer between radiation and atomic transitions." In *Solid-state laser engineering*. Springer *series in optical sciences*. New York: Springer. doi:10.1007/0-387-29338-8_2.

Lakowicz, J.R. (Ed.) 1999. *Principles of fluorescence spectroscopy* (2nd ed.). New York: Kluwer Academic/Plenum Publishers.

Lesniak, W.G., A. Jyoti, M.K. Mishra, N. Louissaint, R. Romero, D.C. Chugani, S. Kannan, and R.M. Kannan. 2013. "Concurrent quantification of tryptophan and its major metabolites." *Analytical Biochemistry* 443(2), 222–231. doi:10.1016/j.ab.2013.09.001.

Mauldin, R.E., T.M. Primus, T.A. Buettgenbach, J.J. Johnston, and G.M. Linz. 2006. "A simple HPLC method for the determination of chlorpyrifos in black oil sunflower seeds." *Journal of Liquid Chromatography & Related Technologies* 29(3): 339–348.

Pelit Arayıcı, P. 2015. *Protective new generation peptide vaccine models: Development of bioconjugates of viral rabies peptides with various adjuvants.* (id) Master Thesis Yildiz Technical University.

Sekar, R.B. and A. Periasamy. 2003. "Fluorescence resonance energy transfer (FRET) microscopy imaging of live cell protein localizations." *The Journal of Cell Biology* 160(5): 629–633. doi:10.1083/jcb.200210140.

Shu, J.Y., B. Panganiban, and T. Xu. 2013. "Peptide-polymer conjugates: From fundamental science to application." *Annual Review of Physical Chemistry* 64(1): 631–657. doi:10.1146/annurev-physchem-040412-110108.

Skoog, D.A., F.J. Holler, and T.A. Nieman. (Eds) 1998. *Principles of instrumental analysis* (5th ed.). Belmont: Brooks/Cole.

Stan Tsai, C. 2007. *Biomacromolecules: Introduction to structure, function and informatics.* USA: Wiley-Liss.

Suchaoin, W., A. Mahmood, K. Netsomboon, and A. Bernkop-Schnürch. 2017. "Zeta-potential-changing nanoparticles conjugated with cell-penetrating peptides for enhanced transfection efficiency." *Nanomedicine* 12(9): 963–975.

Walker, J.M. 2009. *The protein protocols handbook* (3rd ed.). New York: Springer-Verlag LLC.

8 Polymeric Nanoparticle Preparation Methods

Mithat Çelebi

CONTENTS

8.1 INTRODUCTION

Nanoparticles are produced with polymers with a size range of 10–1000 nm. They were first used in the 1970s. The active material in nanoparticles can be dissolved, encapsulated, and/or adsorbed (Çırpanlı 2009; Kulkarni and Rao 2013; Zhong et al. 2017; Lai et al. 2014; Senel et al. 2000). Polymers are used for the preparation of nanoparticles. The nanoparticle formulations were prepared using natural (Essa et al. 2020) and non-biodegradable synthetic polymers such as poly(acrylamide) (Owens et al. 2007) and poly(methyl methacrylate) (Matsuyama and Mishima 2009). The most important disadvantages of non-biodegradable synthetic polymers are the toxicity risks during the synthesis of polymers. The toxic monomers can remain residual in the nanoparticle structure. In the preparation of nanoparticles, surface modification of polymers can be carried out when needed. As the surface alteration with polyethylene glycol (PEG) inhibits recognition by the mononuclear phagocytic system, nanoparticles can stay in the bloodstream for a long time in this way. Polymers such as polyvinylpyrrolidone (PVP), PEG, dextran, chitosan, and surfactants such as oleic acid, sodium salt, and laurylamine are used for coating of nanoparticles to stabilization (Jokerst et al. 2011; H. Wang et al. 2005; Wang, Dijkstra, and Karperien 2016; Desai 2016; Mai Hoa et al. 2009; Petzold and Schwarz 2006; Suk et al. 2016; Şengel Türk, Hasçiçek and Gönül 2007; Gok 2012). Proteins have disadvantages such as the occurrence of antigenic reactions in the utilization of non-biodegradable synthetic polymers. Synthetic polymers which are not biodegradable have toxic risk because none of the monomers is polymerized (Çırpanlı 2009).

There are several methods for production of polymeric nanoparticles such as emulsion solvent diffusion (Esfandyari-Manesh et al. 2020), salting out (Kulkarni and Rao 2013), nanoprecipitaiton (Bordes et al. 2019), and spray drying (Rampino et al. 2013). Polymeric nanoparticles are prepared by two different methods: dispersion of preformed polymer nanoparticles and polymerization reactions from monomers. Preparation of nanoparticles with previously synthesized polymer nanoparticles has been mentioned in this chapter.

8.2 EMULSION SOLVENT DIFFUSION

The emulsion solvent diffusion method is the first method to form polymeric nanoparticles (Rao and Geckeler 2011a). A hydrophobic polymer is mixed in an organic solution using a diffusing agent and high-speed homogenization in a non-solvent (De et al. 2014). The polymer solution is set in vaporizing dissolver agents such as dichloromethane, chloroform, and ethyl acetate to form the emulsion (Rao and Geckeler 2011b; Allemann, Gurny, and Doelker 1993; Anton, Benoit, and Saulnier 2008). The evaporation is carried out by continuous mixing or under reduced pressure in a magnetic stirrer at room temperature (Wang et al. 2016a) (Figure 8.1).

Two main methods are employed for the arrangement of emulsions: the preparing of single emulsion – oil-in-water (o/w) and double emulsion – (water-in-oil)-in-water (w/o)/w. Water is used as the non-solvent in o/w emulsions. The single emulsion method is appropriate for hydrophobic drugs. This method is simple and cost-effective because it eliminates the recycling step and reduces agglomeration. However, it can just be used for a liposoluble structure. The double emulsion method is utilized once the acting material to be encapsulated is hydrophilic (proteins and peptides) (Essa et al. 2020; Derman, Kizilbey, and Mustafaeva Akdeste 2013). Poly-ε-caprolakton (PCL), polylactide (PLA), poly(lactic-co-glycolic acid) (PLGA), and poly(3-hydroxybutyrate) (PHB) are used for preparation of nanoparticles. Dexamethasone (Parde et al. 2017), paclitaxel (Lee and Lim 2017), and itraconazole (Curic, Möschwitzer, Peter, and Fricker 2017) and linezolid (Shaji and Kumbhar 2019) drug molecules were encapsulated using this method (Figure 8.2).

An active substance containing aqueous solution is dispersed in the organic phase solution comprising the polymer using the double solvent diffusion method. W/O

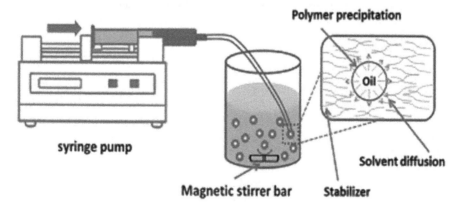

FIGURE 8.1 Preparation of nanoparticles by solvent diffusion method (Wang et al. 2016a).

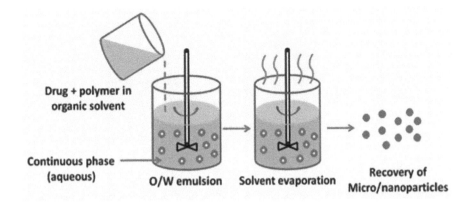

FIGURE 8.2   Preparation of nanoparticles by o/w emulsion evaporation method (Wang et al. 2016a).

emulsion is formed in the outer aqueous phase to give a (w/o)/w binary emulsion. The resulting (w/o)/w emulsion forms droplets of polymer solution containing the active agent solution. After removal of the solvent by diffusion or evaporation, the nanoparticle suspension containing the aqueous solution of the active substance is formed. This suspension is centrifuged to have nanoparticles. After washing with water, the dried nanoparticles are lyophilized (Julienne, Alonso, and Benoit 1992; Yoo et al. 2010; Muthu 2009; Wang et al. 2016a). Preparation of polymeric nanoparticles by the solvent emulsion diffusion method is shown in Table 8.1.

8.3   SALTING OUT

Production of polymeric nanoparticles using the salting out method has also been discovered by Galindo-Rodriguez et al. (Galindo-Rodriguez et al. 2004). Salting out is based on the precipitation of particles. The polymer is dissolved in organic solvents (acetone, etc.) that can be miscible with aqueous solution. The resulting solution is emulsified in electrolytes such as magnesium chloride ($MgCl_2$), calcium chloride ($CaCl_2$), and an aqueous solution containing a surfactant and stabilizer such as Tween 80, polyvinlyprolydone, polyvinyl alcohol, and hydroxyethyl cellulose. Nanoparticles are precipitated due to the organic solvent separated from the aqueous phase under the influence of ions. Nanospheres are obtained by increasing the diffusion of the organic solvent into the aqueous phase. Salting- out is an advantageous method for substances that are sensitive to heat, and electrolyte type and concentration are important parameters for encapsulation efficiency (De et al. 2014). Firstly, the polymer and the active substance are dissolved in an organic solvent such as acetone, and then emulsified in aqueous gel solution of colloidal stabilizers such as PVP containing the salting out agents The o/w emulsion is diluted to a sufficient volume for the solvent diffusion and nanoparticles are obtained by diffusion (Wei et al. 2009; Wang et al. 2016a) (Figure 8.3).

TABLE 8.1
Preparation of Polymeric Nanoparticles by Solvent Emulsion Diffusion Method

Polymeric Nanoparticle	Organic Solvent	Stabilizer	Emulsion	Particle Size	Reference
Linezolid-loaded Eudragit RS 100	Dichloromethane	PVA	Double emission w/o/w	47–119 nm	(Shaji and Kumbhar 2019)
Dexamethasone-loaded ethyl cellulose, Eudragit RS and ethyl cellulose/Eudragit RS	Ethyl acetate	PVA	Single emulsion o/w	105 nm	(Balzus et al. 2017)
itraconazole-loaded poly(butyl cyanoacrylate)	Chloroform	PVA, SDS, Poloxamer 188, Lutrol F68, Sodium cholate and Tween 80	Single emulsion o/w	80 nm	(Curic, Peter, and Fricker 2017)
Poly(lactic-co-glycolic acid) (PLGA)	Dichloromethane (DCM)	Polyvinyl alcohol (PVA)	Single emulsion o/w	106 nm	(Ghitman, Raluca, and Iovu 2017)
Poly-ε-caprolactone (PCL)	DCM	PVA	Double emission solvent evaporation method w/o/w	311–689 nm	(Alex et al. 2016)
Delafloxacin (DFL)-loaded stearic acid (lipid) chitosan	Ethyl acetate	Pluronic 127	Single-emulsion-solvent evaporation technique	299–368 nm	(Anwer et al. 2020)
PLGA	DCM	PVA	Single emulsification and the nanoprecipitation techniques	157–174 nm	(Hernández-Giottonini et al. 2020)
8- chitosan-coated polylactic acid (PLA) or poly(lactic-co-glycolic) acid (PLGA)	Chloroform	PVA	Single emulsion solvent evaporation	816–1194 nm	(Mohammed et al. 2019)
Poly(lactic-co-glycolic acid) (PLGA)	Ethyl acetate	PVA	Solvent evaporation procedure	329 nm	(Miele et al. 2018)

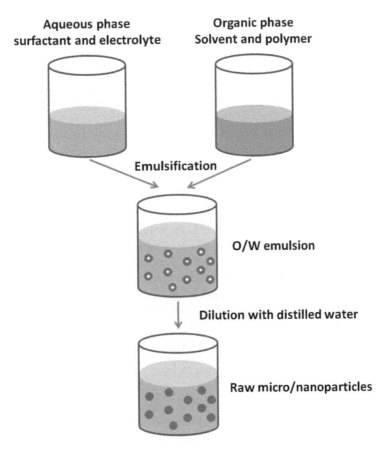

FIGURE 8.3 Preparation of nanoparticles by salting out method (Wang et al. 2016a).

Salting out is similar to the emulsion solvent diffusion method. Salting out is performed at a fixed mixing rate. Polymeric nanoparticles are prepared by taking off solvent and the precipitation agent by using crossflow filtration. This method is economic, scalable, and practical for the temperature-responsive material. Also, it does not need a high temperature. However, nanoparticles require washing (Nasir, Kausar, and Younus 2014; Zohri et al. 2009).

Polyethylene glycol (PEG)lyted liposomes (nanoparticles) are prepared with the extrusion method using ammonium sulfate salting-out agent by Bozo et.al. (Bozó et al. 2016).

8.4 NANOPRECIPITATION

Basically, the polymer collapses on the interface as a result of the displacement of a semipolar solvent in a lipophilic solution that can be mixed with water. Firstly, the polymer is added and then the active substance is dissolved in acetone. This organic solution is added to aqueous solution which is mixed with a magnetic stirrer and contains a stabilizer (surfactant) (Kökcu 2019). In the lipophilic solution, the water is miscible with the water at the interface as a result of the displacement of the semipolar solvent (Fessi et al. 1989; Fessi et al. 1991). This organic solution is injected into the distilled water mixed with the magnetic stirrer and containing the stabilizer. The final volume of suspension is adjusted to the required amount by removing acetone and some water under vacuum (Muthu 2009). In this method, the covalent binding of the active substance to the polymer or the surface of the nanoparticle is (1) before the nanoparticle is prepared and (2) adsorption of the active substance to the polymeric carrier system: different working strategies have been developed, such as the confinement of the active substance in the polymeric matrix, when preparing the nanoparticle (Dreis et al. 2007; Jung, Breitenbach, and Kissel 2000; De et al. 2014)

The nanoprecipitation method is also referred to as the solvent displacement method or the inter-surface precipitation method. This method takes place between two phases, one is organic, and the other is the water phase. In the organic phase the polymer consists of a surfactant and a solvent suitable for the polymer and surfactant, while the water phase consists of water or an aqueous solution. The active substance may be in the water or organic phase, depending on the environment in which it is dissolved. In this technique, synthetic biodegradable polyesters such as polycaprolactone (Ramanujam et al. 2018), poly(lactide) (Martins et al. 2018), poly(lactide-co-glycolide) (Hernández-Giottonini et al. 2020), and poly(alkylcyano acrylate) (Yan et al. 2016) were used. The most important parameters affecting particle formation in this technique are the organic phase drop rate, aqueous phase mixing speed, organic phase adding method, and o/w ratio. (Gok 2012; Mora-Huertas, Fessi, and Elaissari 2010).

Preparation of nanoparticles by the nanoprecipitation method was used by Fessi et al. This method is quick and easy, and nanoparticles are automatically formed in one step. Two immiscible solvents are required for preparation of nanoparticles. It is desired to dissolve the polymer and the active ingredient in the organic solvent. Once the polymer solution is added to the aqueous phase, it is necessary for the organic solvent be removed quickly from the medium. For encapsulation of the active substance, the polymer solution should easily diffuse into the dispersion medium. It is possible to obtain appropriate nanoparticles (100–300 nm) with a low polydispersity index. It can be worked with large group polymers such as PLGA, cellulose derivatives, and PCL in nanosphere preparation. Also, this method does not require an advanced mixing technique, sonication, or high temperature. A surfactant may not always be needed, and most importantly there is no need to use toxic organic solvent, and this technique is more suitable for hydrophobic active substances. Slightly soluble in water, 100% encapsulation efficiency can be achieved with substances dissolved in organic solvents such as ethanol and acetone. In recent years, promising results have been achieved with studies with water-soluble drugs (Çırpanlı 2009). Quérette, Fleury and Sintes-Zydowicza (2019) prepared poly(hdroxy)urethane nanoparticles DMSO solvent and SDS surfactant by the nanoprecipitation method (Quérette, Bordes, and Sintes-Zydowicz 2020). Ding et al. prepared superparamagnetic iron oxide nanoparticles in poly(methylmethacrylate) nanoparticles by the micromixer-assisted nanoprecipitation method (Ding et al. 2018). Hesperidin-diazepam-loaded

PLGA nanoparticles were optimized and developed using the nanoprecipitation method (Dang et al. 2015).

8.5 SPRAY DRYING

The spray drying method has been investigated for years to achieve polymeric nanoparticles by converting to powders by atomization. It is a consistent method. Khan et al. studied it with tramadol hydrochloric acid-loaded chitosan nanoparticles (Nasir, Kausar, and Younus 2014; Khan et al. 2012). Also paclitaxel-loaded polyester nanoparticles with poly(ε-caprolactone) and poly(lactide-co-glycolide) polymer have been studied (López-Gasco et al. 2011; Nasir, Kausar, and Younus 2014). Emami et al. (2014) prepared celecoxib-loaded nanoparticles with the spray drying method. Emami et al. (2014) utilized lactose and mannitol as sugar carriers for the spray drying process (Emami et al. 2014; Nasir, Kausar, and Younus 2014). Kirimlioglu and Ozturk (2020) studied chitosan nanoparticles (Yurtdaş Kırımlıoğlu and Öztürk 2020).

REFERENCES

Alex, A.T., A. Joseph, G. Shavi, J.V. Rao, and N. Udupa. 2016. "Development and evaluation of carboplatin- loaded PCL nanoparticles for intranasal delivery development and evaluation of carboplatin-loaded PCL nanoparticles for intranasal delivery." *Drug Delivery* 23(7): 2144–2153. doi: 10.3109/10717544.2014.948643.

Allemann, E., R. Gurny, and E. Doelker. 1993. "Drug-loaded nanoparticles - preparation methods and drug targeting issues." *European Journal of Pharmaceutics and Biopharmaceutics* 39(5): 173–191.

Anton, N., J.P. Benoit, and P. Saulnier. 2008. "Design and production of nanoparticles formulated from nano-emulsion templates-A review." *Journal of Controlled Release*. https://doi.org/10.1016/j.jconrel.2008.02.007.

Anwer, M.K., M. Iqbal, M.M. Muharram, M. Mohammad, E. Ezzeldin, M.F. Aldawsari, A. Alalaiwe, and F. Imam. 2020. "Development of lipomer nanoparticles for the enhancement of drug release, anti-microbial activity and bioavailability of delafloxacin." *Pharmaceutics*. https://doi.org/10.3390/pharmaceutics12030252.

Balzus, B., F. Feleke, S. Hönzke, C. Gerecke, F. Schumacher, S. Hedtrich, B. Kleuser, and R. Bodmeier. 2017. "European journal of pharmaceutics and biopharmaceutics formulation and ex vivo evaluation of polymeric nanoparticles for controlled delivery of corticosteroids to the skin and the corneal epithelium." *European Journal of Pharmaceutics and Biopharmaceutics* 115: 122–130. doi:10.1016/j.ejpb.2017.02.001.

Bordes, C., M. Nouri, S. Shahriari, G. Pazuki, S. Ding, N. Anton, T.F Vandamme, et al. 2019. "Increase of vanillin partitioning using aqueous two-phase system with promising nanoparticles." *Scientific Reports* 13(10): 1–10. doi:10.1038/s41598-019-56120-8.

Bozó, T., T. Mészáros, J. Mihály, A. Bóta, M.S.Z. Kellermayer, J. Szebeni, and B. Kálmán. 2016. "Aggregation of PEGylated liposomes driven by hydrophobic forces." *Colloids and Surfaces B: Biointerfaces*. https://doi.org/10.1016/j.colsurfb.2016.06.056.

Çırpanlı, Y. 2009. *Development of polymeric and oligosaccaride based nanoparticular formulations containing camptotesi and evaluation of in vitro-in vivo*. PhD. Thesis. Hacettepe Universty, Ankara.

Curic, A., M.J. Peter, and G. Fricker. 2017. "Development and characterization of novel highly-loaded itraconazole poly (butyl cyanoacrylate) polymeric nanoparticles."

European Journal of Pharmaceutics and Biopharmaceutics 114: 175–185. doi:10.1016/j. ejpb.2017.01.014.

Dang, Sç, D. Sharma, G. Philip, R. Gabrani, and J. Ali. 2015. "Dual agents loaded polymeric nanoparticle: Effect of process variables." International Journal of Pharmaceutical Investigation. https://doi.org/10.4103/2230-973x.160853.

De, A., R. Bose, A. Kumar, and S. Mozumdar. 2014. *"Targeted delivery of pesticides using biodegradable polymeric nanoparticles."* In Springer Briefs in Molecular Science. Springer New Delhi Heidelberg New York Dordrecht London.

Derman, S., K. Kizilbey, and Z. Mustafaeva Akdeste. 2013. "Polymeric nanoparticles." *Journal of Engineering and Natural Sciences* 31(May 2014): 109–122.

Desai, K.G. 2016. "Chitosan nanoparticles prepared by ionotropic gelation: An overview of recent advances." *Critical Reviews in Therapeutic Drug Carrier Systems*. https://doi. org/10.1615/CritRevTherDrugCarrierSyst.2016014850.

Ding, S., M.F. Attia, J. Wallyn, C. Taddei, C.A. Serra, N. Anton, M. Kassem, et al. 2018. "Microfluidic-assisted production of size-controlled superparamagnetic iron oxide nanoparticles-loaded poly(methyl methacrylate) nanohybrids." *Langmuir*. https://doi. org/10.1021/acs.langmuir.7b01928.

Dreis, S., F. Rothweiler, M. Michaelis, J. Cinatl, J. Kreuter, and K. Langer. 2007. "Preparation, characterisation and maintenance of drug efficacy of doxorubicin-loaded human serum albumin (HSA) nanoparticles." *International Journal of Pharmaceutics*. https://doi. org/10.1016/j.ijpharm.2007.03.036.

Emami, J., A. Pourmashhadi, H. Sadeghi, J. Varshosaz, and H. Hamishehkar. 2014. "Formulation and optimization of celecoxib-loaded PLGA nanoparticles by the tagu-chi design and their in vitro cytotoxicity for lung cancer therapy." *Pharmaceutical Development and Technology* 1–10. https://doi.org/10.3109/10837450.2014.920360.

Esfandyari-Manesh, M., M. Abdi, A.H. Talasaz, S.M. Ebrahimi, F. Atyabi, and R. Dinarvand. 2020. "S2P peptide-conjugated PLGA-Maleimide-PEG nanoparticles containing imatinib for targeting drug delivery to atherosclerotic plaques." *DARU, Journal of Pharmaceutical Sciences*. https://doi.org/10.1007/s40199-019-00324-w.

Essa, D., P.P.D. Kondiah, Y.E. Choonara, and V. Pillay. 2020. "The design of poly(lactide-co-glycolide) nanocarriers for medical applications." *Frontiers in Bioengineering and Biotechnology* 8(February): 1–20. https://doi.org/10.3389/fbioe.2020.00048.

Fessi, H., F. Puisieux, J.P. Devissaguet, N. Ammoury, and S. Benita. 1989. "Nanocapsule formation by interfacial polymer deposition following solvent displacement." *International Journal of Pharmaceutics*. https://doi.org/10.1016/0378-5173(89)90281-0.

Fessi, H., J.P. Devissaguet, F. Puisieux, and C. Thies. 1991. *Process for the preparation of dispersible colloidal systems of a substance in the form of nanoparticles*. U.S., issued 1991.

Galindo-Rodriguez, S., E. Allémann, H. Fessi, and E. Doelker. 2004. "Physicochemical parameters associated with nanoparticle formation in the salting-out, emulsification-diffusion, and nanoprecipitation methods." *Pharmaceutical Research* 21(8): 1428–1439. doi:10.1023/B:PHAM.0000036917.75634.be.

Ghitman, J., S. Raluca, and H. Iovu. 2017. "Experimental contributions in the synthesis of PLGA nanoparticles with excellent properties for drug delivery : Investigation of key parameters experimental contributions in the synthesis of PLGA nanoparticles with excellent properties for drug delivery." The Journal Scientific Bulletin Series B 79(2): 101–112.

Gok, M.K. 2012. *Preparation of natural and synthetic biocompatible polymeric nanoparticular gene delivery systems and investigation of transfection effect*. Istanbul University, Institute of Science, Chemical Engineering Department, Chemical Technologies Department PhD thesis.

Hernández-Giottonini, K.Y., R.J. Rodríguez-Córdova, C.A. Gutiérrez-Valenzuela, O. Peñuñuri-Miranda, P. Zavala-Rivera, P. Guerrero-Germán, and A. Lucero-Acuña. 2020. "PLGA nanoparticle preparations by emulsification and nanoprecipitation techniques: Effects of formulation parameters." *RSC Advances.* https://doi.org/10.1039/c9ra10857b.

Jokerst, J.V., T. Lobovkina, R.N. Zare, and S.S. Gambhir. 2011. "Nanoparticle PEGylation for imaging and therapy." *Nanomedicine.* https://doi.org/10.2217/nnm.11.19.

Julienne, M.C., M.J. Alonso, and J.P. Benoit. 1992. "Preparation of poly (D, L- lactide/glycolide) nanoparticles of controlled particle size distribution: Application of experimental designs." *Drug Development And Industrial Pharmacy* 18: 1063–1077.

Jung, T., A. Breitenbach, and T. Kissel. 2000. "Sulfobutylated poly(vinyl alcohol)-graft-poly(lactide-co-glycolide)s facilitate the preparation of small negatively charged biodegradable nanospheres." *Journal of Controlled Release* 67(2–3): 157–169. doi:10.1016/S0168-3659(00)00201-7.

Khan, M.S., K. Rohitash, M. Vijaykumar, S.C. Pandey, D. Gowda, F.M. Ahmed, A.R. Sidiqui, and M.S. Khan. 2012. "Development and evaluation of nasal mucoadhesive nanoparticles of an analgesic drug." *Der Pharmacia Lettre* 4(6): 1846–1854.

Kökcu, Y. 2019. "Development of Antitumor and Antioxidant Featured Tripeptide Loaded Polymeric Nanoparticules." M.Sc Thesis. Istanbul University, Istanbul.

Kulkarni, A.A., and P.S. Rao. 2013. *Synthesis of polymeric nanomaterials for biomedical applications. nanomaterials in tissue engineering: Fabrication and applications.* Woodhead Publishing Limited. https://doi.org/10.1533/9780857097231.1.27.

Lai, P., W. Daear, R. Löbenberg, and E.J. Prenner. 2014. "Overview of the preparation of organic polymeric nanoparticles for drug delivery based on gelatine, chitosan, poly(d,l-lactide-co-glycolic acid) and polyalkylcyanoacrylate." *Colloids and Surfaces B: Biointerfaces.* https://doi.org/10.1016/j.colsurfb.2014.03.017.

Lee, E.J. and K.H. Lim. 2017. "Hardly water-soluble drug-loaded gelatin nanoparticles sustaining a slow release: Preparation by novel single-step O / W / O emulsion accompanying solvent diffusion." *Bioprocess and Biosystems Engineering.* https://doi.org/10.1007/s00449-017-1825-8.

López-Gasco, P., I. Iglesias, J. Benedí, R. Lozano, J.M. Teijón, and M.D. Blanco. 2011. "Paclitaxel-loaded polyester nanoparticles prepared by spray-drying technology: In vitro bioactivity evaluation." *Journal of Microencapsulation* 28(5): 417–429. doi:10.31 09/02652048.2011.576785.

Mai, H., L. Thi, T.T. Dung, T.M. Danh, N.H. Duc, and D.M. Chien. 2009. "Preparation and characterization of magnetic nanoparticles coated with polyethylene glycol." *Journal of Physics: Conference Series.* https://doi.org/10.1088/1742-6596/187/1/012048.

Martins, J.P., D. Liu, F. Fontana, M.P.A. Ferreira, A. Correia, S. Valentino, M. Kemell, et al. 2018. "Microfluidic nanoassembly of bioengineered chitosan-modified FcRn-targeted porous silicon nanoparticles @ hypromellose acetate succinate for oral delivery of antidiabetic peptides." *ACS Applied Materials and Interfaces* 10(51): 44354–44367. doi:10.1021/acsami.8b20821.

Matsuyama, K. and K. Mishima. 2009. "Preparation of poly(methyl methacrylate)-TiO2 nanoparticle composites by pseudo-dispersion polymerization of methyl methacrylate in supercritical CO2." *Journal of Supercritical Fluids.* https://doi.org/10.1016/j.supflu.2009.03.001.

Miele, D., S. Rossi, G. Sandri, B. Vigani, M. Sorrenti, P. Giunchedi, F. Ferrari, and M.C. Bonferoni. 2018. "Chitosan oleate salt as an amphiphilic polymer for the surface modification of poly-lactic-glycolic acid (PLGA) nanoparticles. preliminary studies of mucoadhesion and cell interaction properties." *Marine Drugs.* https://doi.org/10.3390/md16110447.

Mohammed, M., H. Mansell, A. Shoker, K.M. Wasan, and E.K. Wasan. 2019. "Development and in vitro characterization of chitosan-coated polymeric nanoparticles for oral delivery and sustained release of the immunosuppressant drug mycophenolate mofetil." *Drug Development and Industrial Pharmacy.* https://doi.org/10.1080/03639045.2018. 1518455.

Mora-Huertas, C.E., H. Fessi, and A. Elaissari. 2010. "Polymer-based nanocapsules for drug delivery." *International Journal of Pharmaceutics.* https://doi.org/10.1016/j. ijpharm.2009.10.018.

Muthu, M.S. 2009. "Nanoparticles based on PLGA and its co-polymer: An overview." *Asian Journal of Pharmaceutics.* https://doi.org/10.4103/0973-8398.59948.

Nasir, A., A. Kausar, and A. Younus. 2014. "A review on preparation, properties and applications of polymeric nanoparticle-based materials a review on preparation, properties and applications of polymeric nanoparticle-based materials." *Polymer-Plastics Technology and Engineering* 2015: 37–41. doi:10.1080/03602559.2014.958780.

Owens, D.E., Y. Jian, J.E. Fang, B.V. Slaughter, Y.H. Chen, and N.A. Peppas. 2007. "Thermally responsive swelling properties of polyacrylamide/poly(acrylic acid) interpenetrating polymer network nanoparticles." *Macromolecules.* https://doi.org/10.1021/ma071089x.

Parde, C., B. Balzus, F. Feleke, S. Hönzke, C. Gerecke, F. Schumacher, S. Hedtrich, et al. 2017. "Development of lipomer nanoparticles for the enhancement of drug release, anti-microbial activity and bioavailability of delafloxacin." *European Journal of Pharmaceutics and Biopharmaceutics* 10. https://doi.org/10.3109/10717544.2014.948 643.

Petzold, G. and S. Schwarz. 2006. "Dye removal from solutions and sludges by using polyelectrolytes and polyelectrolyte-surfactant complexes." *Separation and Purification Technology* 51(3): 318–324. doi:10.1016/j.seppur.2006.02.016.

Quérette, T., C. Bordes, and N. Sintes-Zydowicz. 2020. "Non-isocyanate polyurethane nanoprecipitation: Toward an optimized preparation of poly(hydroxy)urethane nanoparticles." *Colloids and Surfaces A: Physicochemical and Engineering Aspects.* https://doi.org/10.1016/j.colsurfa.2019.124371.

Quérette, T., E. Fleury, and N. Sintes-Zydowicza. 2019. "Non-isocyanate polyurethane nanoparticles prepared by nanoprecipitation." *European Polymer Journal* 114: 434–445. doi:10.1016/j.eurpolymj.2019.03.006.

Ramanujam, R., B. Sundaram, G. Janarthanan, E. Devendran, M. Venkadasalam, and M.C.J. Milton. 2018. "Biodegradable polycaprolactone nanoparticles based drug delivery systems: A short review." *Biosciences, Biotechnology Research Asia.* https://doi.org/10.13005/bbra/2676.

Rampino, A., M. Borgogna, P. Blasi, B. Bellich, and A Cesàro. 2013. "Chitosan nanoparticles: Preparation, size evolution and stability." *International Journal of Pharmaceutics.* https://doi.org/10.1016/j.ijpharm.2013.07.034.

Rao, J.P. and K.E. Geckeler. 2011a. "Polymer nanoparticles: Preparation techniques and size-control parameters." *Progress in Polymer Science (Oxford).* https://doi.org/10.1016/j .progpolymsci.2011.01.001.

Rao, J.P. and K.E. Geckeler. 2011b. "Progress in polymer science polymer nanoparticles : Preparation techniques and size-control parameters." *Progress in Polymer Science* 36(7): 887–913. doi:10.1016/j.progpolymsci.2011.01.001.

Senel, S., M.J. Kremer, S. Kaş, P.W. Wertz, a.a. Hincal, and C.a. Squier. 2000. "Enhancing effect of Chitosan on peptide drug delivery across Buccal Mucosa." *Biomaterials* 21(20): 2067–2071. doi:10.1016/s0142-9612(00)00134-4.

Shaji, J. and M. Kumbhar. 2019. "Formulation and characterization of linezolid loaded eudragit Rs 100 polymeric nanoparticles." *IJPSR* 10(4): 1944–1952. https://doi.org/10.1 3040/IJPSR.0975-8232.10(4).1944-52.

Suk, J.S., Q. Xu, N. Kim, J. Hanes, and L.M. Ensign. 2016. "PEGylation as a strategy for improving nanoparticle-based drug and gene delivery." *Advanced Drug Delivery Reviews.* https://doi.org/10.1016/j.addr.2015.09.012.

Şengel Türk, C. T., C. Hasçiçek, and N. Gönül. 2007. "Nanoparticulate Drug Delivery Systems for Targeting the Drugs to the Brain." *Journal of Neurolagicial Sciences[Turkish]* 24(3): 254–263.

Wang, H., X. Qiao, J. Chen, X. Wang, and S. Ding. 2005. "Mechanisms of PVP in the preparation of silver nanoparticles." *Materials Chemistry and Physics.* https://doi.org/10.1016/j.matchemphys.2005.05.005.

Wang, R., P.J. Dijkstra, and M. Karperien. 2016. "Dextran." *Biomaterials from Nature for Advanced Devices and Therapies.* https://doi.org/10.1002/9781119126218.ch18.

Wang, Y., P. Li, T. Truong, D. Tran, J. Zhang, and L. Kong. 2016a. "Manufacturing techniques and surface engineering of polymer based nanoparticles for targeted drug delivery to cancer." *Nanomaterials* 6(2): 1–7. https://doi.org/10.3390/nano6020026.

Wei, X.W., C.Y. Gong, M.L. Gou, S.Z. Fu, Q.F. Guo, S. Shi, F. Luo, G. Guo, L.Y. Qiu, and Z.Y. Qian. 2009. "Biodegradable poly(ε-caprolactone)-poly(ethylene glycol) copolymers as drug delivery system." *International Journal of Pharmaceutics* 381(1): 1–18. doi:10.1016/j.ijpharm.2009.07.033.

Yan, L, X. Cui, T. Harada, S.F. Lincoln, S. Dai, and T.W. Kee. 2016. "Generation of fluorescent and stable conjugated polymer nanoparticles with hydrophobically modified poly(acrylate)S." *Macromolecules.* https://doi.org/10.1021/acs.macromol.6b02002.

Yoo, M.K., S.K. Kang, J.H. Choi, I.K. Park, H.S. Na, H.C. Lee, and E.B. Kim, et al. 2010. "Targeted delivery of chitosan nanoparticles to peyer's patch using M cell-homing peptide selected by phage display technique." *Biomaterials.* https://doi.org/10.1016/j.biomaterials.2010.06.059.

Yurtdaş Kırımlıoğlu, G. and A.A. Öztürk. 2020. "Levocetirizine dihydrochloride-loaded chitosan nanoparticles: Formulation and in vitro evaluation." *Turkish Journal of Pharmacetical Science* 17(1): 27–35. doi:10.4274/tjps.galenos.2018.34392.

Zhong, T., Y. Jiao, L. Guo, J. Ding, Z. Nie, L. Tan, and R. Huang. 2017. "Investigations on porous PLA composite scaffolds with amphiphilic block PLA −b -PEG to enhance the carrying property for hydrophilic drugs of excess dose." *Journal of Applied Polymer Science* 134(8): 1–7. doi:10.1002/app.44489.

Zohri, M., T. Gazori, S.S. Mirdamadi, A. Asadi, and I. Haririan. 2009. "Polymeric nanoparticles: Production, applications and advantage." *Journal of Nanotechnology* 3(1).

9 Peptides in Brain Disorders

Hüseyin Ünübol and Gökben Hızlı Sayar

CONTENTS

> *One of the ways to harm the patient is to leave the patient untreated.*
>
> **Professor Dr. Nevzat Tarhan**

9.1 INTRODUCTION

The brain is known as a communication center made up of billions of nerve cells (neurons). Neural networks allow messages to be transmitted to different structures within the spinal column and peripheral nervous system. These neural networks coordinate and regulate everything we feel, think, and do, that is, control almost all functions. The human brain is the most complex organ in the body. The brain is about 1300 grams of gray and white matter at the very center of all activities. It is essential for many jobs, for example driving a car, enjoying food, bringing out a work of art, and enjoying the daily activities we need. Briefly, the brain regulates basic body functions, undertakes the function of interpreting and reacting to events and shaping emotions, thoughts, and behaviors. The brain consists of many parts that work in teams. Different brain regions are responsible for the coordination and administration of specific functions (Tarhan and Nurmedov 2011). Peptides have the ability to affect important brain regions required for life-sustaining functions, so there are studies on peptide drug and peptide-polymer vaccines in brain diseases, and the facts about the benefits of peptides are revealed. In this chapter, explanations about the use of peptides in common brain diseases such as addiction disorder, Alzheimer's disease, Parkinson's disease, multiple sclerosis, and psychiatric diseases will be made and the importance of peptides will be emphasized.

9.2   HOW DOES THE BRAIN COMMUNICATE?

From neuron to neuron, nerve cells in the brain receive and transmit messages in the form of electrical impulses. After the nerve cell receives the message, it allows it to be sent to other neurons. Neurotransmitters are chemical precursors of the brain. Messages between neurons are carried through these chemicals. They transmit messages between neurons.

Receptors are the chemical receptors of the brain and connect to a specific site on the recipient cell called a neurotransmitter receptor.

The neurotransmitter and its receptor act as keys and locks. In this specific mechanism, each receptor transmits the appropriate message only after interacting with the correct neurotransmitter type. Carriers signal brain chemical converters. Carriers are located in a neurotransmitter-secreting cell and recycle these neurotransmitters (e.g. return them to the secreted cell), thus interrupting the neuronal signal. Other substances, such as amphetamine or cocaine, can cause a considerable amount of natural neurotransmitter secretion of nerve cells or prevent normal recycling of these brain chemicals. Because of this glitch it is highly amplified, so the message that has been raised at the end disrupts the communication channels. The difference in effect can be compared to the difference between whispering into the ear and shouting into a microphone.

To understand brain disorders dopaminergic, GABAergic, opioid, cholinergic, serotonergic, and other pathway receptors and the effects of genetic differences on carriers are very important. Functional disorders of these pathways cause brain disorders such as schizophrenia, ADHD, impulse control disorders, addiction, Parkinson's and Alzheimer's disease.

9.3   ADDICTIVE DISORDER

Studies in alcohol, smoking, sex, gambling, and drug-metabolizing enzymes such as dependence, dopaminergic, GABAergic, opioid, cholinergic, serotonergic, and other pathway receptors, and the effects of genetic differences on carriers were investigated. Dopamine is a very important pathway because of understanding the mechanism of some brain disorders including addiction.

9.3.1   THE BRAIN (DOPAMINE) PATHWAYS

First of all, brain regions such as the frontal cortex, accumulator nucleus, and ventral tegmental are important for the natural reward mechanisms perceived by the brain, such as food, music, and art. Typically, there is an increase in dopamine during the response to natural rewards such as food. But the other (smoking, alcohol, drugs, etc.) targets the brain's reward system by installing dopamine in addictive substances. The final support for this information comes from many functional magnetic resonance imaging (fMRI) studies. In these studies on humans, it was found that alcohol, nicotine, cocaine, chocolate, and opiate craving is induced by a reminder of increased metabolic activity in the anterior cingulate gyrus and other frontal lobe areas of the brain (Tarhan and Nurmedov 2011).

9.3.1.1 What Is Dopamine?

Dopamine is a chemical that plays an important role in hormone regulation of the central nervous system. Gender and its effect on growth hormones are known. Dopamine secretion in the brain of a depressed person decreases with serotonin. There is a horizontal relationship between dopamine and the chemical serotonin, which is related to the sense of happiness, which is related to dopamine and happiness. Lack of dopamine actually means lack of reward. The functions of the pathways carrying *dopamine* to other parts of the brain are described below:

a) Nigrostriatal pathway: It extends to the dorsal area of the substantia sigra and the striatum. The combination of complex movements with cognitive perception, sensory-motor coordination, and initiation of movement are related to the dopamine neurotransmitter herein.

b) Mesolimbic pathway: It extends from the ventral tegmental area (VTA) to the ventral aspect of the striatum, including the amygdala, hippocampus, and orbitofrontal cortex. Because it includes the limbic system, it has emotional, cognitive, and sensorimotor functions, and as a result, pleasant feelings are associated with the reward.

c) Mesocortical pathway: The anterior cingulate from the VTA extends into the parietal and temporal areas, including the dorsal striatum, the prefrontal cortex, and the thalamus. It plays a role in functions such as memory.

d) Tuberoinfundibular pathway: It extends from the hypothalamus to the pituitary. There are also primary effects on prolactin level and other hormonal functions.

In sleepless individuals, the brain increases dopamine secretion as a defense (the presynaptic receptor sensitivity is reduced, negative feedback is released, and dopamine is secreted). For this reason, insomnia can reduce depression, while mania can trigger mental illness. The same mechanism plays an important role by increasing the release of this chemical related to the sense of pleasure in substance addiction. Symptoms such as shivering when substance use is discontinued are similar to those seen in Parkinson's patients. The Parkinson's dopamine system is associated with neuron damage. Studies show that, with the chronic use of alcohol in the brain, dopaminergic system activity is slowing down. Dopamine release increased at the beginning of alcohol use while the decrease in long-term use indicates the effect of alcohol on the histotoxicity.

9.4 ATTENTION DEFICIT HYPERACTIVITY DISORDER (ADHD)

The tasks of the frontal lobe are the alignment of incoming information, associating current experiences with past experiences, controlling behaviors, suppressing inappropriate reactions, and organizing and planning for the future. These are also called executive functions of the frontal lobe. The principal of the executive functions is the ability to initiate, sustain, inhibit, and distract attention. Therefore, dysfunction in the frontal lobe can lead to impairment of attention, impulse control, and/or cognitive activity.

9.4.1 How Are Movements Planned?

The prefrontal cortex is the region that briefly converts the senses: attention, perception, perceptual analysis, abstract thinking, and social behavior from all lobes into behavior and works in cooperation with the amygdala and thalamus. Glutamatergic and dopaminergic activity are communicated to each other by providing communication between neurons (Bozzi and Borrelli 2006).

MOTOR CORTEX. The region where all accounts are made and decisions made before a move is made (STOP AND THINK BEFORE YOU DO!). Then select the appropriate movement of synaptic neurons (information bearing) from the PREMOTOR CORTEX. After deciding on the appropriate motion, it also enters the field of engine movement (motor vehicle field). Managerial functions: motor planning, distracting, changing cognitive sets, monitoring, and adapting behavior through attention and processing memory, are associated with dopaminergic activity. Patients with ADHD have generally evolved to include three subtypes as diagnostic criteria: predominantly careless subtype (most common in females), predominantly hyperactive/impulsive subtype (most commonly in males), and united subspecies (Tarhan 2016) (Figure 9.1).

ADHD. Because it is defined as a neurocognitive disease characterized by hyperactivity, carelessness, working memory defects, and impulsivity, the use of peptide drugs makes a significant contribution to the development of treatment modulation of the neurocognitive approach. Although the etiology of ADHD is not fully known, the disorder of the dopaminergic system is highly affected by the effects of current ADHD therapies. In the most common treatment mechanism known for this disease, there are stimulants (e.g. methylphenidate) that increase synaptic dopamine by blocking the dopamine transporter (DAT) directly (Polanczyk et al. 2007, 2015). Although these pharmacological agents are effective, they are associated with various side effects, including risks for future substance use disorders in patients with ADHD.

In a study to elute a protein complex resulting from the interaction between DAT and the dopamine D2 receptor (D2R) there was a use of an interference-inducing peptide (TAT-DATNT) on Sprague-Dawley (SD) rats and it was determined that locomotor behavior was induced with the peptides that were used (TAT-DATNT). Furthermore, using in vivo microdialysis and high-performance liquid chromatography, it was found that the degradation of D2R-DAT increases the level of extracellular dopamine. More importantly, the involved TAT-DATNT peptide spontaneously

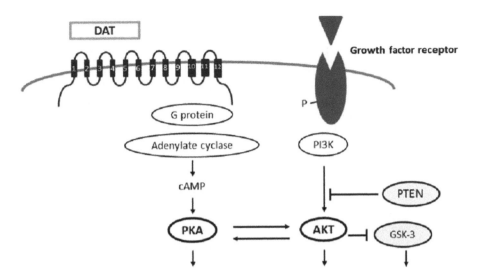

FIGURE 9.1 The importance of DAT (dopamine transporter) in conducting cellular activation in the postsynaptic membrane and the importance of working in cooperation with growth factors (Kitagishi et al. 2015).

changes the hyperactivity in the spontaneous hypertensive rat animal model (SHR). This peptide offers a different way to regulate the activity of DAT and dopaminergic neurotransmission, and a potential target region for future development of ADHD therapies (Lai et al. 2018).

There are reports, in contrast to methylphenidate at low doses and in high doses, that SHR failed to offer the same therapeutic effects on hyperactivity in rats (Cao et al. 2012).

A similar study showed that a higher dose of TAT-DATNT (4.0 nmol) in SHR rats was tested with different dose effects, and showed a U-shaped dose-response curve on hyperactivity in SHR, but when given much a higher dose, methylphenidate on the contrary, it increases excessive dopaminergic neurotransmission and fulfills the stimulatory effects observed in SHR rats. (Lai et al. 2018). Wistar Kyoto (WKY) rats reported higher DAT levels in the striatum of SHR rats than in WKY rats at two weeks of age when compared to a control strain for SHR rats (Watanabe et al. 1997). There was no significant difference in D2R or DAT levels between SHR and WKY rats in general. TAT-DATNT at the same dose did not have an effect on locomotor activity in WKY rats, but the dopaminergic effect was found in SD rats. On the other hand, similar to the effects of TAT-DATNT, the effects of methylphenidate on WKY rats were dose-dependent (Umehara et al. 2013; Yang et al. 2006).

9.5 MULTIPLE SCLEROSIS DISORDER

Myelin oligodendrocyte glycoprotein (MOG) is a protein specifically expressed on the surface of oligodendrocytes and myelin in the central nervous system and is of great importance. MOG has been identified and studied as a putative candidate auto-antibody target in MS. An immunogenic MOG35-55 peptide (MEVGWYRSPFSRVVHLYRNGK) (Heidari et al. 2019; Kamisli et al. 2018) with solid phase peptide synthesis method is used for vaccination in experimental autoimmune encephalomyelitis (EAE) type of MS model. Poly (N-vinyl-2-pyrrolidone-co-acrylic acid) polymer will be synthesized by the method described earlier (Shibata, Nakagawa, and Tsutsumi 2005; Kamada et al. 2003; Kodaira et al. 2004) and polylacrylate-glycolic acid (PLGA) nanoparticles will be synthesized by the method described earlier (Çalman et al. 2018). The nanoparticle-peptide and bipolymer-peptide conjugation method were described earlier (Acar 2006; Hermanson 1996; Mansuroğlu 2007; Kızılbey 2012). If this peptide can be used with all these biopolymers, an immune response could be high. (EAYKAAEKA YAAKEAAKEAAKAKAEKKAAYAKAKAAKYEKKAKKAAAEYKKK) an approved peptide drug is also delivered by the solid phase peptide synthesis method. The drug name is glatiramer and is from the DrugBank pharmaco-informatic database program.

9.6 PARKINSON'S DISEASE

Parkinson's disease (PD) is known to be one of the most common neurodegenerative brain diseases in the world. PD is characterized by different effects in the patient, for example, it is a chronic neurodegenerative disorder characterized by the loss of dopaminergic neurons in the substantia nigra, the loss of striatal dopamine levels, and the resulting extrapyramidal motor dysfunction. The differential diagnosis of PD is difficult to detect especially in the early stage of the disease because PD is genetically and pathologically diverse. This is a factor that indicates that PD has different causes of disease diagnosis. For example, alpha synuclein (α-syn) aggregate exists in certain patients, while Lewy bodies might be missing or the two pathologies can be watched remembering dynamic changes for dopamine discharge and striatal content, α-syn pathology, deficiencies in engine works that are influenced in pre-show, and show periods of PD, irritation, and biochemical and sub-atomic changes like those seen in PD. α-syn aggregate is very important for the disease. An immunogenic peptide with the sequence of CGGVDPDN (Games et al. 2013) is developed with solid phase peptide synthesis method as a vaccine in PD for this aggregate. So, it can be said that peptides can be neuroprotection for the PD, such as the TFP5 peptide, FITCGGGKEAFWDRCLSVINLMSSKMLQINAYARAARRAARR, SCP peptide FITCGGGGGGFWDRCLSGKGKMSSKGGGINAYARAARRAARR, and TP5 peptide KEAFWDRCLSVINLMSSKMLQINAYARAARRAARR are depleted in neuroinflamation and apoptosis. In a study 1-methyl-4-phenyl-1,2,3,6-tetrahydropyridine (MPTP/MPP) was used for PD in a mouse model (Binukumar et al. 2015). Also PD01A is a Phase 1 epitope vaccine in experiment – a synthetic α-syn mimicking peptide-polymer based on α-syn aggregate by inducing an immune response that generates antibodies specifically against it and is a potential cure for PD (Romero-Ramos et al. 2014). Frequently vaccines are developed for T cell response but immunogenicity of the epitope vaccines composed of three different α-syn peptides fused with a foreign universal T cell epitope but three peptide-based epitope vaccines are given below, composed of different B cell epitopes such as Syn85-99 (AGSIAAATGFVKKDQ), α-syn109-126 (QEGILEDMPVDPDNEAYE), α-syn126-140 (EMPSEEGYQDYEPEA), and P30 (FNNFTVSFWLRVPKVSASHLE) peptides (Ghochikyan et al. 2014). As a result, peptides can have different immune cell response in brain disorders.

9.7   OTHER PSYCHIATRIC DISORDERS

Peptides are not used for depression, bipolar disorders, borderline disorders, and other specific psychiatric disorders but there are studies for schizophrenia. The pathophysiology of schizophrenia is related to imbalances in the dopaminergic system, but also includes neurochemical changes in other neurotransmitter systems that may be indirectly related to serotonergic, glutamatergic, central cholinergic, and gamma-aminobutyric systems. Antipsychotic drugs (APDs) are widely used for disease-positive symptoms. Classical APDs are designed to correct neurochemical imbalances and are therefore based on dopamine (DA) D2 receptor antagonists. Despite the effectiveness of treatment with APDs, 30% of patients do not deal with schizophrenia symptomatology and 60% continue to have positive symptoms. POP peptide is an 81 kDa monomeric serine peptidase and may be effective for the cognitive impairment of schizophrenia. The interaction of POP with other cells, independent of its catalytic activity, can be observed by effects on neuron growth, synaptic plasticity, and protein secretion (López et al. 2013). So, peptides can also be used for cell growth, cell penetrating, and protein expression in cognitive diseases. As is known, peptides are also useful in protecting cell biological agents such as microtubule activity. The benefit of protecting biological agents is very important in maintaining cell stability during diseases. There is an imbalance in the expression of ADNP/ADNP2 in the brain that can affect the progression of the disease in the schizophrenia model that exists in the brain tissue associated with schizophrenia and in animal models. Among the traits of these genes are the regulation of interacting microtubules, so their impairment can lead to the formation of the schizophrenia model. Its activity depends on the neuroprotective protein (ADNP) which contains a small peptide motif as NAPVSIPQ sequence that provides potent neuroprotection for tau pathology, neuronal cell death as well as social and cognitive dysfunctions. (Gozes 2011). Especially in amyloid beta pathology, DAEFRHDSGY (Asp - Ala - Glu - Phe - Arg - His - Asp - Ser - Gly - Tyr) peptide, Wang et al. synthesized peptide immunogens, A1-14 peptide immunogens for UBITh® AD immunotherapeutic vaccine by using automated solid-phase synthesis for Alzheimer's disease. Wang's group holds two immunogenic peptide patents which have high responder rates, strong on-target immunogenicity, and a potential of cognition improvement, composition as vaccines for the prevention and treatment of Alzheimer's disease (Wang et al. 2017). In Table 9.1 studies have shown that peptides may be useful in neuronal diseases. As described in Chapter 13, peptides have beneficial properties, especially antioxidant peptides, and their therapeutic properties in brain diseases and their therapeutic properties in patients should be investigated.

TABLE 9.1

Peptides That Are Used in Neuronal Diseases

Peptides	Function	References
PACAP	Acts as a regulatory peptide stimulates adenylate cyclase; helps improve cognition	Nonaka et al. 2002; Rat et al. 2011; Nonaka et al. 2012
Exendin	It acts as an agonist for GLP-1 receptors, which causes increased insulin release; it helps to lower blood sugar levels	During et al. 2003; Banks, During, and Niehoff 2004
β-Sheet Breaker	It is seen as therapeutic candidate for the treatment of AD	Bieler and Soto 2004
Albumin	It helps transport molecules for targeted drug delivery, improves the stability of substrates and changes delivery patterns	Morales 2007; Migliore et al. 2010; Falcone et al. 2014
Vasoactive Intestinal Peptides (VIP)	Used in the pathogenesis of large NDs including AD, HD, and PD	White et al. 2010; Morell, Souza-Moreira, and Gonzalez-Rey 2012
GALP	Helps regulate the feeding and reduce body weight	Nonaka et al. 2008; Shiba et al. 2010; Shioda et al. 2011
Leptin	It is an anorexigenic satiety signaling factor, which helps to reduce weight and food intake	Novakovic et al. 2009; Schulz et al. 2012
Peptides Aβ (16-20) KLVFF	Peptide inhibitor for AD	Funke and Willbold 2012
HKQLPFFEED	Inhibitor in AD	Lin et al. 2014
RYYAAFFARR	Act as an inhibitor in AD	Liu et al. 2014
GV1001 is a novel 16-amino-acid peptide	Oxidative stress causes the death of neural stem cells, this peptide functions by exerting antioxidant effects, increasing survival signals and decreasing death signals	Hyun-Hee et al. 2019
AP90	Inhibits Ab aggregation especially in AD	Kellock et al. 2016
IGLMVG-NH2	Act as an inhibitor in AD	Bansal et al. 2016
Nonhemolytic 11-residue peptide (NAVRWSLMRPF) from the E-protein sequence of SARS coronavirus	Act as an inhibitor in AD	Ghosh et al. 2017
PABA-functionalized peptide dendrimers	Act as Antioxidants with cationic and neutral radicals scavenging potency. Also, it acts as new compounds with marked selectivity against human melanoma cell or glutamate-stressed CGC neurons	Sowinska et al. 2019

Abreviations: Alzheimer's disease (AD); cerebral giant cell (CGC); glucagon-like peptide-1(GLP1); Huntington's disease (HD); neuronal disease (ND); Parkinson's disease (PD)

So as a result, according to the literature, brain disorders need to be studied more at the molecular mechanism level. In this molecular mechanism, peptides can play important roles and be useful for future treatments, but further investigation is needed.

REFERENCES

Acar, S. 2006. *Peptide protein covalent conjugation.* Istanbul: YTU Institute of Science and Technology, Bioengineering Program.

Banks, W.A., M.J. During, and M.L. Niehoff. 2004. "Brain uptake of the glucagon-like peptide-1 antagonist exendin(9–39) after intranasal administration." *Journal of Pharmacology Experimental Therapeutics* 309(2): 469–475. doi:10.1124/jpet.103.063222.

Bansal, S., I.K. Maurya, N. Yadav, C.K. Thota, V. Kumar, K. Tikoo, V.S. Chauhan, and R. Jain. 2016. "C-terminal fragment, Ab32–37, analogues protect against ab aggregation induced toxicity." *ACS Chemical Neuroscience* 7: 615–623. doi:10.1021/acschemneuro.6b00006.

Bieler, S. and C. Soto. 2004. "Beta-sheet breakers for Alzheimer's disease therapy." *Current Drug Targets* 5(6): 553–558. doi:10.2174/1389450043345290.

Binukumar, B., V. Shukla, N.D. Amin, P. Grant, M. Bhaskar, S. Skuntz, ... H.C. Pant. 2015. "Peptide TFP5/TP5 derived from Cdk5 activator P35 provides neuroprotection in the MPTP model of Parkinson's disease." *Molecular Biology of the Cell* 26(24): 4478–4491.

Bozzi, Y. and E. Borrelli. 2006. "Dopamine in neurotoxicity and neuroprotection: What do D2 receptors have to do with it?" *Trends in Neurosciences* 29(3): 167–174.

Çalman, F., P. Pelit Arayıcı, H.K. Büyükbayraktar, M. Karahan, Z. Mustafaeva, and R. Katsarava. 2018. "Development of vaccine prototype against zika virus disease of peptide-loaded PLGA nanoparticles and evaluation of cytotoxicity." *International Journal of Peptide Research and Therapeutics* 25: 1057–1063.

Cao, A.H., L. Yu, Y.W. Wang, J.M. Wang, L.J. Yang, and G.F. Lei. 2012. "Effects of methylphenidate on attentional set-shifting in a genetic model of attention-deficit/hyperactivity disorder." *Behavioral and Brain Functions* 8(1): 10.

During, M.J., L. Cao, D.S. Zuzga, L.S. Francis, H.L. Fitzsimons, X. Jiao, R.I. Bland, et al. 2003. "Glucagon-like peptide-1 receptor is involved in learning and neuroprotection." *Nature Medicine* 9(9): 1173–1179. doi:10.1038/nm919.

Falcone, J.A., T.S. Salameh, X. Yi, B.J. Cordy, W.G. Mortell, A.V. Kabanov, and W.A. Banks. 2014. "Intranasal administration as a route for drug delivery to the brain: Evidence for a unique pathway for albumin." *Journal of Pharmacology and Experimental Therapeutics* 351(1): 54–60. doi:10.1124/jpet.114.216705.

Funke, S.A. and D. Willbold. 2012. "Peptides for therapy and diagnosis of Alzheimer's disease." *Current Pharmaceutical Desingn* 18(6): 755–767. doi:10.2174/138161212799277752.

Games, D., P. Seubert, E. Rockenstein, C. Patrick, M. Trejo, K. Ubhi, and B. Ettle. 2013. "Axonopathy in an α-Synuclein transgenic model of lewy body disease is associated with extensive accumulation of C-terminal–truncated α-synuclein." *The American Journal of Pathology* 182(3): 940–953.

Ghochikyan, A., I. Petrushina, H. Davtyan, A. Hovakimyan, T., Saing, A. Davtyan, and M.G. Agadjanyan. 2014. "Immunogenicity of epitope vaccines targeting different B cell antigenic determinants of human α-synuclein: Feasibility study." *Neuroscience Letter* 560: 86–91. doi:10.1016/j.neulet.2013.12.028

Ghosh, A., N. Pradhan, S. Bera, A. Datta, J. Krishnamoorthy, N.R. Jana, and A. Bhunia. 2017. "Inhibition and degradation of amyloid b (Ab40) fibrillation by designed small peptide: A combined spectroscopy, microscopy, and cell toxicity study." *ACS Chemical Neuroscience* 8: 718–722. doi:10.1021/acschemneuro.6b00349.

Gozes, I. 2011. "Microtubules, schizophrenia and cognitive behavior: Preclinical development of davunetide (NAP) as a peptide-drug candidate." *Peptides* 32(2): 428–431.

Heidari, M., M. Ghodusi Borujeni, S. Kabirian Abyaneh, and P. Rezaei. 2019. "The effect of spiritual care on perceived stress and mental health among the elderlies living in nursing home." *Journal of Religion and Health*, 58(4): 1328–1339. doi:10.1007/s10943-019-00782-1.

Hermanson, G.T. 1996. *Bioconjugate techniques*. San Diego: Academic Press, 297–364.

Hyun-Hee, P., Y. Hyun-Jung, K. Sangjae, K. Gabseok, C. Na-Young, L. Eun-Hye, L.Y. Joo, Y. Moon-Young, L. Kyu-Yong, and K. Seong-Ho. 2019. "Neural stem cells injured by oxidative stress can be rejuvenated by GV1001, a novel peptide, through scavenging free radicals and enhancing survival signals." *Neurotoxicology*. https://doi.org/10.1016/j.neuro.2016.05.022.

Kamada, H., Y. Tsutsumi, K. Sato-Kamada, Y. Yamamoto, Y. Yoshioka, T. Okamoto, S. Nakagawa, S. Nagata, and T. Mayumi. 2003. "Synthesis of a poly(vinylpyrrolidone-co-dimethyl maleic anhydride) co-polymer and its application for renal drug targeting" *Nature Biotechnology* 21(4): 399–404. doi:10.1038/nbt798.

Kamışlı, S., O. Çiftçi, A. Taşlıdere, N. Başak Türkmen, and C. Ozcan. 2018. "The beneficial effects of 18 beta-glycyrrhetinic acid on the experimental autoimmune encephalomyelitis (EAE) in C57BL/6 mouse model." *Immunopharmacology and Immunotoxicology* 40: 344–352.

Kellock, J., G. Hopping, B. Caughey, and V. Daggett. 2016. "Peptidescomposed of alternating L- and D-amino acids inhibit amyloidogenesis in three distinct amyloid systems independent of sequence." *The Journal of Molecular Biology* 428: 2317–2328. doi:10.1016/j.jmb.2016.03.013.

Kızılbey, K. 2012. *Melanoma Ny-Eso-1 antigens 157–167, 155–163 peptide sequences synthesis and development of polyelectrolytes and bioconjugates*. Ph.D., YTU Graduate School of Science, Bioengineering Program, Istanbul.

Kodaira, H., Y. Tsutsumi, Y. Yoshioka, H. Kamada, Y. Kaneda, Y. Yamamoto, and S. Tsunoda. 2004. "The targeting of anionized polyvinylpyrrolidone to the renal system." *Biomaterials* 25(18): 4309–4315.

Lai, T.K.Y., P. Su, H. Zhang, and F. Liu. 2018. "Development of a peptide targeting dopamine transporter to improve ADHD-like deficits." *Molecular Brain* 11(1): 66.

Lin, L.X., X.Y. Bo, Y.Z. Tan, F.X Sun, M. Song, J. Zhao, Z.H. Ma, M. Li, K.Y. Zheng, and S.M. Xu. 2014. "Feasibility of bsheet breaker peptide H102 treatment for Alzheimer's disease based on bamyloidhypothesis." *PLoSOne* 9: e112052. doi:10.1371/journal.pone.0112052.

Liu, J., W. Wang, Q. Zhang, S. Zhang, and Z. Yuan. 2014. "Study on the efficiency and interaction mechanism of adecapeptide inhibitör of b-amyloid aggregation." *Biomacromolecules* 15: 931–939. doi:10.1021/bm401795e.

López, A., L. Mendieta, R. Prades, S. Royo, T. Tarragó, and E. Giralt. 2013. "Peptide POP inhibitors for the treatment of the cognitive symptoms of schizophrenia." *Future Medicinal Chemistry* 5(13): 1509–1523.

Mansuroğlu, B. 2007 *Development of biohibrid constructed and bioconvenient polymer-peptide conjugates*. Ph.D. Thesis, YTU Institute of Science and Technology, Bioengineering Program, Istanbul.

Migliore, M.M., T.K. Vyas, R.B. Campbell, M.M. Amiji, and B.L. Waszczak. 2010. "Brain delivery of proteins by the intranasal route of administration: A comparison of cationic liposomes versus aqueous solution formulations." *Journal of Pharmaceutical Science* 99(4): 1745–1761. doi:10.1002/jps.21939.

Morales, J. 2007. "Defining the role of insulin detemir in basal insulin therapy." *Drugs* 67(17): 2557–2584. doi:10.2165/00003495-200767170-00007.

Morell, M., L. Souza-Moreira, and E. Gonzalez-Rey. 2012. " VIP in neurological diseases: More than a neuropeptide." *Endocrine, Metabolic and Immune Disorders - Drug. Targets* 12(4): 323–332. doi:10.2174/187153012803832549.

Nonaka, N., W.A. Banks, H. Mizushima, S. Shioda, and J.E. Morley. 2002. "Regional differences in PACAP transport across the blood–brain barrier in mice: A possible influence of strain, amyloid β protein, and age." *Peptides* 23(12): 2197–2202. doi:10.1016/S0196-9781(02)00248-6.

Nonaka, N., S.A. Farr, H. Kageyama, S. Shioda, and W.A. Banks. 2008. "Delivery of galanin-like peptide to the brain: Targeting with intranasal delivery and cyclodextrins." *Journal of Pharmacology Exprimental Therapeutics* 325(2): 513–519. doi:10.1124/jpet.107.132381.

Nonaka, N., S.A. Farr, T. Nakamachi, J.E. Morley, M. Nakamura, S. Shioda, and W.A. Banks. 2012. "Intranasal administration of PACAP: Uptake by brain and regional brain targeting with cyclodextrins." *Peptides* 36(2): 168–175. doi:10.1016/j.peptides.2012.05.021.

Novakovic, Z.M., M.C. Leinung, D.W. Lee, and P. Grasso. 2009. "Intranasal administration of mouse [D-Leu-4]OB3, a synthetic peptide amide with leptin-like activity, enhances total uptake and bioavailability in Swiss Webster mice when compared to intraperitoneal, subcutaneous, and intramuscular delivery sys- tems." *Regulatory Peptides* 154(1–3):107–111. doi:10.1016/j.regpep.2009.01.002.

Polanczyk, G.V., M.S. de Lima, B.L. Horta, J. Biedermanand, and L.A. Rohde. 2007. "The worldwide prevalence of ADHD: A systematic review and metaregression analysis." *The American Journal of Psychiatry* 164: 942–948. doi:10.1176/ajp.2007.164.6.942.

Polanczyk G.V., G.A. Salum, L.S., Sugaya, A. Caye, and L.A. Rohde. 2015. "Annual research review: A meta-analysis of the worldwide prevalence of mental disorders in children and adolescents." *Journal of Child Psychology and Psychiatry* 56(3): 345–365.

Rat, D., U. Schmitt, F. Tippmann, I. Dewachter, C. Theunis, E. Wieczerzak, R. Postina, F. van Leuven, F. Fahrenholz, and E. Kojro. 2011. "Neuropeptide pituitary adenylate cyclase-activating polypeptide (PACAP) slows down Alzheimer's disease-like pathology in amyloid precursor protein-transgenic mice." *FASEB Journal: Official Publication of the Federation of American Societies for Experimental Biology* 25(9): 3208–3218. doi:10.1096/fj.10-180133.

Romero-Ramos, M., M. von Euler Chelpin, and V. Sanchez-Guajardo. 2014. "Vaccination strategies for Parkinson disease." *Human Vaccines & Immunotherapeutics* 10(4): 852–867.

Schulz, C., K. Paulus, O. Jöhren, and H. Lehnert. 2012. "Intranasal leptin reduces appetite and induces weight loss in rats with diet-induced obesity (DIO)." *Endocrinology* 153(1): 143–153. doi:10.1210/en.2011-1586.

Shiba, K., H. Kageyama, F. Takenoya, and S. Shioda. 2010. "Galanin-like peptide and the regulation of feeding behavior and energy metabolism." *FEBS Journal* 277(24): 5006–5013. doi:10.1111/j.1742-4658.2010.07933.x.

Shibata, H., S. Nakagawa, and Y. Tsutsumi. 2005. "Optimization of protein therapies by polymer-conjugation as an effective DDS." *Molecules* 10: 162–180.

Shioda, S., H. Kageyama, F. Takenoya, and K. Shiba. 2011. "Galanin-like peptide: A key player in the homeostatic regulation of feeding and energy metabolism?" *International Journal of Obesity (London)* 35(5): 619–628. doi:10.1038/ijo.2010.202.

Sowinska, M., M. Morawiak, M. Bochyńska-Czyż, A.W. Lipkowski, E. Ziemińska, B. Zabłocka, and Z. Urbanczyk-Lipkowska. 2019. "Molecular antioxidant properties and in vitro cell toxicity of the p-aminobenzoic acid (PABA) functionalized peptide dendrimers §." *Biomolecules* 9: 89.

Tarhan, N. 2016. *ADHD symptoms*. http://www.interhaber.com. Accesing date: 13.2.2020.

Tarhan, N. and S. Nurmedov. 2011. *Addiction*. İstanbul: Timaş Publication.

Umehara, M., Y. Ago, T. Kawanai, K. Fujita, N. Hiramatsu, K. Takuma, and T. Matsuda. 2013. " Methylphenidate and venlafaxine attenuate locomotion in spontaneously hypertensive rats, an animal model of attention-deficit/hyperactivity disorder, through alpha2-adrenoceptor activation." *Behavioural Pharmacology* 24(4):328–331. doi:10.1097/FBP.0b013e3283633648.

Wang, C.Y., P.N. Wang, M.J. Chiu, C.L. Finstad, F. Lin, S. Lynn, and P.A. Frohna. 2017. "UB-311, a novel UBITh ® amyloid β peptide vaccine for mild Alzheimer's disease." *Alzheimer's & Dementia: Translational Research & Clinical Interventions* 3(2): 262–272.

Watanabe, Y., M. Fujita, Y. Ito, T. Okada, H. Kusuoka, and T. Nishimura. 1997. "Brain dopamine transporter in spontaneously hypertensive rats." *The Journal of Nuclear Medicine* 38(3):470–4.

White, C.M., S. Ji, H. Cai, S. Maudsley, and B. Martin. 2010. "Therapeutic potential of vasoactive intestinal peptide and its receptors in neurological disorders." *CNS and Neurological Disorders - Drug Targets* 9(5): 661–666. doi:10.2174/187152710793361595.

Yang P.B., A.C. Swann, and N. Dafny. 2006. "Dose-response characteristics of methylphenidate on locomotor behavior and on sensory evoked potentials recorded from the VTA, NAc, and PFC in freely behaving rats." *Behavioral Brain Function* 2: 3.

10 Pseudo-Proteins and Related Synthetic Amino Acid-Based Polymers Promising for Constructing Artificial Vaccines

Ramaz Katsarava

CONTENTS

10.1 INTRODUCTION: BIODEGRADABLE POLYMERS COMPOSED OF α-AMINO ACIDS

It is impossible to imagine modern medicine without applications of various natural or artificial materials including polymers. Among the polymers, so-called biodegradable polymers (BPs) play one of the key roles. These kinds of polymers are widely used as resorbable surgical materials and therapeutic systems (various drug delivery and drug-eluting devices).

The BPs could be broken down in the physiological environment after their functions are fulfilled, resulting in small-sized debris which can be cleared from the body via metabolic means. To be fragmented into small-sized breakdown products, the BPs should contain chemical bonds in the backbones which are easily cleavable either enzymatically or chemically with reasonable rates. Normally these chemical bonds are cleaved via either hydrolysis or (rarely) a redox mechanism. The polymers degradable via a redox mechanism could contain suitable bonds (e.g. reduction-sensitive S-S disulfide links), whereas the BPs which degrade via a hydrolytic mechanism should contain hydrolytically labile bonds such as highly polarized C-C bonds (e.g. in polycyanoacrylates), ester, ortho-ester, and anhydride bonds, as well as amide, urethane, or urea bonds; the latter are also subjected to hydrolysis but with much lower rates. The BPs which contain only amide (peptide) bonds undergo mostly enzyme-catalyzed (specific) hydrolysis, e.g. hydrolytic degradation of collagen is catalyzed by collagenase. The anhydride bonds are cleaved predominantly via chemical (non-specific) hydrolysis whereas the ester bonds could be subjected to both enzymatic and chemical hydrolysis.

The BPs are classified as naturally occurring and synthetic ones (Table 10.1). Proteins, polysaccharides, nucleic acids, and bacterial polyesters pertain to the naturally occurring BPs. Among them collagen and albumin are the most popular for medical uses, such as resorbable surgical materials and drug delivery systems.

Collagen is widely used as various surgical biomaterials – bone grafts, scaffolds in tissue regeneration in forms of sponges, thin sheets, or gels, in reconstructive surgery, and in wound care as the artificial skin substitutes to manage severe burns and wounds, etc. (Muthukumar et al. 2018).

The therapeutic applications of collagen are, for example, drug delivery shields in ophthalmology, medicated sponges for burns/wounds, mini-pellets, and tablets for protein delivery, gel formulation in combination with liposomes for sustained drug delivery, controlling material for transdermal delivery, microspheres for drug delivery, nanoparticles for drug and gene delivery, etc. (Chak et al. 2013).

Another protein such as albumin is also highly applicable for therapeutic purposes. Its potential as a carrier for various molecules, proteins, or haptens was discussed in Chapter 5.

This wide biomedical applicability of proteins is due to special properties of proteins such as *a high affinity with tissues* (a tissue compatibility) owing to NH-CO

TABLE 10.1
Main Types of BPs

Naturally Occurring	Synthetic
• Proteins	• Polyesters
• Polysaccharides	• Polyamides
• Hyaluronic acid	• Poly(ester amide)s
• Nucleic acids	• Poly(orto esters)
• Bacterial polyesters (Poly-β-hydroxybutyrates)	• Polyanhydrides
	• Polyphosphasenes
	• Amino acid-based polymers (pseudo-proteins and related polymers)

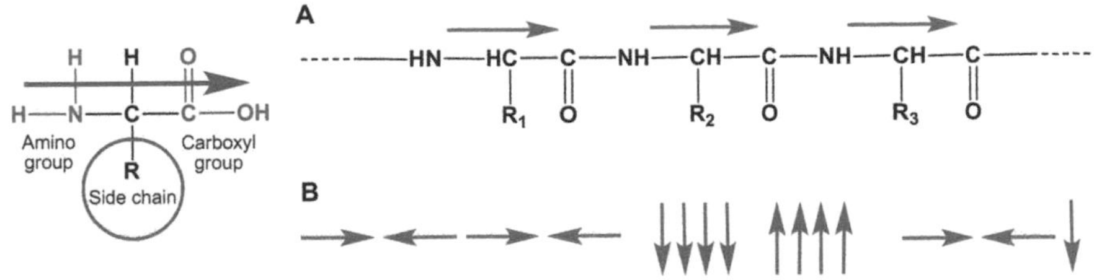

FIGURE 10.1 Possible orientations of AAs in macromolecules: **A** – "head-to-tail" orientation (proteinaceous molecular architecture), **B** – another type of orientation (non-proteinaceous molecular architecture).

bonds, and *releasing α-amino acids* upon biodegradation that means a nutritive potential for cells, and as a result, acceleration of tissue regeneration.

However, proteins as biomaterial have some serious limitations such as batch-to-batch variation, risk of disease transmission, a rather narrow range of material properties, and immune rejection. The last two limitations are connected with molecular architecture of collagen (proteins in whole). It is known that α-amino acids (AAs) in a protein molecule have so-called "head-to-tail" orientation as it is depicted in Figure 10.1A (assuming that the α-amino group is a "head" and the α-carboxyl group is a "tail"). This natural proteinaceous (i.e. protein like) architecture of macromolecules containing only amide (peptide) groups is easily recognizable by the immune system of the body. This can cause immune incompatibility of the biomaterial with the body tissues.

Furthermore, the presence of only peptide (NH-CO) bonds in a protein molecule significantly limits their material properties. For these reasons, the synthetic AA-based polymers (AABPs) having non-proteinaceous molecular architecture (Figure 10.1B) seem to be more promising for biomedical applications. Such polymers can contain, along with NH-CO bonds, additional hetero-bonds like ester, urea, urethane, etc. in the backbones thereby increasing the range of material properties and further decreasing the similarity with proteins providing in that way low to zero immunogenicity of the BPs.

Four basic families of synthetic AABPs such as poly(amino acid)s, pseudo-poly(amino acid)s, polydepsipeptides, and pseudo-proteins have been designed for the applications as biodegradable biomaterials.

All these polymers similar to proteins reveal a high tissue compatibility and release AAs upon biodegradation, and at the same time they are free of the limitations typical for proteins discussed above.

10.2 POLY(AMINO ACID)S

The A-B type polymers – poly(amino acids)s (PAAs) having conventional proteinaceous "head-to-tail" molecular architecture – have been synthesized and studied by many famous chemists (Khuphe and Thornton 2018) and explored as potential

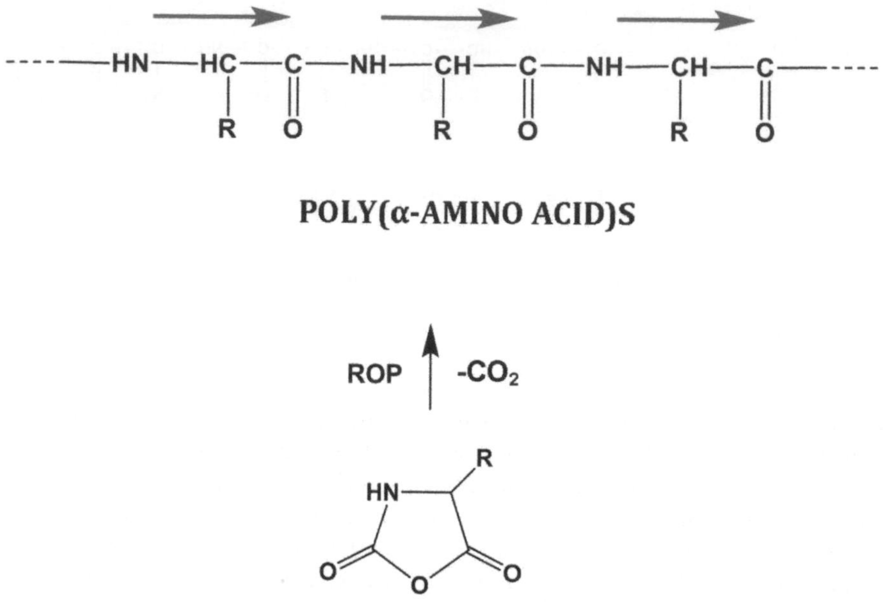

FIGURE 10.2 The synthesis of poly(α-amino acid)s via ROP of NCAs. NCAs.

new biomaterials starting from about 1970 (Katchalski 1974). The PAAs could be obtained via AB-type step-growth polymerization (solution polycondensation) of activated hetero-bifunctional monomers) (Katchalski and Sela 1958), or by Merrifield solid phase peptide synthesis (Khuphe and Thornton 2018) similar to the synthesis of peptide epitopes (see Chapters 4 and 5). However, this strategy of the PAA synthesis is complex and expensive.

The rational and commonly used synthesis of PAAs consists in the ring-opening polymerization (ROP) of suitable monomers – N-carboxy-α-amino acid anhydrides (NCAs). These readily undergo polymerization, with carbon dioxide evolution yielding the corresponding PAAs (Khuphe and Thornton 2018) (Figure 10.2).

Despite some prospects of the synthetically made PAAs as potential biomaterials, the studies revealed that most of the PAAs could not be considered as suitable biomedical materials due to their immunogenicity stipulated by proteinaceous molecular architecture (Katchalski 1974). So far, only a small number of poly(γ-substituted glutamates) and copolymers thereof (Sidman et al. 1983) have been identified as promising candidate materials for biomedical applications. These polymers could be considered as promising carriers for constructing artificial vaccines.

10.3 PSEUDO-POLY(AMINO ACID)S

More promising than PAAs, as biodegradable biomaterials, the so-called "pseudo-poly(amino acid)s" – the AABPs, backbones of which are constructed by utilizing the side-chain functional groups of suitably modified trifunctional α-L-amino acids or dipeptides made of them (Kobauri et al. 2019). Accordingly, pseudo-poly(amino acid)s (PPAAs) have non-proteinaceous molecular architecture (Figure 10.1B) and differ from conventional PAAs which have proteinaceous molecular architecture (Figure 10.1A). Such an approach offers the opportunity to create polymers from naturally occurring metabolites (AAs) but without some of disadvantages of conventional PAAs resulting from the repeating peptide bonds (see above). Homopolymerization or copolymerization of AAs with non-AA monomers has been achieved through a variety of reactions, leading to PPAAs of various classes – polyesters (PEs), polyamides (PAs), polyureas (PUs), and polyurethanes (PURs).

For synthesizing PPAA-PEs two naturally occurring hydroxyl AAs – *trans*-4-hydroxy-proline and serine were exploited.

Thermal polycondensation of N-acyl *trans*-4-hydroxy-proline (Figure 10.3) resulted in the PPAA-PEs of valuable material properties (Kwon and Langer 1989). The length of the acyl substituents influenced both physical-chemical characteristics and molecular weights of the resulting PPAA-PEs – the highest M_w 42.39 kDa was achieved in the case of the longest chain acyl substituent – N-palmitoyl substituted *trans*-4-hydroxy-proline. These kinds of PPAA-PEs are suitable for constructing artificial vaccines – epitope-loaded micro- and nanoparticles.

N-benzyloxycarbonyl(Z)-protected hydroxyproline was used for synthesizing functional PPAA-PE with free amino groups – poly(*trans*-4-hydroxy-L-proline ester) (Lim, Choi, and Park 1999), Figure 10.4:

PPAA-PE with useful material properties on the basis of L-serine was obtained by ROP of N-protected-L-serine β-lactone (Figure 10.5). After removing

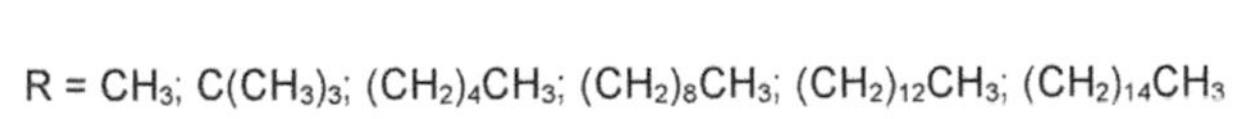

PPAAs - polyesters

$R = CH_3; C(CH_3)_3; (CH_2)_4CH_3; (CH_2)_8CH_3; (CH_2)_{12}CH_3; (CH_2)_{14}CH_3$

FIGURE 10.3 Synthesis of the PPAAs-PEs by thermal polycondensation of N-substituted trans-4-hydroxy-proline methyl ester.

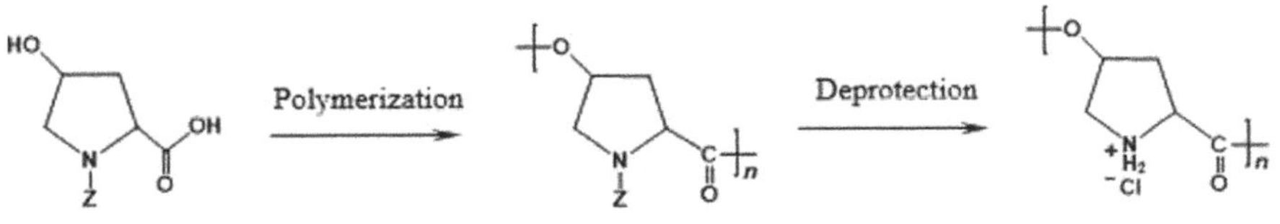

FIGURE 10.4 Synthesis of the PPAAs-PE with free amino groups by thermal polycondensation of N-benzyloxycarbonyl(Z)-protected hydroxyproline.

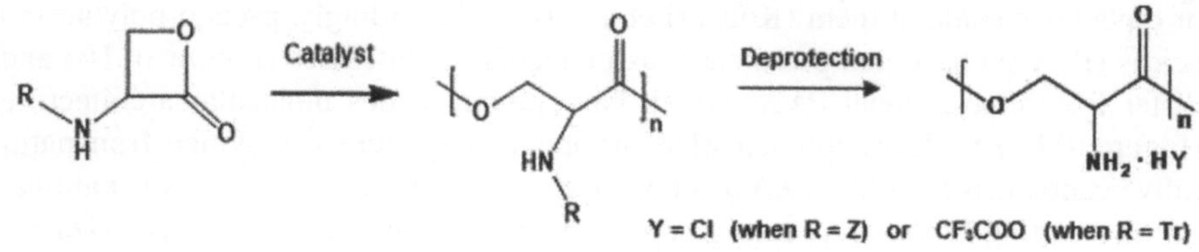

FIGURE 10.5 Synthesis of the PEs by ROP of N-protected serine.

benzyloxycarbonyl (**Z**) (Zhou and Kohn 1990) or triphenylmethyl (Trityl, **Tr**) (Fietier, Le Borgne, and Spassky 1990) protections, functional PPAA – poly(L-serine ester) containing free lateral amino groups was obtained (Figure 10.5).

A series of polymers – PAs, Pus, and PEURs were synthesized on the basis of L-lysine esters used as diamines (Zavradashvili et al. 2013; Katsarava, Kantaria, and Kobauri 2019). Among the lysine-based polymers one of the most attractive is water-soluble poly-(L-lysine citramide) (Couffin-Hoarau, Boustta, and Vert 2001), entirely composed of naturally occurring building blocks. Synthesis of this highly functional PPAA-PA is given in Figure 10.6.

The functional polymers depicted in Figures 10.4– 10.6 are highly promising carriers for covalent conjugation of various biologicals including epitopes.

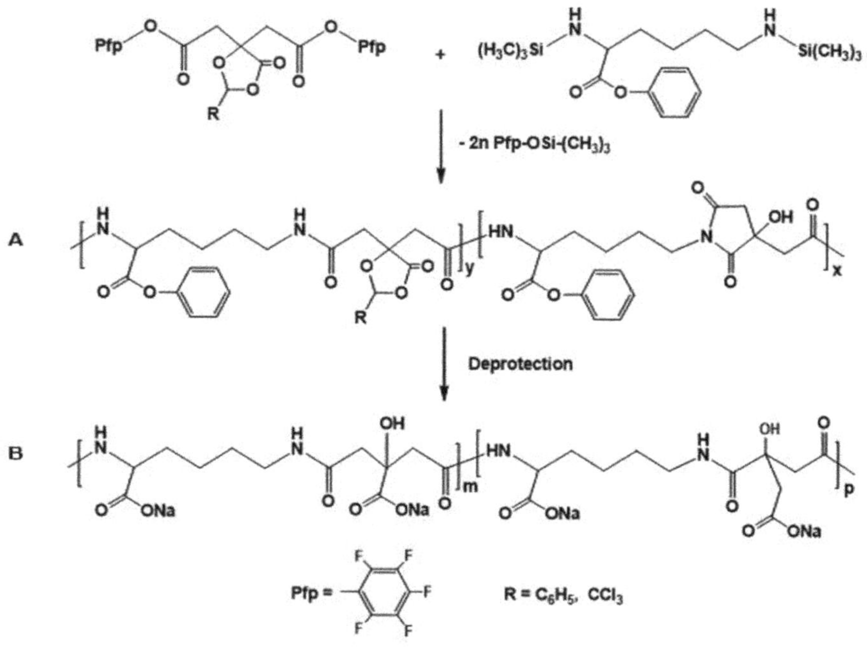

FIGURE 10.6 Synthesis of functional PPAA-PAs composed of lysine and citric acid.

10.4 POLYDEPSIPEPTIDES

The next representatives of the synthetic AABPs are polydepsipeptides (PDPs) – an important class of biodegradable biomaterials completely composed of naturally occurring ("physiological") building blocks such as α-amino acids and α-hydroxy acids (Feng et al. 2010). The PDPs, similar to PAAs above, could be synthesized from preformed linear di- or higher activated hetero-bifunctional depsipeptides via step-growth polymerization (solution polycondensation) (Stewart 1969). However, the rational and commonly used synthesis of the PDPs consists in ROP of cyclic depsipeptides – morpholine-2,5-diones (Helder et al. 1985) as depicted in Figure 10.7.

Depsipeptide-based polymers and copolymers can be designed either without or with different pendant functional groups including carboxyl, hydroxyl, amine, and thiol functions by varying α-hydroxy acids and α-amino acids (i.e. by varying R_1 and R_2 substituents of 2,5-dioxomorpholine rings, Figure 10.7). Though the AAs are inserted in the backbones of PDPs in a "head-to-tail" manner, they are interleaved with ester bonds that provide non-proteinaceous molecular architecture less recognizable by the immune system.

Biodegradable PDPs are of interest for constructing synthetic vaccines using various technological approaches.

10.5 PSEUDO-PROTEINS

10.5.1 SYNTHESIS OF THE PSEUDO-PROTEINS

Highly promising representatives of the synthetic AABPs, having valuable material properties and suitable for biomedical applications, were first reported in 1994 (Arabuli et al. 1994). Along with AAs the main "bricks" for constructing these polymers were physiological, non-toxic, cheap, and readily available organic compounds such as diols and dicarboxylic acids. We used the name **pseudo-proteins (PPs)** for these polymers to distinguish them from other classes of the synthetic AABPs (Katsarava, Kantaria, and Kobauri 2019).

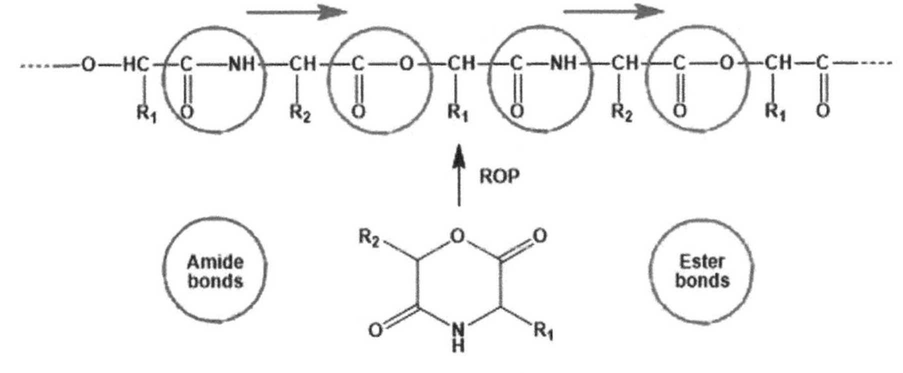

FIGURE 10.7 Synthesis of polydepsipeptides by ROP of cyclic depsipetides.

Key bis-nucleophilic monomers. One of the main advantages of the PPs over other classes of the synthetic AABPs consist in key bis-nucleophilic monomers – diamine-diesters. They are basically synthesized as stable di-p-toluenesulfonic acid salts (TDADEs, Figure 10.8) using a simple and cost-effective procedure – direct thermal condensation of cheap and readily available components such as AAs, diols in refluxed cyclohexane (Katsarava et al. 2016) in the presence of p-toluenesulfonic acid monohydrate (Figure 10.8) which serves as both amino group protector and the condensation reaction catalyst (Katsarava and Gomurashvili 2011). After cooling the reaction mixture to room temperature, the TDADE is almost quantitatively precipitated (yield up to 95%) and is separated from the reaction mixture by simple filtration. The used organic solvent (cyclohexane) can be recycled by distillation and reused. The TDADEs made of hydrophobic AAs are purified by recrystallization from water.

Additional merits of the TDADE-monomers are stability upon storage, highly active terminal amino groups for successful chain growth along with the designed-in non-proteinaceous ("head-to-head") orientation of amino acids, enzyme-specific lateral groups R, and labile (hydrolysable) ester bonds (Figure 10.9) which goes after the step-growth polymerization (SGP) into macromolecules made of them. The TDADEs are the universal bis-nucleophilic monomers which, after interacting with various bis-electrophiles (Figure 10.10), lead to the synthesis of three basic classes of the PPs such as poly(ester amide)s (PEAs), poly(ester urethane)s (PEURs), and poly(ester urea)s (PEUs), see Figure 10.9.

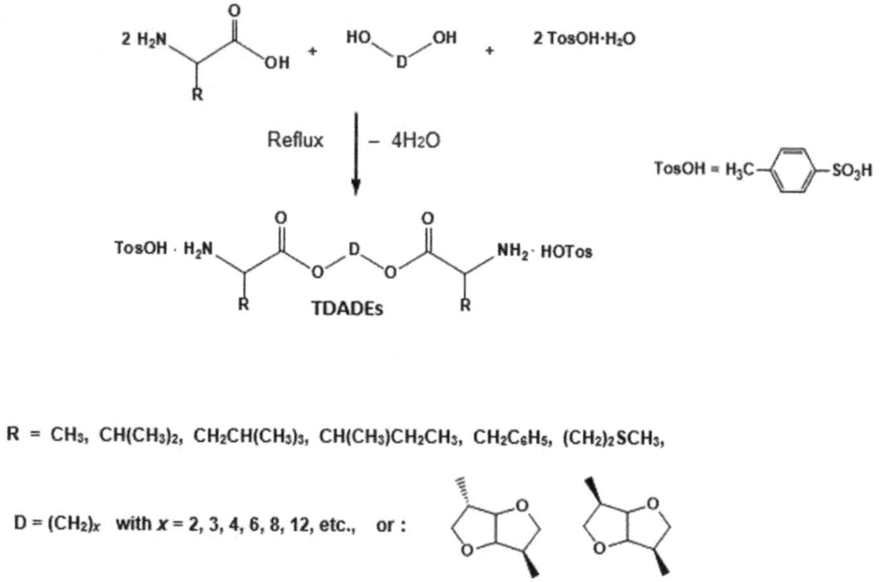
2 H2N
OH
R
HO
OH
D
2 TosOH·H2O
Reflux
– 4H2O
TosOH = H3C–SO3H
TosOH · H2N
NH2 · HOTos
TDADEs
R = CH3, CH(CH3)2, CH2CH(CH3)2, CH(CH3)CH2CH3, CH2C6H5, (CH2)2SCH3,
D = (CH2)x with x = 2, 3, 4, 6, 8, 12, etc., or :

FIGURE 10.8 Synthesis of diamine-diester monomers (TDADEs) by direct thermal condensation of AAs with fatty diols.

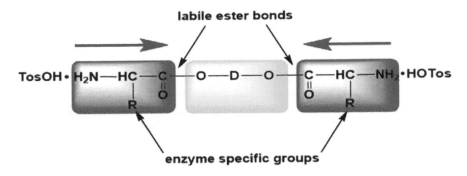

FIGURE 10.9   The structure of DADEs – key monomers for synthesizing PPs.

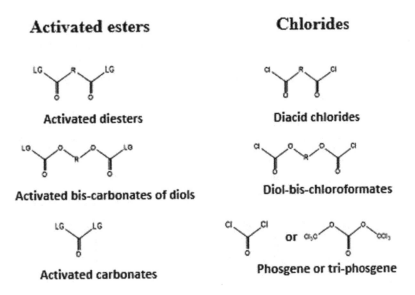

FIGURE 10.10   Key bis-electrophilic monomers. The key bis-electrophilic monomers for synthesizing PPs – counter-partners of the TDADEs are activated diesters and chlorides of various classes and LG.

Many diacid chlorides, phosgenes (as solutions in organic solvents), and triphos-genes are purchasable products, whereas diol bis-chloroformates, activated diesters of diacids, activated bis-carbonates of diols, and activated carbonates are custom made.

Through varying the bis-electrophilic monomers, it is possible to construct dif-ferent classes of the PPs. Diacid chlorides and activated diesters of diacids are used for synthesizing the PEAs, diol-*bis*-chloroformates, and activated bis-carbonates of diols – the PEURs, phosgene/triphosgene and activated carbonates – the PEUs. Two types of SGP were exploited for synthesizing the PPs – interfacial polycondensation (IP) using diacid chlorides, and solution active polycondensation (SAP) using acti-vated diesters of diacids as bis-electrophilic monomers.

These syntheses are shown schematically in Figure 10.11 for the PP-PEAs. The other classes of the PPs could be synthesized similarly – the PEURs via IP using diol-*bis*-chloroformates and via SAP using activated bis-carbonates of diols, the PEUs via IP using phosgene or thriphosgene and via SAP using activated carbonates (Katsarava and Gomurashvili 2011).

Three classes of the PPs developed are composed of tunable building blocks (Figure 10.12) – PEAs of three blocks such as AA, diol, and dicarboxylic acid, PEURs of three blocks – AA, and two diols, and PEUs of two blocks – AA and diol. In case of PEURs and PEUs the third building block – carbonic acid – is invariable.

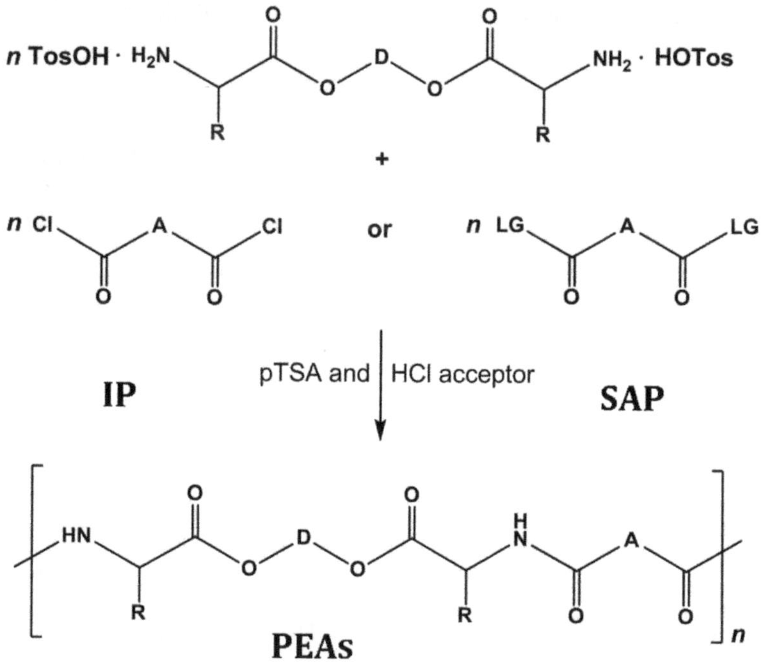

FIGURE 10.11 The synthesis of PP–PEAs via interfacial polycondensation (IP) and solu-tion active polycondensation (SAP).

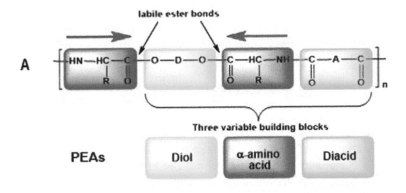

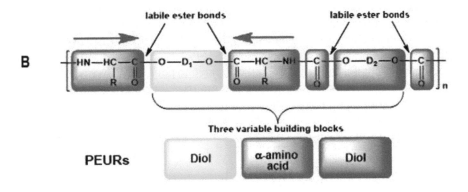

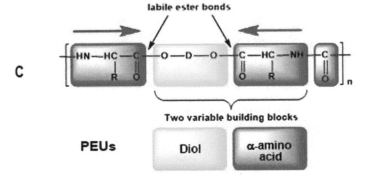

FIGURE 10.12   General structure of the various classes of PPs: **A** – PEAs, **B** – PEURs, **C** – PEUs.

A variety of building blocks, polymers classes and synthetic methods discussed above, allow tuning of the structure and material properties of the PPs in the widest range. As expected, based on the non-proteinaceous architecture, PPs show low to zero immunogenicity (Katsarava, Kulikova, and Puiggalí 2016).

10.5.2 THE BASIC PROPERTIES OF THE PPs

The main merits of the PPs are high yields, cheapness, and stability (long shelf-life) of the starting AA-based monomers (TDADEs), various chemical bonds built into the polymeric backbones, and virtually unlimited synthetic possibilities. As carriers of epitopes when constructing artificial vaccines, the PPs possess several advantages over the classical polyesters used for the same purposes. It is obvious that the three classes of PPs (depicted in Figure 10.12) along with a huge availability of variable building blocks and a possibility to synthesize copolymers containing fragments of different classes, including ones bearing lateral functional groups, allow the design of carriers with widely tunable chemical, biochemical, physical-chemical, and mechanical properties (Katsarava and Gomurashvili 2011; Zavradashvili et al. 2017). The PPs of all classes contain hydrophilic links (NHCO) which provide a high affinity with tissues. After biodegradation, the PPs release weaker acidic products in much lower quantities (per unit weight) compared to poly-α-hydroxy acids and from this point of view could be considered as more friendly systems for cells and epitopes of a natural origin. For example, most PPs of the PEAs classes (PP-PEAs) after ultimate hydrolysis (Figure 10.13) release neutral (zwitterionic) AAs and diols, and

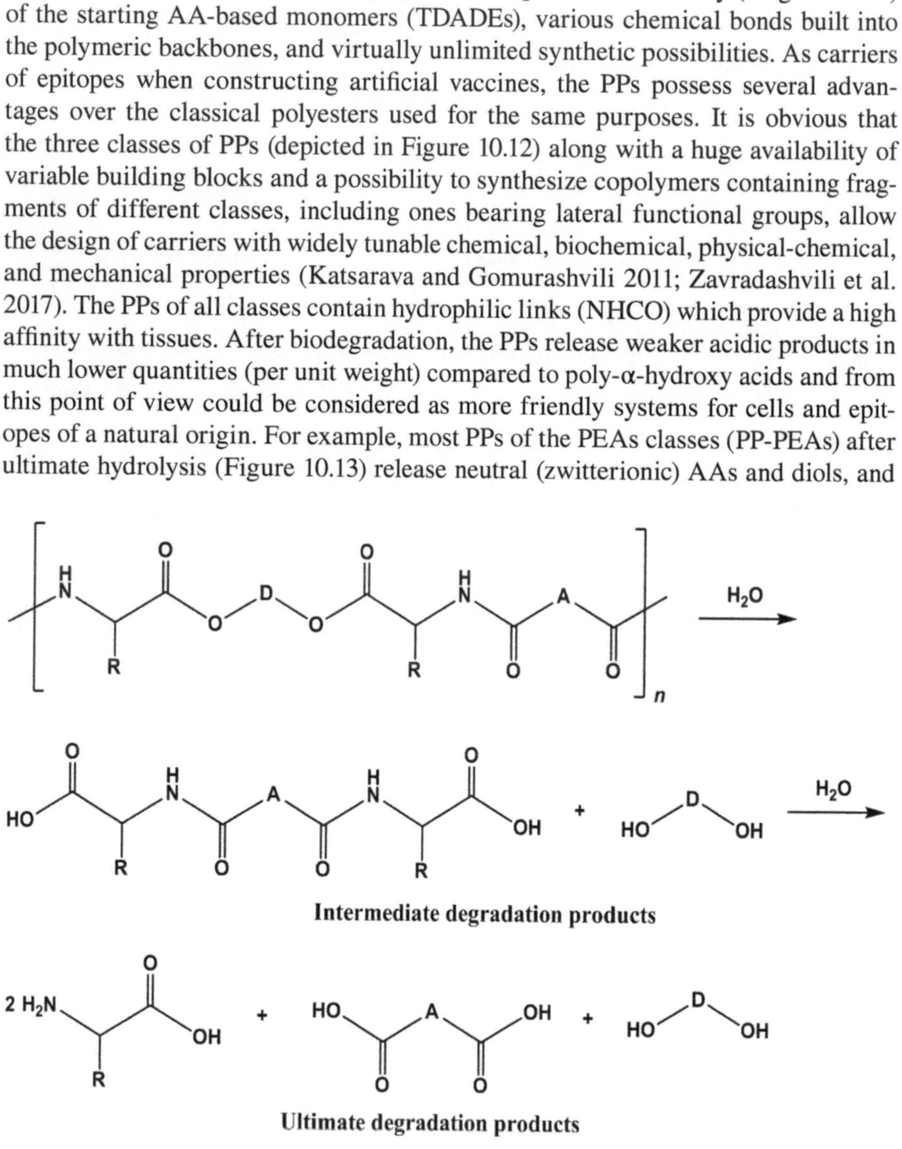

Intermediate degradation products

Ultimate degradation products

FIGURE 10.13   Hydrolytic degradation of PP-PEAs.

relatively week fatty diacids HOOC-A-COOH (e.g. sebacic acid, $A = (CH_2)_8$, with 4.72 and 5.45 pKa). Ultimate biodegradation products of other two classes of the PPs – PEURs and PEUs are normal metabolites such as AAs and CO_2, and neutral and readily metabolizable diols HO-D-OH (as it is shown for PP-PEUs in Figure 10.14).

Interestingly, even acidic products, released after ester-bond's hydrolysis of PP-PEUs before the ultimate biodegradation, did not cause inflammation that was ascribed to the self-buffering property of PEU preventing significant pH drop during the degradation process (Stakleff et al. 2013). Many PPs are amorphous and that provides smooth biodegradation of devices made of them. Other advantages of PPs over aliphatic polyesters, especially over the polyesters obtained via ROP are (Katsarava and Gomurashvili 2011; Katsarava, Kulikova, and Puiggalí 2016):

- Polycondensation synthesis without using toxic catalysts
- Synthesis under atmospheric conditions at moderate temperatures (20–80°C)
- Tunable hydrophobicity/hydrophilicity balance
- Higher affinity (owing to NHCO bonds) and better compatibility with tissues
- Solubility in common organic solvents (ethanol, isopropanol, THF, acetone, etc.)
- Longer shelf-life

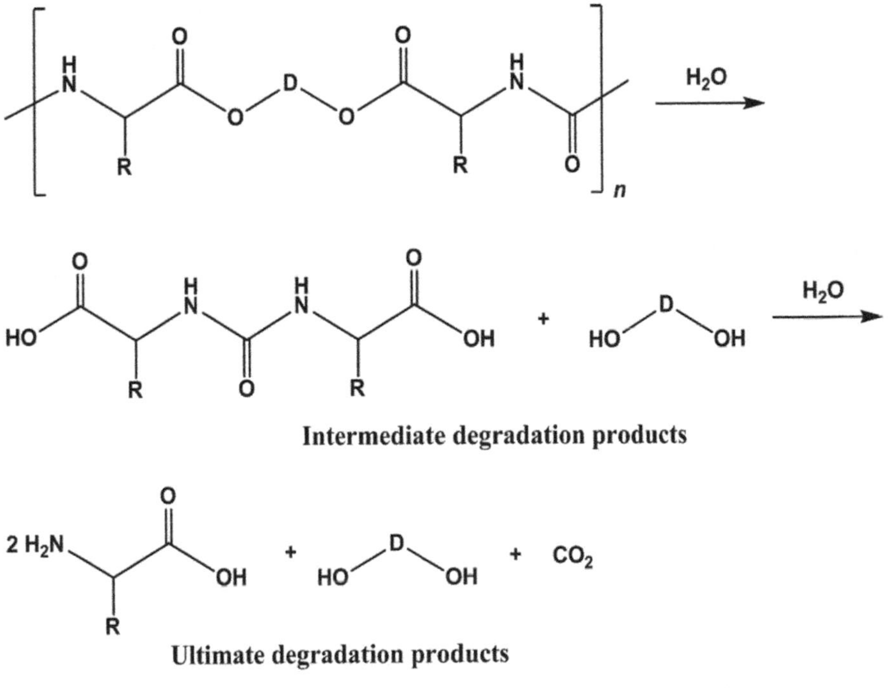

Intermediate degradation products

Ultimate degradation products

FIGURE 10.14   Hydrolytic degradation of PP-PEUs.

The molecular weights of the PPs are reasonably high and vary within the range 24,000–167,000 Da (Mw), polydispersity – within 1.20–1.81. As expected, based on the non-proteinaceous architecture, PPs show low to zero immunogenicity.

10.5.3 FUNCTIONAL PPS

The PPs above are of regular type, i.e. they do not contain lateral functional groups and are mostly suitable for physical targeting (Morachis, Mahmoud, and Almutairi 2012) biologicals by fabricating nanovehicles (nanoparticles, NPs). The scope of applications of PPs can be substantially expanded by designing functional ones. This can be achieved (i) by combination of TDADEs with functionalized TDADEs or with other types of functional co-monomers, or (ii) by synthesizing various active pre-polymers with subsequent functionalization by means of polymer-analogous reactions (Zavradashvili et al. 2017). For example, PPs containing free lateral functional -COOH groups (PP-PEA polyacids), suitable for coupling bioactive agents, were synthesized by copolymerization of regular TDADEs with di-p-toluenesulfonic acid salt of lysine benzyl ester under the conditions of SAP with subsequent deprotection of the obtained polymeric benzyl esters by selective catalytic hydrogenolysis using Pd catalyst (Figure 10.15).

Functional PPs containing unsaturated double bonds were synthesized by SAP (Chkhaidze et al. 2011) of TDADEs with di-p-nitrophenyl fumarate (NF), as depicted in Figure 10.16.

Analogously, epoxy-functionalized PPs (epoxy-PP-PEAs) by SAP (Zavradashvili et al. 2013) of TDADEs with di-p-nitrophenyl *trans*-epoxy-succinate (N*t*ES) were synthesized, as depicted in Figure 10.17.

The obtained unsaturated and epoxy PP-PEAs can be used for covalent attachment of various drugs and bioactive compounds via activated double bonds (through

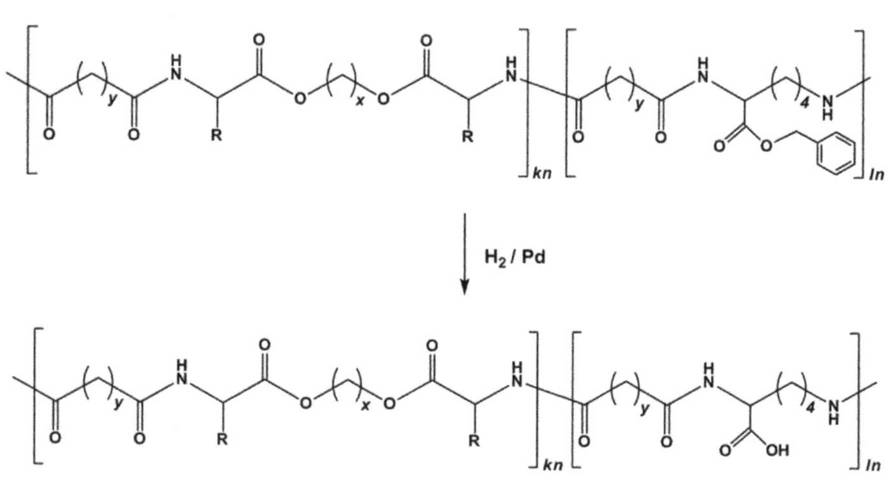

FIGURE 10.15 Synthesis of PP-PEA polyacids.

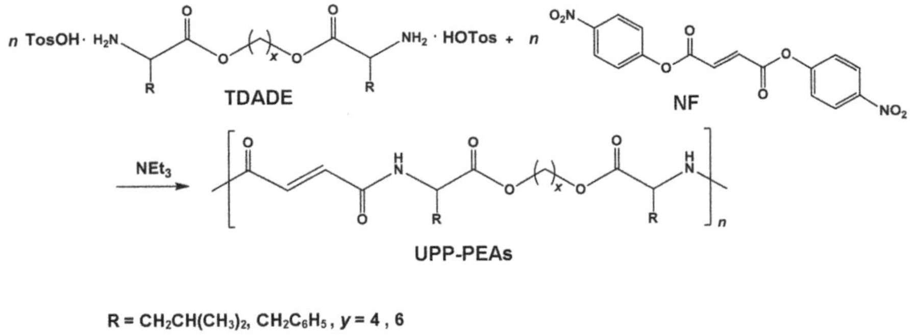

FIGURE 10.16 Synthesis of unsaturated PP-PEAs composed of fumaric acid.

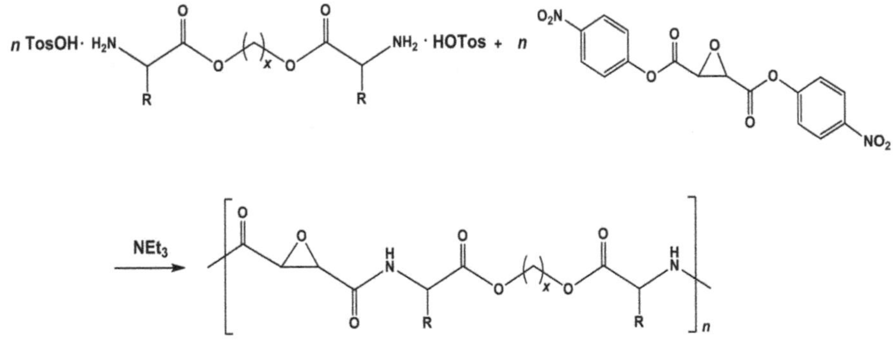

FIGURE 10.17 Synthesis of functional PP-PEA composed of epoxy-succinic acid.

sulfhydyl groups) or epoxy groups (through amino groups). These functional PPs are capable of being subjected to further functionalization by transforming them into polyacids, polyamines, and polyols by interaction of hetero-bifunctional compounds (Zavradashvili et al. 2017). For example, unsaturated PP-PEA can be transformed into suitable derivatives by interaction with hetero-bifunctional PEG-derivatives such as thiol-PEG-amine (Figure 10.18, FG = NH$_2$) or thiol-PEG-carboxymethyl (Figure 10.18, FG = COOH). Analogously, epoxy-PP-PEAs can be functionalized by interaction with hetero-bifunctional PEG-derivatives such amine-PEG-carboxy-methyl (Figure 10.19, FG = COOH), amine-PEG-thiol (Figure 10.19, FG = NH$_2$) or amine-PEG-hydroxyl (Figure 10.19, FG = OH). After modifying with hetero-bifunctional PEG-derivatives the polymers become water soluble and contain functional groups at attached lateral PEG-substituents remote form the polymeric backbones. Such kinds of the functionalized PPs have a high potential for the application as carriers for the conjugation of epitopes of various origin.

The PEG-modified PPs are also promising as surfactants for fabricating NPs with simultaneous PEGylation of their surface. PEG decreases the affinity of plasma

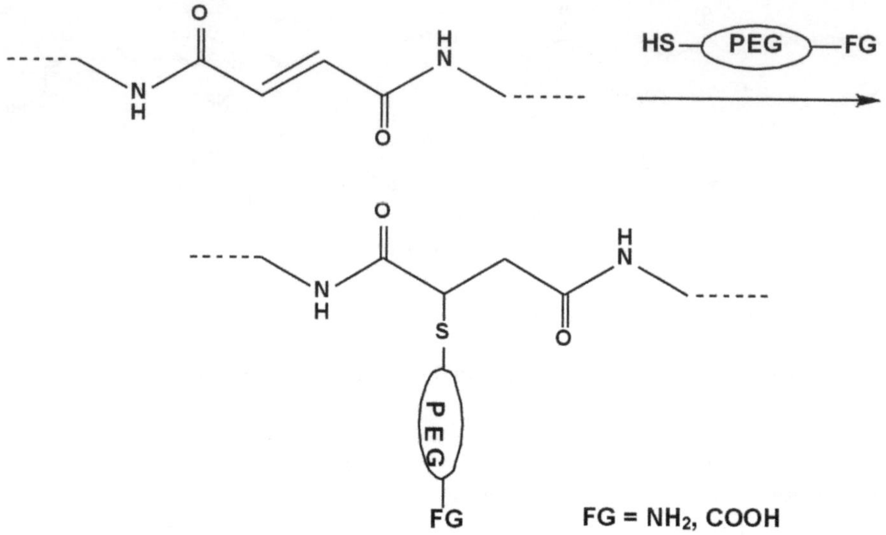

FIGURE 10.18 Functionalization of unsaturated PP-PEAs with hetero-bifunctional PEG-derivatives.

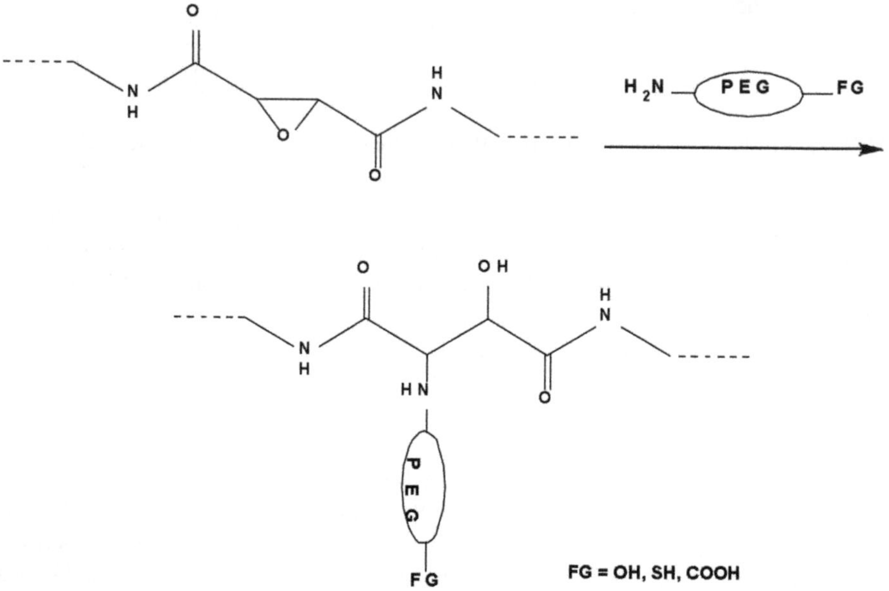

FIGURE 10.19 Functionalization of epoxy-PP-PEAs with hetero-bifunctional PEG-derivatives.

proteins (opsonins) for adsorption on NPs – long chains of PEG form a random cloud around the NPs thereby preventing absorption of opsonins and in that way suppressing phagocytosis. Along with the protection of the NPs from phagocytosis the PEGylation substantially increases the bioavailability of NPs (Laurent and Yahia 2013). It has to be noted that functional groups (FG in Figures 10.18 and 10.19) of the PEG cloud (functionalized cloud – Figure 10.20) can be used for the conjugation of various markers, vectors, etc. to the surface of NPs.

It is synthesized on the basis of AA arginine (Memanishvili et al. 2014). Most of the cationic PPs were soluble in water, showed an excellent cell compatibility and are promising as gene transfection agents. The cationic PP made of more hydrophobic monomers such as, e.g. PEA **8R6** composed of sebacic acid and 1,6-hexanediol (Figure 10.21) is insoluble in water and is of interest for fabricating positively charged NPs.

10.5.4 Nanoparticles Made of PPs

A systematic study of fabricating PP-based NPs by cost-effective polymer deposition/solvent displacement (nanoprecipitation) method was done by Katsarava and

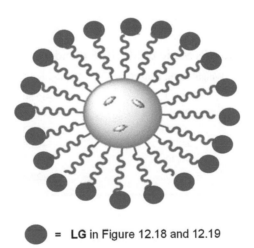

● = **LG** in Figure 12.18 and 12.19

FIGURE 10.20 NP with functionalized PEG cloud.

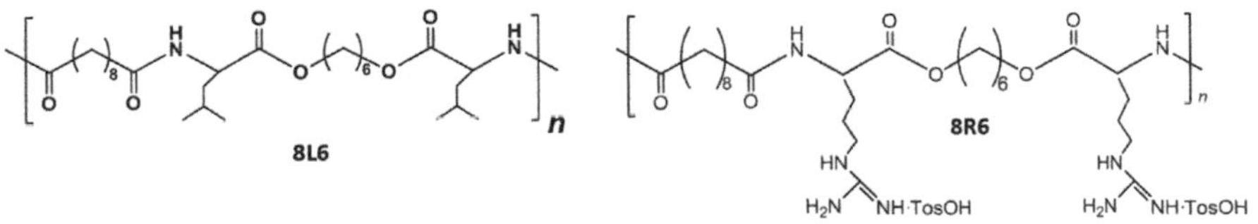

8L6

8R6

FIGURE 10.21 PPs used for preparing NPs: regular PEA 8L6 and arginine-based cationic PEA 8R6.

co-workers (Kantaria et al. 2016). Four regular and one cationic PPs, eight surfactants, and four organic solvents miscible with water were tested for the fabrication of the NPs. It was found that depending on the nature of the PPs, organic solvents, surfactants, and the fabrication conditions the size (mean particle diameter) of the NPs could be tuned within 42 ÷ 398 nm, the zeta-potential within −12.5 ÷ +28 mV. The study revealed that in terms of particle size, stability, and biocompatibility the best are: as a polymer – the PP-PEA composed of sebacic acid (**8**), L-leucine (**L**), 1,6-hexanediol (**6**) labeled as **8L6** (Figure 10.21), as a solvent (organic phase) – dimethyl sulfoxide (DMSO), and as a surfactant – Tween 20. Pure **8L6** formed negatively charged NPs with zeta-potential (ZP) −12 mv (Table 10.2), whereas its blending with cationic PP **8R6** (Figure 10.21) resulted in positively charged NPs with ZP within +(20–28) mv depending on the **8R6** content in the blend. The positive surface charge (positive zeta-potential) of NPs is favorable for penetration through cells and biological barriers – it is known that a positive charge helps with the NPs' adhesion to the surface of cells and stimulates penetration into the cells via endocytosis

The NPs showed a high stability upon storage over 90 days (Table 10.3).

TABLE 10.2
The ZP of the NPs Made of 8L6/8R6 Blends

Polymer	Weight Ratio 8L6/8R6	Zeta-potential (mV) ± SD
8L6	100/0	−12.50 ± 1.99
8L6/8R6	95/5	+19.6 ± 3.6
	90/10	+23.2 ± 4.7
	80/20	+25.8 ± 2.7
	70/30	+27.2 ± 3.2
	50/50	+28.0 ± 4.9

TABLE 10.3
The Stability of the Prepared NPs Upon Storage at 4–5°C

Polymer	Time			
	Freshly prepared	After 30 days	After 60 days	After 90 days
	MPD (nm) ± SD [PDI ± SD]			
8L6	70.1 ± 2.3	72.2 ± 1.3	70.4 ± 1.9	71.8 ± 2.3
	[0.188 ± 0.002]	[0.181 ± 0.006]	[0.179 ± 0.005]	[0.178 ± 0.009]
8L6/8R6 (70/30)	125.7 ± 4.3	118.4 ± 5.1	121.9 ± 3.8	119.2 ± 4.1
	[0.221 ± 0.014]	[0.229 ± 0.012]	[0.219 ± 0.009]	[0.218 ± 0.006]

10.6 CONCLUSION

The PPs made by step growth polymerization of TDADEs with various bis-electrophilic counter-partners have a high potential as carriers in the design of artificial vaccines – as both vehicles (nanocontainers) and water soluble functional polymers for covalent conjugations of epitopes.

REFERENCES

Arabuli, N., G. Tsitlanadze, L. Edilashvili, D. Kharadze, T. Goguadze, V. Beridze, Z. Gomurashvili, and R. Katsarava. 1994. "Heterochain polymers based on natural α-amino acids. Synthesis and enzymatic hydrolysis of regular poly(ester amide)s based on bis-(Lphenylalanine) α-ω-alkylene diesters and adipic acid." *Macromolecular Chemistry and Physics* 195: 2279–2289.

Chak V., D. Kumar, and S. Visht. 2013."A review on collagen based drug delivery systems." *International Journal of Pharmacy Teaching & Practices* 4: 811–820.

Chkhaidze, E., D. Tugushi, D. Kharadze, Z. Gomurashvili, C.C. Chu, and R. Katsarava. 2011. "New unsaturated biodegradable poly(ester amide)s composed of fumaric acid, L-leucine and α,ω-alkylene diols." *Journal of Macromolecular Science, Part A, Pure and Applied Chemistry* 48: 544–555.

Couffin-Hoarau, A.C., M. Boustta, and M. Vert. 2001. "Enlarging the library of poly-(L-lysine citramide) polyelectrolytic drug carriers." *Journal of Polymer Science: Part A: Polymer Chemistry* 39: 3475–3484.

Feng Y., J. Lu M. Behl, and A. Lendlein. 2010. "Progress in depsipeptide-based biomaterials." *Macromolecular Bioscience* 10: 1008–1021.

Fietier, I., A. Le Borgne, and N. Spassky. 1990. "Synthesis of functional polyesters derived from serine." *Polymer Bulletin* 24: 349–353.

Helder, J., F.E. Kohn, S. Sato, J.W. van den Berg, and J. Feijen. 1985. "Synthesis of poly[oxyethylidenecarbonylimino-(2-oxoethylene)] [poly(glycine-D,L-lactic acid)] by ring opening polymerization." *Macromolecular Chemistry Rapid Communications* 6: 9–14.

Kantaria, T., T. Kantaria, S. Kobauri, M. Ksovreli, T. Kachlishvili, N. Kulikova, D. Tugushi, and R. Katsarava. 2016. "Biodegradable nanoparticles made of amino acid based ester polymers: Preparation, characterization, and in vitro biocompatibility study." Applied Science 6: 444.

Katchalski, E. 1974. *Poly(amino acids): Achievements and prospects.* New York: Wiley.

Katchalski, E. and M. Sela. 1958. "Synthesis and chemical properties of poly-α-amino acids." *Advances in Protein Chemistry* 13: 243–492.

Katsarava, R. and Z. Gomurashvili. 2011. "Biodegradable polymers composed of naturally occurring α-amino acids. In Lendlein, A., and Sisson, A. (Eds), *Handbook of biodegradable polymers - isolation, synthesis, characterization and applications.* Wiley-VCH, Verlag GmbH & Co. KGaA, 107–131.

Katsarava, R., N. Kulikova, and J. Puiggalí. 2016. "Amino acid based biodegradable polymers – promising materials for the applications in regenerative medicine (Review)." *Journal of Regenerative Medicine* 1: 12.

Katsarava, R., D. Tugushi, V.Beridze, and N. Tawil. 2016. *Composition comprising a polymer and a boactive agent and method of preparing thereof.* US 2016/0375139 A1.

Katsarava, R., T. Kantaria, and S. Kobauri. 2019. "Pseudo-proteins and related synthetic amino acid based polymers." *Journal of Materials Education,* in press.

Khuphe M., and P.D. Thornton. "Poly(amino acids)." Chapter 7 In Parambath, A. (Ed.), *Engineering of biomaterials for drug delivery systems.* Woodhead Publishing, p.199.

Kobauri, S., T. Kantaria, N. Kupatadze, N. Kutsiava, D. Tugushi, and R. Katsarava. 2019. "Volume pseudo-proteins: A new family of biodegradable polymers for sophisticated biomedical applications." *Nano Technology & Nano Science Journal (NTNS)* 1(1): 37–42.

Kwon, H.Y. and R. Langer. 1989. "Pseudopoly(amino acids): A study of the synthesis and characterization of poly(*trans*-4-hydroxy-N-acyl-L-proline esters)." *Macromolecules* 22: 3250–3255.

Laurent, S. and L.H. Yahia. 2013. *Protein corona: Applications and challenges, in: B. Martinac (Ed.), protein-nanoparticle interactions.* Berlin: Springer-Verlag.

Lim, Y., Y.H. Choi, and J. Park. 1999. "A self-destroying polycationic polymer: Biodegradable poly(4-hydroxy-L-proline ester)." *Journal of the American Chemical Society* 12: 5633–5639.

Memanishvili, T., N. Zavradashvili, N. Kupatadze, D. Tugushi, M. Gverdtsiteli, V.P. Torchilin, C. Wandrey, L. Baldi, S.S. Manoli, and R. Katsarava. 2014. "Arginine-based biodegradable ether-ester polymers of low cytotoxicity as potential gene carriers." *Biomacromolecules* 15: 2839–2848.

Morachis, J.M., E.A. Mahmoud, and A. Almutairi. 2012. "Physical and chemical strategies for therapeutic delivery by using polymeric nanoparticles." *Pharmacological Reviews* 64: 505–519.

Muthukumar, T., G. Sreekumar, T.P. Sastry, and M. Chamundeeswari. 2018. "Collagen as a potential biomaterial in biomedical applications." *Reviews on Advanced Materials Science* 53: 29–39.

Sidman, K.R., W.D. Steber, A.D. Schwope, and G.R. Schnaper. 1983. "Controlled release of macromolecules and pharmaceuticals from synthetic polypeptides based on glutamic acid." *Biopolymers* 22: 547–556.

Stakleff K.S., F. Lin, L.A. Smith Callahan, M.B. Wade, A. Esterle, J. Miller, and M.L. Becker. 2013. "Resorbable, amino acid-based poly(ester urea)s crosslinked with osteogenic growth peptide with enhanced mechanical properties and bioactivity." *Acta Biomaterialia* 9: 5132–5142.

Stewart, F.H.C. 1969. "Synthesis of polydepsipeptides with regularly repeating unit sequences." *Australian Journal of Chemistry* 22: 1291–1298.

Zavradashvili N., G. Jokhadze M. Gverdtsiteli G. Otinashvili N. Kupatadze Z. Gomurashvili D. Tugushi, and R. Katsarava. 2013. "Amino acid based epoxy-poly(Ester Amide)s - a new class of functional biodegradable polymers: Synthesis and chemical transformations." *Journal of Macromolecular Science, Part A, Pure and Applied Chemistry* 50: 449–465.

Zavradashvili N., G. Jokhadze M. Gverdtsiteli, D. Tugushi, and R. Katsarava. 2017. "Biodegradable functional polymers composed of naturally occurring amino acids (Review)." *Research and Reviews in Polymer* 8: 105–128.

Zhou, Q.X. and J. Kohn. 1990. "Preparation of poly(L-serine ester): A structural analogue of conventional poly(L-serine)." *Macromolecules* 23: 3399–3406.

11 Peptide Vaccines in Cancers

Öznur Özge Özcan, Rümeysa Rabia Kocatürk, and Fadime Canbolat

CONTENTS

11.1 INTRODUCTION

The reason why peptide-based vaccines are used therapeutically in cancers is to reveal immune responses to cancer through antigenic epitopes obtained from tumor antigens. Peptide-based vaccines are synthesizable, highly stable entities and there is remarkable oncogenic potential. For the characteristics of peptide vaccines in preventing tumor proliferation, pharmacokinetic parameters should be determined and understood in terms of their biological distribution. In this chapter, the therapeutic aspects of peptide-based cancer vaccines, how they are obtained, and the inclusion of new technologies, especially the interesting aspects of polymeric carriers, will be discussed. Although peptides are very small molecules, successful results have been obtained in in vitro trials, but their potential cannot be adequately preserved in in vivo results and phase studies. For this reason, the advantages and disadvantages of the therapeutic effects of these vaccines in terms of immunotherapy and their use with nanotechnology open the way to discuss new generation technologies in this chapter.

11.2 IMMUNOTHERAPY IN CANCER

The 2018 Nobel Physiology and Medicine Award was given to pioneers in the field of cancer immunotherapy, as they firmly confirmed the molecular bases adapted to the adaptive immune system to combat cancers. The immune system has been clinically validated to combat a patient's unique tumor (Goldberg 2019). Although immunotherapy contains molecular strategy (Motzer et al. 2018) and cytotoxic chemotherapy agents targeting cancer cells, it shows superior results compared to conventional treatments (Reck et al. 2016). Normally, the main task of our immune system is to protect our body against foreign infections and diseased pathogens and cancers. The immune responses created by our body are divided into two, as B and T lymphocytes, and humoral immunity and cellular immunity mediated by these products (Bielinska et al. 2014). The humoral immune system eliminates harmful pathogens and toxins through antibodies produced by B cells, while the cellular immune system allows the inhibition of intracellular microbes by activation of antigen presenting cells (APCs) and proliferation of T cells (shown in Figure 11.4). The cellular immune system responds faster to pathogens and other harmful conditions than the humoral immune system (Ponte et al. 2010; Tan and Coussens 2007; Tsiantoulas et al. 2014). Our natural and adaptive immune systems play an important role not only against pathogens and infections, but also in anticancer immune responses (Figure 11.1) (Dranoff 2004; Lakshmi et al. 2013). Our natural immune cells can produce the signals needed to stimulate responses from T and B cells for anticancer stimulation (Woo, Corrales, and Gajewski 2015). The adaptive immune system has B cells, CD8+ cytotoxic T cells, and CD4+ helper T cells. APCs, on the other hand, recognize foreign antigens in terms of anticancer functions and serve as a bridge between naive T cells and the innate and adaptive immune system (Binder 2014).

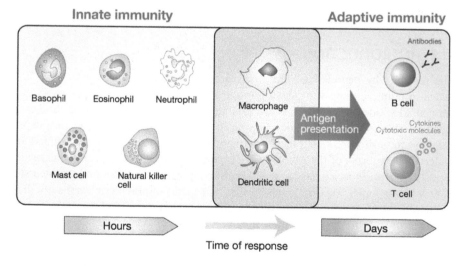

FIGURE 11.1 Description of cells in cancer's innate and adaptive immune system (Yamauchi and Moroishi 2019; Zhang and Chen 2018).

Malignant tumors commonly cause death in the world, and are the second highest cause of mortality (Yong et al. 2012). Progress has been made for decades in traditional cancer treatments, mainly through the removal of tumor tissue by an operation, use of chemotherapy drugs, and radiation treatments. While these treatment options were not successful for the treatment of every cancer patient, they achieved success in certain cancer groups. In this case, it indicates that there are no adequate or optimal treatments for tumors (Yang et al. 2015) that are not accessible by surgery or it also resulted in the formation of resistant tumor tissues that do not allow chemotherapy and radiation. Therefore, immunotherapy methods, which can have fewer side effects and prevent metastasis compared to traditional treatments, have brought innovation to the field of cancer treatment (in Figure 11.1) (Zhang and Chen 2018).

It was targeted to activate the immune system while planning the first immune-related treatments in cancer immunotherapy. Cancer cells (cytokines, interleukins, interferons, etc.) were seen with no specific target with immunotherapies. Antigens originating from the molecular basis of the cancer cell and monoclonal antibodies (MAB) were discovered thanks to these antigens, which paved the way for more research. However, adverse conditions were encountered, such as cancer cells showing immune resistance, escaping the immune response, or suppressing the immune system (Konsoulova 2015).

Immunotherapy is of great importance not only with the progression and treatment of cancer, but also in revealing the relationship between the immune system and the molecular mechanisms of cancer. The immune system is characterized by a large number of interconnected cells and their tissues throughout the body. Basically, three immune cells complete this integrity: 1) lymphocytes (T cells, B cells, and natural killer (NK) cells), 2) myeloid cells (macrophages, dendritic cells, and antigen-presenting cells), and 3) granulocytes (neutrophils, basophils, and eosinophils) (shown in Figure 11.2) (Lesterhuis et al. 2011). The general immune system produces specific antibodies and protects us from harmful foreign pathogens. These antibodies bind to targeted antigens, creating an immune response of antigen-containing cells to destroy them (Konsoulova 2015). The innate immune system produces granulocytes, macrophages, dendritic cells, mast cells, and NK cells, which act as a primary protective mechanism in situations triggered by mutation of cells as a result of stress caused by any reason. Then, in the second step, by triggering the adaptive immune response by acting as a bridge function of APC cells, cancer cells are attempted to be destroyed by specifically producing immune cells including B and T immune cells, CD 4 and CD8 expressing T-lymphocytes (Disis 2010). Healthy cells in our body begin to spread rapidly by multiplying uncontrollably with genetic or epigenetic negative effects.

Such genetic effects stimulate the natural immune system that reacts immediately. NK cells are distinguished from other immune cells by the fact that inhibitory and stimulating molecules play an important role in complex expression processes (Medzhitov 2001). Major histocompatibility complex (MHC) inhibitor receptors play a crucial role in ensuring the activation of NK cells in the natural cytotoxicity of the innate antitumor response. Cytotoxicity receptors are expressed on the surface of the NK cell. Expression of these receptors is stimulated on the surface of

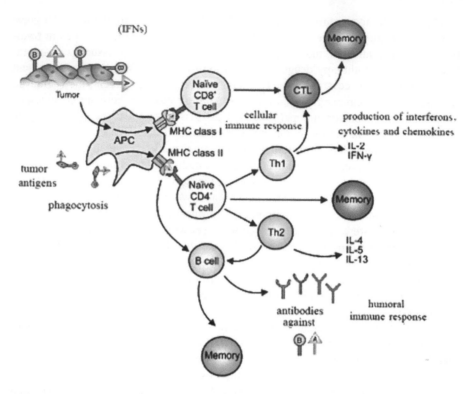

FIGURE 11.2 Representation of the innate and adaptive immune response (humoral and cellular T and B cell response). Tumor cell proteins APCs act as a bridge between these two systems. APCs from T lymphocyte cells help T cells proliferate and mature into antibody-producing plasma cells. In peptide-based vaccines, Dentritic cells can present cross-presentation epitopes to CD8 + T cells by antigen uptake (Konsoulava 2015).

tumor cells (Moretta et al. 2001). Binding of MHC-I-related ligands to these surfaces provides tumor cytotoxicity and IFN-y production, activation of cytotoxic NK cells, T cells, and CD8 T cells (Bauer et al. 1999; Kumar et al. 1995). With this receptor stimulation inside the cell, it activates macrophages and dendritic cells by releasing cytoplasmic stress molecules out of the cell. With this activation, interleukin production, which is a cellular immune response, begins and then the adaptive immune system is activated (Medzhitov 2001; Diefenbach et al. 2000).

The adaptive immune response continues in the "second step" as an immune response. Dendritic cells present cancer antigens to MHC class I and II, thereby developing T and B lymphocytes in the tumor-specific cellular response by dendritic cells and antibody formation in the humoral response. APCs and macrophages of the humoral response compose against antigens formed together with dendritic cells. Antigen proteins are cleaved to peptide sequences by phagocytosis or receptor endocytosis. They produce cytokines as MHC class II immune responses and contribute to more autoimmune responses in case of insufficiency. (Steinman 2012). It uses "immune control points" which are endogenous cross-presentations that control the activation of increased T cells after the breakdown of antigens. If the immune system works properly, our body is effectively protected from various foreign pathogens (Josefowicz et al. 2009; Shevach 2002).

Cancer immunotherapy has gained a lot of attention around the world due to the fact that immune checkpoint inhibitors are being produced and studied, and their recent development (Wada et al. 2016). The immune system plays a dual role in cancer: 1) it inhibits the progression of the tumor by killing cancer cells or stopping their growth and 2) facilitates tumor growth by selecting tumor cells that are more likely to survive in an immunocompetent host (Schreiber, Old, and Smyth 2011). As the immune system understood biological expression of cancerous cell destruction, stopping, and tumor development, the existence of "cancer immuno-regulation'" and immunotherapy methods has been inevitable. There are two types of immunotherapy methods in cancer: vaccination and cell transfer therapy. Cancer vaccines are based on active immunotherapy to induce an anti-tumor immune response when immunized with the content of tumor antigens, while cell transfer therapies are based on passive stimulating therapies. It is aimed at actively destroying cancer cells due to the penetration and increase of autologous lymphocytes in tumor cells (Melief et al. 2015; Morgan et al. 2006). In this context, it is now technologically possible to design T cells specific for tumor antigens against cancer and for this, cell transfer therapies can be developed by including genetic engineering technologies. More recently, a new method of immunotherapy, also known as immune checkpoint blockade, has been designed in the clinic (Porter et al. 2011).

Briefly, a variety of cancer immunotherapies have been developed for cancer treatments, including immune checkpoint blockage therapy, cytokine therapy, cancer vaccine, and chimeric antigen receptor (CAR)-T cell therapy and many have been studied (Riley et al. 2019). Monoclonal and polyclonal antibodies, antibodies, cytokines, cellular immunity, and vaccines used in cancer immunotherapy have become increasingly successful therapeutic agents in both preclinical models and clinical trials, particularly in the treatment of solid cancers. Histologically, these alternative methods of immunotherapy are used to reduce the size of solid cancer and in combination treatments with chemotherapy (Baxevanis et al. 2009). Another aim of tumor immunotherapy is to enable individuals with a high family history of cancer to gain immunity against target tumors, actively or passively (Kakimi et al. 2016). With the increasing advances in cancer immunotherapy, it has become possible to control both natural and adaptive immune cells (illustrated in Figure 11.1) and their responses on tumor cells (Zhang and Chen 2018).

Cancer vaccines found among all these methods are usually produced in the form of tumor-associated antigens and immune adjuvants. The most important advantage of cancer vaccines is that they contain antigens that can produce a cancer-specific immunological response and can provide long-term immune memory to inhibit cancer recurrence (Baxevanis et al. 2009). Another advantage of vaccines is the ability to combine immunotherapy with traditional treatments, which may be the most effective strategy for the patient's benefit (Zitvogel et al. 2008). In addition, vaccination against cancer-specific antigens can make the tumor more susceptible before applying chemotherapy, which can provide the most effective treatment (Ramakrishnan,

Antonia, and Gabrilovich 2008). Technological advances in the use of vaccines in cancer immunotherapy are addressed in many ways including DNA or RNA vaccines, recombinant viruses that express tumor genes, genetically engineered tumor cells or dendritic cells (DC), monoclonal antibodies, and synthetic peptides that represent the abundance of tumor immunogenic epitopes (van der Burg et al. 2006).

Of new approaches to develop clinical methods in immunotherapy, only two approaches, preventive and therapeutic, are used in vaccines (Fink 2018). Most of these immunotherapy methods mentioned have various safety and efficacy challenges. Especially for cancer vaccines, the protective and limited therapeutic benefits, complex production processes, and determination of uncertain effective dosages, clinical results that create insufficient antitumor immune responses are encountered (Cicchelero, de Rooster, and Sanders 2014) (Figure 11.3).

There are many immunotherapeutic modalities in cancer immunotherapy, including by the United States Food and Drug Administration (FDA) and European Medicine Agency approval. Such modalities include many strategic methods to increase the effectiveness of cancer immunotherapy and clinical methods that try to transform them into effective cancer therapies including FDA-approved cytokines, immunomodulatory mAbs, chimeric antigen receptors, aptamers, blocking tumor-induced immune suppression, blocking activity in cancer cells, modulation of tumor necrosis factor-inducing ligand (TRAIL) receptors, lymphodepletion, in vitro-activated Natural killer T (NKT) cells and their clinical application, modulation of

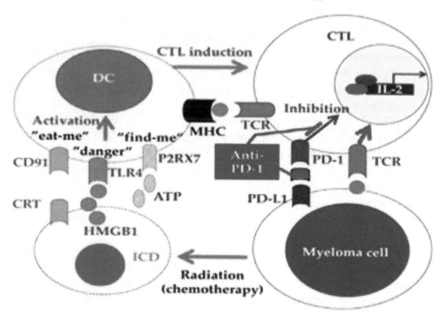

FIGURE 11.3 The presentation of traditional therapies in cancer such as radiotherapy and chemotherapy in below. Demonstration of the effect of immunotherapy on cancer cells is shown. In particular, immune checkpoint blockage inhibitors (e.g. Anti PD-1) can trigger IL-2 cells as a systemic antitumor immunological response (Tamura et al. 2019).

CTLA-4 and PD-1 checkpoints (Marin-Acevedo et al. 2018; David et al. 2011), other vaccinations, adjuvants and multiple peptide vaccines for cancer immunotherapy (Baxevanis et al. 2009). This chapter focuses on synthetic peptide vaccines. In the field of immunotherapy, peptide vaccines can stimulate T cells, in particular and has been specific. It took more than 20 years to develop peptide-based vaccines to treat growing tumors (Rosenberg, Yang, and Restifo 2004). A vaccine developed specifically for tumor antigens may have wide application and use in preventing metastasis of many different types of cancer. Cancer cells growing uncontrollably in cancer patients mask carboxyrate antigens by adding carbohydrate moieties to avoid an autoimmune response, and consequently eliminating threatening tumor cells. In fact, this mechanism plays a major role in the recurrence of cancer after successful treatments (Lazoura and Apostolopoulos 2005).

11.3 ANTITUMOR IMMUNE RESPONSE OF PEPTIDE VACCINES

The therapeutic strategies of peptide vaccines in the fight against cancers are to control metastatic and local tumor cells. As an alternative to monoclonal antibodies intended to lead the cell to controlled apoptosis in passive immunotherapy, the active immunotherapy method developed by tumor-specific antigens (TSA) and tumor-associated antigens (TAA) have been developed. Peptides consisting of many amino acids and showing antitumor properties are antigenic epitopes developed from TAA or TSA against many types of cancer (Nezafat et al. 2014; Buhrman et al. 2013; Lazoura and Apostolopoulos 2005). Peptide vaccines are used to stimulate T cells against specific expressed antigens from cancer cells (Milani et al. 2013). The advantage of peptide vaccines compared to conventional treatments is their low toxicity, significantly increasing the survival rate and life span of healthy cells in cancer patients (Asahara et al. 2013; Hui et al. 2013; Tanaka et al. 2013).

Proteins expressed in tumor cells have the potential to elicit immune responses to antigens and create long-term memory (Kruger, Greten, and Korangy 2007; Lage 2014). Each antigen is tumor specific. Large epitopes of overexpressed proteins between different tumor antigen types continue to be extensively investigated for cancer vaccines due to their potential to elicit CD8+ and CD4+ T cell immune responses. Compared to full amino acid length protein antigens that require cellular internalization and many modification processes, peptides bind to MHC molecules to produce more effective cancer vaccines with high stability and cellular uptake (Pol et al. 2015). The function of peptide vaccines is related on three stages, improving an immune response against cancer. By therapeutic application of the peptide vaccine, the DCs around the cancer cells turn into mature DCs to initiate this immune response (Mellman, Coukos, and Dranoff 2011). Peptides systemically delivered by the differentiation of these DC cells are introduced into the cell by MHC class I and II molecules. Binding of MHC class I to peptides and stimulating cytotoxic T lymphocytes in general is the target of peptide vaccines (Flutter and Gao 2004). Peptides are first delivered to the endoplasmic reticulum by an antigen-specific transport into the cell (Neefjes et al. 2011). The peptide inside the cell binds to MHC class I and from there it is transmitted to the Golgi. The peptide-MHC I linkage delivered the Golgi to cell membrane interacts with T cell receptors to stimulate CD8+ T cells (Van Kaer 2002). MHC class II cells work differently molecularly than MHC I class. Peptides which are belong to MHC class II can stimulate CD4+ T helper cells. MHC class II peptides may have longer amino acid sequences than MHC I stimuli (Rosenberg, Yang, and Restifo 2004). Peptides are taken into the cell by antigen-presenting phagolysosomes to enter the cell. It combines with MHCI class II invariant chain and is called MHC class II compartment (MIIC). MIICs are delivered into the phagolysosomes. Inside the phagolysosome, li is separated from MIIC and bound with peptides (Neefjes et al. 2011). Leukocyte antigen (HLA)-DM functions to bind li and peptide complexes to MHC class II outside the cell membrane. After the extracellular peptides combine with MHC class II, CD4+ T helper cells are stimulated, and peptides are identified. Schematic representation of all these processes is given in Figure 11.4). In the next steps, peptide-carrying DCs come to lymphoid tissues and generate T cell immune response such as CD4+ T helper cells and CD8+ T cytotoxic cells (Callan et al. 2000). T cells that are stimulated at the last stage come to the tumor environment and cause a cytotoxic response. In particular, CD 8 cytotoxic T lymphocytes (CTL) cells secrete CD4+ T helper 1 cells (Th1), IFN-y, TNF-a and interleukin-2 (IL-2) and several cytokines. It has been found to have

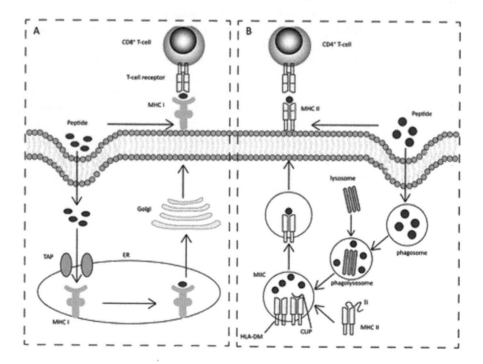

FIGURE 11.4 Introduction of peptide-based vaccines from about immune cells MHC I or MHC II pathway. Peptides which are of high solubility can bind directly to MHC molecules on the cell membrane of DCs. Depending on amino acid sequences the peptide sequence consists of peptide-based vaccine uptake (Yang et al. 2015).

strong effects throughout the tumor environment (Bos and Sherman 2010; Dosset et al. 2012; Kennedy and Celis 2008).

Antigen presentation is a difficult-to-understand process involving a range of events, ranging from an undefined pathogen or cell biological activity to binding to large histocompatibility complex proteins (MHC) class I (MHC-I) or class II (MHC-II) (Comber and Philip 2014; Klug et al. 2009). Since the response against pathogens occurs in the cell as antigen presentation, their cytotoxic response is MHC-I antibody and T helper response is MHC-II (Parmiani et al. 2014), especially immune response against tumor involved with the other T cell units (Reinherz, Keskin, and Rainhol 2014). The peptides presented by MHC-I against tumors can range from 3 to 30 amino acids in length (Cerezo et al. 2015).

11.4 KEY CONCEPTS OF PEPTIDE SELECTION IN IMMUNE RESPONSES

Peptide selection is very important for the immune response and cancer pathogenesis that will occur when applying peptide vaccines in tumor cells. Peptide epitopes are two types in terms of tumor antigens such as TSA and TAA. TSAs are generally associated with tumor cells from viral infection or mutation. More commonly preferred TSAs are found in malignant and healthy cells compared to TAAs (Wei et al. 2011).

11.4.1 TSA-Derived Peptides as Immunogens

Peptides from TSA from viral oncogene products show strong immune responses. Viral oncogenes do not show activity in normal cells, so viral oncoproteins are not produced. However, TSA-derived peptides are potential vaccine candidates as these proteins must be produced to maintain malignant status in cancer development. Various viral oncoproteins have been preferred as vaccine systems to create T cell immune response especially in cancer types caused by viral effects. Liao et al. studied type 16 E5 peptides developed for cervical cancer caused by papillomavirus. The result of this study, tumor growth was inhibited, and the vaccine caused strong T cell immune response (Liao et al. 2013). Many specific TSA-derived peptides have been identified for human tumors, and our body does not have full tolerance for these peptides. Thus, they have amino acid sequences (9–13 amino acid lengths) recognized by T cell receptors through MHC molecules (Novellino, Castelli, and Parmiani 2005). The antigenic recombinant peptide of the Epstein-Barr virus has shown a low immune response in the phase studies of nasopharyngeal carcinomas, although T cells are used to generate immune responses and safety (Hui et al. 2013). Peptides by T cells turned out to be tumor-specific within the scope of HLA class I and/or HLA-Class II (Melero et al. 2014; Parmiani et al. 2002).

11.4.2 TAA-Derived Peptides as Immunogens

Apart from viral effects, other TAA-derived peptides are also associated with the presence of antigens present in many proteins of mutagenic genes produced by tumor cells. In the production of new epitopes, instead of detecting a number of mutated antigens in mouse and human systems also creates different discussions on the characterization steps in TAA-derived peptide vaccines (Palucka, Banchereau, and Mellman 2010). In recent years, the detection of TAA-derived peptides, which caused high immune response against both mutation and deviral pathogens, with proteomic and genomic studies conducted with DNA sequencing developments, leads to further development of peptide vaccines (Parmiani et al. 2007). The advantage of genomic analysis in this direction is to provide a better understanding of the effects of nucleotide changes, deletion and insertion mutations, inversions, and translocations on antigen protein, especially in mutagenic genes. Mutation determinations differ according to tumor types. For example, in tumors such as melanoma, immune systems have the potential to create new antigens (Johnson et al. 2014; Wölfel et al. 1995). As a result, TAA-derived peptides are detected as differentiation antigens, tumors caused by microorganisms, excessive protein synthesis resulting from overexpressing genes and global tumor antigens. An example of antigen differentiation is the $gp100_{209-2M}$ peptide vaccine administered on melanoma patients between stage 1 and 3, which induces CD8+ T cells (Hu et al. (2013, 2014); Schwartzentruber et al. 2011; Walker et al. 2009). Overexpressed antigens include overexpression of HER protein 2, which is common in many types of cancer, such as breast, ovarian, and lung. In the combined treatment of HER-2/neu protein-derived peptides with granulocyte – macrophage colony stimulation factor, patients' strong and HER2-specific T cell immunological effect has been demonstrated (Disis et al. 2002; Obara et al. 2018). In global tumor antigens, telomerase expressions appear to increase by 90–95% in patients with tumors. Telomerase peptide vaccines have been shown to be effective in T cell immunity in these patients. (Bernhardt et al. 2006). Brunsvig et al. demonstrated intradermal GV1001 injections (hTERT: 611-626) peptide vaccine had immunogenic, safe, and stimulated strong specific immune responses, especially in lung cancer patients (Brunsvig et al. 2006, 2011).

11.4.3 Long Peptides Vaccine

It is known that long peptides, which are considered to be longer than 13 amino acids, have better penetration with MHC molecules and have higher immunity than other short peptides due to their higher affinity. In addition, HPV long peptides are known to show a higher T cell response in humans (Welters et al. 2008). However, in general, when peptide vaccines are designed, TAA-derived peptides, which are highly compatible with HLA alleles and have short amino acid sequences, are selected from antigens in many HLA subtypes (Thundimadathil 2012). Long peptide chains are also synthesized because these short chains cause HLA restrictions and may reduce immune tolerance by presenting T and B cells that are not APC capable. Selected antigenic epitopes are also available from in silico and in vitro experiments, but the peptides obtained by these methods should be TAA-derived peptides (Hajighahramani et al. 2017) and optimally associated with MHC cells on HLA alleles (Khazaie, Bonertz, and Beckhove 2009).

The important advantages of peptide vaccines over other vaccines are that they are easy to synthesize, show high chemical stability and do not contain oncogene-stimulating properties. Although peptide vaccines are not currently available on the market, clinical studies are still ongoing. Peptides can offer comfortable and simple ways to tackle the complex causes of the molecular biological infrastructure of tumor antigens in particular (De Paula Peres et al. 2015; Melief and van der Burg 2008). There are some conditions that can limit the therapeutic effect of the peptide vaccine (Parmiani et al. 2014). First of all, tumor cells can lower the activity of MHC molecules or disable the immune response effect of antigens. Another reason is that peptide vaccines are limited to HLA class I and II, and consequently clinical trials are still moderate. Most peptide-based vaccines target a limited number of MHC class I peptide sequences to respond to CTL. The most troublesome disadvantage may increase the PD-L1 activity and express levels of surface ligands acting on these receptors on PD-1-mediated surfaces (Yang et al. 2015). In the application of peptide-based vaccines, approaches to eliminate these disadvantages can be followed by the combined administration of immune-modulating antibodies with peptide vaccines, the use of effective adjuvants, the selection of HLA class II-restricted helper epitopes, the combination of low and high affinity peptides. Another approach is combination of peptides with nanotechnological areas so they are called nanodrug delivery systems (Chen, Chen and Liu 2019) and to design and discover peptides with improved technological bioinformatics programs (Löffer et al. 2016).

11.5 NANOTECHNOLOGICAL APPROACH IN IMMUNOTHERAPY

The therapeutic index is a very important parameter to improve prognosis in diseases and especially cancer patients. The drug doses given in cancer pathophysiology are of great importance when determining the therapeutic effect because a very small percentage acts on leukocytes. With the analysis of these values, treatment approaches for breaking the local immune tolerance without disrupting the immune tolerance of the patients systemically significantly change the good results from the disease. In this context, the nanotechnology and bioengineering fields, which develop new treatment approaches in cancer technologically, offer great advantages to cancer immunotherapy (Jeanbart and Swartz 2015).

Agents are developed to obtain more effective anti-tumor immune responses with various biomaterials including nanoparticles or microparticles, scaffolds, implants, and cell-based platforms to develop effective therapeutic strategies (Goldberg 2015; Gu and Mooney 2016; Jeanbart and Swartz 2015; Weber and Mulé 2015). In order to increase cancer treatment, new nanotechnologically local treatments have been developed supported by various biomaterials with radiotherapy, phototherapy, and even chemotherapies. Such local treatments are also used to increase the effectiveness of immunogenic agents. Combined with primary cancer cell death and immune adjuvants initiated by these nanotechnological therapies, they have high potential to activate strong systemic anti-tumor responses. At the same time, when combined with immunotherapeutic agents, it can release TAA that can function as "tumor vaccines" on a cellular basis without systemic elimination (Chen et al. 2016).

11.5.1 Polymeric Carriers to Enhance Cancer Immunotherapy

Nanoparticles, which are frequently used among polymeric carriers, have been very effective in the birth of the nanomedicine. They have many advantages compared to traditional methods in cancer immunotherapy and vaccination. Nanoparticles (NPs) can play an effective role in cancer immunotherapy as a means of tumor protection and therapeutic delivery by protecting the active substances. It may increase the stability and solubility of active substances which are used in immunotherapy. It can also prolong drug half-life by avoiding pharmacokinetic elimination after the active ingredients (especially peptides) are introduced into the circulation. If the therapeutic method of immunotherapy is used, it can minimize the side effects such as the negative effect of healthy cells by showing direct targeting cancer cells (Özcan and Karahan 2019). Since such NPs are biodegradable carriers in biological fluids, they can be loaded with therapeutic compounds to achieve concentrated local drug delivery with their sustained release potential (Chow and Ho 2013). The most functional advantage of NPs is that they can only be used with targeting systems, where they can be directed to the cancer cell or tumor microenvironment. These targeting structures are the presence of various ligands (e.g. antibodies, aptamers, or peptide epitopes) that can interact with molecules originating from tumor cells, including receptors located on the surface of cancer cells (Goldberg 2015; Özcan and Karahan 2019; Wang et al. 2014). Rather than conventional treatments with high cytotoxic effects given to tumor cells, administration of immunostimulatory agents may be a more effective option to eradicate tumors. Although applications for systemic use with conventional treatments for tumors are still difficult, it has been found that their use with immunological agents can be quite effective on the immune system (Özcan and Karahan 2019; Sunshine et al. 2014). On the other hand, using drugs of nanometer or micrometer size NPs with small molecules, oligonucleotides, immunomodulatory compounds, etc. (shown in Figure 11.5) can allow them to pass various cellular or textural biological barriers more easily and accumulate in tumor tissues (Dave et al. 2014; Mai et al. 2014; Mu et al. 2016; Shen et al. 2013; Xu et al. 2016).

While NPs can offer effective solutions in conventional drug administration, they can also offer effective solutions in the field of cancer immunotherapy. There are still difficulties in cancer immunotherapy, especially because of the poor immunogenicity of cancer vaccines and because immunotherapies can come up with unwanted side effects (Özcan and Karahan 2019). Many different biological and physicochemical activities of NPs which are especially approved or still studied, are versatile systems that can overcome these problems and cancer immunotherapy (Fan and Moon 2015). Also, despite the important efficacy of cancer vaccines, traditional vaccination approaches have been insufficient on the immune response to achieve successful treatments or complete protection from tumor vaccines (Rosenberg, Yang, and Restifo 2004) (Figure 11.6).

In particular, new strategies are needed to successfully transmit tumor antigens and adjuvant combinations by APCs and to get sufficient immune response for destroying tumor cells (He et al. 2015; Fan et al. 2013; Kumar and Mohammad 2011; Van der Meel et al. 2013). Indeed, polymeric NPs have many opportunities

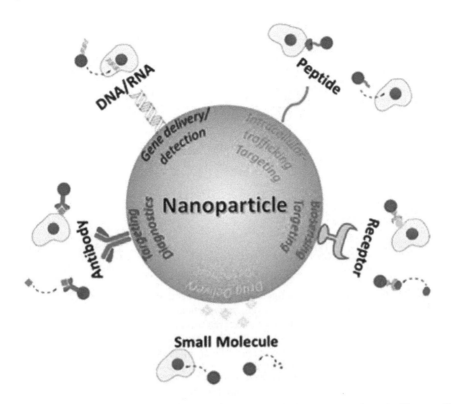

FIGURE 11.5 Targeting tumor cells with NPs conjugation methods (given in Chapter 5) with nucleic acids, peptide epitopes, cancer cell-derived receptors and antibodies (Graczyk et al. 2020).

for cancer immunotherapy. Among these, higher immune response can be achieved by administering antigens (such as pathogen peptide epitopes) and adjuvants (using NPs) as vaccines. In addition, higher immune activity can be achieved in the tumor microenvironment by targeting specific cells such as administration of antibodies, APCs, or dendritic cells with NPs and more specifically to counteract most of the immuno-inhibitory effects of tumors (Hagan, Medik, and Wang 2018). NPs are synthetic structures with nanometer scale dimensions and properties that can be combined with molecular structures (usually derived from polymers, lipids or metal, and metal alloys).

11.5.2 DIFFERENT TYPES OF POLYMERIC CARRIERS FOR CANCER IMMUNOTHERAPY

The production of high specific immune responses against cancer depends on the antigen uptake of APCs, especially DCs. Efficient uptake by DCs depends on the important antigen characteristics of size, shape, and surface charge because the key part in the immune response formation is the ability to adequately communicate with adaptive immune cells, such as CD8 T cells, which are bound to the lymph nodes, by

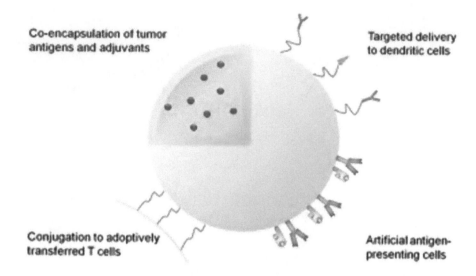

Co-encapsulation of tumor antigens and adjuvants

Targeted delivery to dendritic cells

Conjugation to adoptively transferred T cells

Artificial antigen-presenting cells

FIGURE 11.6 Tumor-specific antigens (antibodies, peptide epitopes, etc.) and adjuvant formulas can be encapsulated or conjugated to nanoparticles. NPs surfaces can be targeted with specific receptors, ligands and antibodies to target tumor cells. In addition, T cell immunotherapy can be performed by transfecting T cells to the NPs surface. Thus, NPs formulations can be taken with dendritic cells to increase the stimulation of MHC class molecules (Fan and Moon 2015).

proper activation of DCs (Bachmann and Jennings 2010; Bousso and Robey 2003; Fehres et al. 2014). The use of NPs in peptide-based vaccines associated with antigen presentation is an optimally designed antigen delivery system. It can be seen that more effective vaccines can be developed on the tumor with high DC cellular uptake and high antigen presentation using NP delivery systems (Bertrand et al. 2014).

In recent years, cancer vaccines against the tumor, antigens (Reyes et al. 2013; Reed, Orr, and Fox 2013), and adjuvant combinations are used with NPs [e.g. poly(lactide-co-glycolide) (PLG), poly(lactic-co-glycolic acid) (PLGA), gold NPs], liposomes, etc. to increase vaccine effectiveness and also virus-like particles (VLP) has gained considerable interest as cancer vaccine platforms to induce antigen-specific immune responses against cancerous cells (Jiang et al. 2005; Krishnamachari et al. 2011; Park et al. 2013; Schwendener 2014; Zhao et al. 2014). VLPs (shown in Chapter 13) are NPs (20–100 nm) derived from various viruses that have been inactivated so that they no longer have copying features. By offering linear and conformational antigens to APCs, both MHC I and II can result in B cell activation and cross-presentation, thus activating CD4+ and CD8+ T cells. VLPs can be designed to target immune cells, increase vaccine efficacy, or be loaded with immunogenic ligands by site-directed mutagenesis or bioconjugation (Buonaguro et al. 2011). When VLPs are used with vaccines, they provide the induction of APCs by increasing the antigen delivery in the lymph nodes that DC cells can reach. As antigen presentation increases, endocytosis continues to provide antigens to adaptive immune cells by stimulating the efficacy of vaccines (Manolova et al. 2008; Molino et al.

2013, 2017). In addition, encapsulation of peptide vaccines developed specifically against cancer with dendrimers has shown that higher T immune response, higher antibody response, and peptides can remain longer in body fluids than in free antigen yield (Lu et al. 2015). However, there is lack of studies about dendrimer NPs to deliver cancer peptide vaccines.

11.5.2.1 PLGA

The permeability and retention effect (EPR) caused by the increase of molecules accumulated in the tumor tissue through capillaries (extravasation pathway) has opened the way for targeting NPs, especially against solid tumors. This scientific way and targeted NP studies have attracted attention with the use of liposomal doxorubicin (Doxil) and albumin-bound paclitaxel (Abraxane) drugs, which have opened the nanomedicine field with the approval of the FDA (Fang, Nakamura, and Maeda 2011; Micha et al. 2006; Barenholz 2012). Recent research has shown that the use of NPs (including PLGA) has a more immune-enhancing effect not only in drug studies, but also with antigens (peptides, epitopes, etc.) compared to conventional administration with vaccine adjuvants. In particular, encapsulation of cancer-developed peptide vaccines with NPs (methods were shown in Chapter 8) showed that higher T immune response, higher antibody response and peptides could remain longer in body fluids than degradation compared to free antigen yield (Akagi, Shima, and Akashi 2011; Xiang et al. 2015). The importance of DC uptake was mentioned for peptide antigens to generate T cell response. NPs (especially PLGA) provide great advantages in DC uptake in terms of size, surface load, hydrophobicity (Benne et al. 2016). It can be seen that NPs (especially PLGA) can be produced in the size of approximately 20 ~ 200 nm. NPs can be effectively internalized by DCs by means of clathrin and caveola-dependent endocytosis pathways (Bachmann and Jennings 2010). Another advantage of PLGA NPs is that when used with antigens, the antigen is released in a controlled manner to prevent the antigen from being detected by a foreign body and being phagocyted by antibodies in vivo studies (Jahan et al. 2018).

PLGA is also FDA approved for humans, biocompatible, and polymeric NPs. Highly stable in PLGA saline buffer, it is very suitable for skin and intramuscular injections in cancer peptide vaccine applications (Kim, Griffith, and Panyam 2019). Transition-like receptor (TLR) 7/8 agonists are among the important adjuvant groups in cancer vaccines to demonstrate strong T cell responses by promoting DC uptake and are also a citcon enhancer in cellular immune response. Although TLR7 agonists were successful in in vitro cancer vaccination studies, they resulted in their accumulation at the injection site in preclinical and clinical studies. As a solution, although it is thought that by applying subcutaneous and intramuscular applications, TLR7 and TLR8 agonists will be provided with advantage in terms of transmission to DCs that can produce the desired immune response. It has been used with PLGA polymeric NPs that can overcome these problems such as enhancing the DC uptake and lymphatic drainage; PLGA NPs overcome these problems by providing high clearance protection for adjuvant formulations (Silva et al. 2013). Since peptides identified by cancer type and biology are not sufficient to increase T cell stimulation, an immunostimulant is also needed to increase the immunogenicity of the vaccine.

For this reason, to increase the immunogenicity of peptide vaccines in cancer, IL-12, granulocyte-macrophage colony stimulating factor (GM-CSF), and TLRs agonists have been actively studied as vaccine adjuvants so far (Lehner et al. 2007; Simons and Sacks 2006; Napolitani et al. 2005).

11.5.2.2 Liposomes

Liposomes have been investigated in the field of cancer research by encapsulating various active substances against cancer, including drugs, protein, vectors, and nucleic acids. Due to their physicochemical properties (size, surface, composition), liposomes are among the colloidal NP types that can be easily modulated to increase the recognition and antigen uptake by DC cells. They are particularly advantageous for peptide antigens that show hydrophilic properties. In previous studies, liposomal vectors were used with peptide epitopes in combination with adjuvant for different vaccine applications (Heurtault et al. 2009; Kakhi et al. 2016; Roth et al. 2005; Thomann et al. 2011). Liposomes are powerful tools that transmit antigen presentation to APCs, deliver locally to tissues, retain, and transfer to cells (Lee et al. 2015).

In in vivo vaccine studies using liposomes, it has been reported that intake of peptide antigens increases the uptake by DC cells and induces cell-mediated anticancer immune response at the vaccination site (Gregoriadis 1994; Hansen et al. 2012; Nordly et al. 2011; Zaks et al. 2006; Varypataki et al. 2017; Zong et al. 2016). PLGA, liposomes, and cationic lipid NPs have been proven to offer Trp2 antigens but PLGA polymers show slow release rates and long-term immune responses compared to liposomes (Demento et al. 2012). Encapsulation of peptide vaccines developed specifically against cancer with liposomes has shown that higher T immune response, higher antibody response and peptides can remain longer in body fluids than the free antigen yield (Chen and Huang 2008; Vasievich et al. 2012). E75 (human epidermal growth factor 2 (HER2)/neu 369-377) "KIFGSLAFL" is an immunogenic peptide antigen consisting of nine amino acids and derived from the HER2 protein overexpressed due to the overwork of the HER2 receptor. E75 (neu 369-377) HER2 has been studied in early-stages breast cancer patients with phase III studies (Schneble et al. 2014). Although clinical data was obtained, E75 breast cancer peptide vaccine should be repeated to maintain immunity, CD8 CTL (cytotoxic T lymphocytes) was observed in dose repeats (Benavides et al. 2009; Hueman et al. 2007; Mittendorf et al. 2008; Murray et al. 2002). Liposomes are easy to produce, biocompatible, and reliable nanoformulations play an important role in vaccine development. According to a study by Arab et al., when the peptide vaccine derived from E75 HER-2/neu was used with liposome NPs it was found to show a strong immune response to the tumor in vivo (Arab et al. 2018). A peptide from the HA influenza virus protein as a CD4 epitope and a peptide from the HPV16 E7 oncoprotein as a CD8 epitope in conjunction with liposome NPs were studied in vivo against human papillomavirus-induced tumors. The results showed a specific immune response and an antitumor activity against lung cancer (Jacoberger-Foissac et al. 2019). The liposome NP formulation created with the HTM4SF5 peptide, which is in the B cell epitope category, stopped the growth in the tumor with specific IgG production after implantation of hepatocellular carcinoma (HCC) cells and given to the experimental animals observed

(Kwon et al. 2012). In the study by Lai et al., in vivo studies of the nanostructured vaccine formulation of TRP2180-188 peptide epitope-loaded liposomes with CpG oligodeoxynucleotide adjuvants were performed. According to the results obtained, the liposomal formulation has been shown to significantly reduce the number of myeloid-derived suppressor cells (MDSCs) and regulatory T cells, while at the same time increasing the number of activated T cells, tumor antigen-specific CD8+ cytotoxic T cells, and interferon-producing cells. When tumor tissues were examined, it was observed that it stopped tumor angiogenesis and tumor cell proliferation and also increased the apoptotic effect (Lai et al. 2018). The epitope peptide (pGPC3) derived from GPC3 is shown in vivo antitumor effect of the formulation created with liposome surfaces (pGPC3-liposome) and peptide-specific CTLs (Iwama et al. 2016). As a result, liposomal vaccines created with cancer-associated peptide antigens may create an innovative vaccine strategy for cancers in the future.

11.6 FUTURE PERSPECTIVE

The target of peptide-based vaccines used for cancers is based on the strategy of stimulating the immune system of patients to recognize cancer cells and eliminating them before they multiply. The given peptide antigens must have the potential to be taken up by the patient's DCs and to activate APC. Thus, peptide-based cancer vaccines can activate the immune system, which has the potential to destroy tumor tissues, and prevent metastasis. Co-administration with NPs due to problems caused by the weak immunogenicity of peptides shows both lower toxicity and higher immunological properties. The use of NPs until the peptide antigens are internalized by the cells can prevent the body from perceiving the antigen peptide as foreign and breaking it down with proteases. In addition, NP systems can contribute to more successful uptake of antigens by cells. As a result, improvements in the stability and permeability of peptide drugs or vaccines with formulations developed (such as NP systems) with studies to be supported in the future may lead to more use of peptide materials in clinical cancer immunotherapies and cheaper and easier-to-produce immunotherapeutics. In addition, further studies of metabolomic, proteomics, and genomic screening of peptide-based cancer vaccines in the body may pave the way for more frequent use of bioactive peptides produced by post-translational modifications or non-ribosomal synthesis. In particular, these studies need urgent study on cancer peptide vaccines supplied with NP systems. In addition, the pharmacokinetic studies of peptide antigens for the use of peptide-based cancer vaccines in the clinic and the difficulties in how they show biological distribution biologically, the parameters determining the doses of the peptides, and the parameters determining the therapeutic indexes should be studied.

REFERENCES

Akagi, T., F. Shima and M. Akashi. 2011. "Intracellular degradation and distribution of proteinencapsulated amphiphilic poly(amino acid) nanoparticles." *Biomaterials* 32(21): 4959–4967. doi:10.1016/j.biomaterials.2011.03.049.

Arab, A., J. Behravan, A. Razazan, Z. Gholizadeh, A.R. Nikpoor, N. Barati, F. Mosaffa, A. Badiee, and M.R. Jaafari. 2018. "A nano-liposome vaccine carrying E75, a HER-2/ neu-derived peptide, exhibits significant antitumour activity in mice." *Journal of Drug Targeting* 26(4): 365–372. doi:10.1080/1061186X.2017.1387788.

Asahara, S., K. Takeda, K. Yamao, H. Maguchi, and H. Yamaue. 2013. "Phase I/II clinical trial using HLA-A24-restricted peptide vaccine derived from KIF20A for patients with advanced pancreatic cancer." *Journal of Translational Medicine* 11(1): 291. doi:10.1186/1479-5876-11-291.

Bachmann, M.F. and G.T. Jennings. 2010. "Vaccine delivery: A matter of size, geometry, kinetics and molecular patterns." *Nature Reviews Immunology* 10(11): 787–796. doi:10.1038/nri2868.

Barenholz, Y. 2012. "Doxil® - The first FDA-APPROVED NANO-DRUG: Lessons learned." *Journal of Controlled Release* 160(2): 117–134. doi:10.1016/j.jconrel.2012.03.020.

Bauer, S., V. Groh, J. Wu, A. Steinle, J.H. Phillips, L.L. Lanier, and T. Spies. 1999. "Activation of NK cells and T cells by NKG2D, a receptor for stress-inducible MICA." *Science Magzine* 285(5428): 727–729. doi:10.1126/science.285.5428.727.

Baxevanis, C.N., S.A. Perez, and M Papamichail. 2009. "Cancer imunotherapy." *Critical Reviews in Clinical Laboratory Sciences* 46(4): 167–189. doi:10.1080/10408360902937809

Benavides, L.C., J.D. Gates, M.G. Carmichael, R. Patil, J.P. Holmes, M.T. Hueman, E.A. Mittendorf, et al. 2009. "The impact of HER2/neu expression level on response to the E75 vaccine: From US military cancer institute clinical trials group study I-01 and I-02." *Clinical Cancer Research* 15(8):2895–904. doi:10.1158/1078-0432.CCR-08-1126.

Benne, N., J. Van Duijn, J. Kuiper, W. Jiskoot, and B. Slütter. 2016. "Orchestrating immune responses: How size, shape and rigidity affect the immunogenicity of particulate vaccines." *Journal of Controlled Release* 234: 124–134. doi:10.1016/j.jconrel.2016.05.033.

Bernhardt, S.L., M.K. Gjertsen, S. Trachsel, M. Møller, J.A. Eriksen, M. Meo, T. Buanes, and G. Gaudernac. 2006. "Telomerase peptide vaccination of patients with non-resectable pancreatic cancer: A dose escalating phase I/II study." *British Journal of Cancer* 95, 1475–1482. doi:10.1038/sj.bjc.6603437.

Bertrand, N., J. Wu, X. Xu, N. Kamaly, and O.C. Farokhzad. 2014. "Cancer nanotechnology: The Impact of passive and active targeting in the era of modern cancer biology." *Advanced Drug Delivery Reviews* 66: 2–25. doi:10.1016/j.addr.2013.11.009.

Bielinska, A.U., P.E. Makidon, K.W. Janczak, L.P. Blanco, B. Swanson, D.M. Smith, and T. Pham et al. 2014. "Distinct pathways of humoral and cellular immunity induced with the mucosal administration of a nanoemulsion adjuvant." The Journal of Immunology doi:10.4049/jimmunol.1301424.

Binder, R.J. 2014. "Functions of heat shock proteins in pathways of the innate and adaptive immune system." *Journal of Immunology* 193(12): 5765–5771. doi:10.4049/ jimmunol.1401417.

Bos, R. and L.A. Sherman. 2010. "CD4+ T-cell help in the tumor milieu is required for recruitment and cytolytic function of CD8+ T lymphocytes." *Cancer Research* 70(21): 8368–8377. doi:10.1158/0008-5472.CAN-10-1322.

Bousso, P. and E. Robey. 2003. "Dynamics of CD8 + T cell priming by dendritic cells in intact lymph nodes." *Nature Immunology* 4(6): 579–585. doi:10.1038/ni928.

Brunsvig, P.F., S. Aamdal, M.K. Gjertsen, G. Kvalheim, C.J. Markowski-Grimsrud, I. Sve, M. Dyrhaug, et al. 2006. "Telomerase peptide vaccination: A phase I/II study in patients with non-small cell lung cancer." *Cancer Immunology, Immunotherapy* 55(12): 1553–1564. doi:10.1007/s00262-006-0145-7.

Brunsvig, P.F., J.A. Kyte, C. Kersten, S. Sundstrøm, M. Møller, M. Nyakas, G.L. Hansen, G. Gaudernack, and S. Aamdal. 2011. "Telomerase peptide vaccination in NSCLC: A phase II trial in stage III patients vaccinated after chemoradiotherapy and an 8-year update on a phase I/II trial." *Clinical Cancer Research* 17(21): 6847–6857. doi:10.1158/1078-0432.CCR-11-1385.

Buhrman, J.D., K.R. Jordan, D.J. Munson, B.L. Moore, J.W. Kappler, and J.E.Slansky. 2013. "Improving antigenic peptide vaccines for cancer immunotherapy using a dominant tumorspecific T cell receptor." *Journal of Biological Chemistry* 288(46): 33213–33225. doi:10.1074/jbc.M113.509554.

Buonaguro, L., M. Tagliamonte, M.L. Tornesello, and F.M. Buonaguro. 2011. "Developments in virus-like particle-based vaccines for infectious diseases and cancer." *Expert Review of Vaccines* 10(11): 1569–1583. doi:10.1586/erv.11.135.

Callan, M.F., C. Fazou, H. Yang, T. Rostron, K. Poon, C. Hatton, and A.J. McMichael. 2000. "CD8(+) T-cell selection, function, and death in the primary immune response in vivo." *Journal of Clinical Investigation* 106(10): 1251–1261. doi:10.1172/JCI10590.

Cerezo, D., M.J. Peña, M. Mijares, G. Martínez, I. Blanca, and J.B. De Sanctis. 2015. "Peptide vaccines for cancer therapy." *Recent Patents on Inflammation & Allergy Drug Discovery* 9(1): 38–45. doi:10.2174/1872213x09666150131141953.

Chen, Q., L. Xu, C. Liang, C. Wang, R. Peng, and Z. Liu. 2016. "Photothermal therapy with immune-adjuvant nanoparticles together with checkpoint blockade for effective cancer immunotherapy." *Nature Communications* 7: 13193. doi:10.1038/ncomms13193.

Chen, Q., M. Chen, and Z. Liu. 2019. "Local biomaterials-assisted cancer immunotherapy to trigger systemic antitumor responses." *Chemical Society Reviews* 48(22): 5506–5526. doi:10.1039/c9cs00271e.

Chen, W. and L. Huang. 2008. "Induction of cytotoxic T-lymphocytes and antitumor activity by a liposomal lipopeptide vaccine." *Molecular Pharmaceutics* 5(3): 464–471. doi:10.1021/mp700126c.

Chow, E.K.H. and D. Ho. 2013. "Cancer nanomedicine: From drug delivery to imaging." *Science Translational Medicine* 5(216): 216rv4. doi:10.1126/scitranslmed.3005872.

Cicchelero, L., H de Rooster, and N.N. Sanders. 2014. "Various ways to improve whole cancer cell vaccines." *Expert Review of Vaccines* 13(6): 721–735. doi:10.1586/14760584.2014.911093.

Comber, J.J. and R. Philip. 2014. "MHC class I antigen presentation and implications for developing a new generation of therapeutic vaccines." *Therapeutic Advances in Vaccines and Immunotherapy* 2(3): 77–89. doi:10.1177/2051013614525375.

Dave, B., S. Granados-Principal, R. Zhu, S. Benz, S. Rabizadeh, P. Soon-Shiong, and K.D. Yu. 2014. "Targeting RPL39 and MLF2 reduces tumor initiation and metastasis in breast cancer by inhibiting nitric oxide synthase signaling." *Proceedings of the National Academy of Sciences of the United States of America* 111(24): 8838–8843. doi:10.1073/pnas.1320769111.

David, L.P., L.L. Bruce, K. Michael, B. Adam, and H.J. Carl. 2011. "Chimeric antigen receptor-modified T cells in chronic lymphoid leukemia." *The New England Journal of Medicine* 365(8): 725–733. doi:10.1056/NEJMoa1103849.

Demento, S.L., W. Cui, J.M. Criscione, E. Stern, J. Tulipan, S.M. Kaech, and T.M. Fahmy. 2012. "Role of sustained antigen release from nanoparticle vaccines in shaping the T cell memory phenotype." *Biomaterials* 33(19): 4957–4964. doi:10.1016/j.biomaterials.2012.03.041.

De Paula Peres, L., F.A.C. da Luz, B. dos Anjos Pultz, P.C. Brígido, R.A. de Araújo, L.R. Goulart, and M.J.B. Silva 2015. "Peptide vaccines in breast cancer: The immunological basis for clinical response." *Biotechnology Advances* 33(8): 1868–1877. doi:10.1016/j.biotechadv.2015.10.013.

Diefenbach, A., A.M. Jamieson, S.D. Liu, N. Shastri, and D.H. Raulet. 2000. "Ligands for the murine NKG2D receptor: Expression by tumor cells and activation of NK cells and macrophages." *Nature Immunology* 1(2): 119–126. doi:10.1038/77793.

Disis, M.L. 2010. "Immune regulation of cancer." *Journal of Clinical Oncology* 28(29): 4531–4538. doi: 10.1200/JCO.2009.27.2146.

Disis, M.L., T.A. Gooley, K. Rinn, D. Davis, M. Piepkorn, M.A. Cheever, K.L. Knutson, and K. Schiffman. 2002. "Generation of T-cell immunity to the HER-2/neu protein after active immunization with HER-2/neu peptide-based vaccines." *Journal of Clinical Oncology* 20(11): 2624–2632. doi:10.1200/JCO.2002.06.171.

Dosset, M., Y. Godet, C. Vauchy,L. Beziadu, Y.C. Lone, C. Sedlik, C.Liard et al. 2012. "Universal cancer peptide-based therapeutic vaccine breaks tolerance against telomerase and eradicates established tumor." *Clinical Cancer Research* 18(22): 6284–6295. doi:10.1158/1078-0432.CCR-12-0896.

Dranoff, G. 2004. "Cytokines in cancer pathogenesis and cancer therapy." *Nature Reviews Cancer* 4(1): 11–22. doi:10.1038/nrc1252.

Fan, Y., and J.J. Moon. 2015. "Nanoparticle drug delivery systems designed to improve cancer vaccines and immunotherapy." *Vaccines* 3(3): 662–685. doi:10.3390/vaccines3030662.

Fan, Y., W. Du, B. He, F. Fu, L. Yuan, H. Wu, and W. Dai. 2013. "The reduction of tumor interstitial fluid pressure by liposomal imatinib and its effect on combination therapy with liposomal doxorubicin." *Biomaterials* 34(9): 2277–2288. doi:10.1016/j.biomaterials.2012.12.012.

Fang, J., H. Nakamura, and H. Maeda. 2011. "The EPR effect: Unique features of tumor blood vessels for drug delivery, factors involved, and limitations and augmentation of the effect." *Advanced Drug Delivery Reviews* 63(3): 136–151. doi:10.1016/j.addr.2010.04.009.

Fehres, C.M., W.W.J. Unger, J.J. Garcia-Vallejo, and Y. van Kooyk. 2014. "Understanding the biology of antigen cross-presentation for the design of vaccines against cancer." *Frontiers in Immunology* 5: 1–10. doi:10.3389/fimmu.2014.00149.

Fink, P.J. 2018. "The cancer immunotherapy revolution: Mechanistic insights." *The Journal of Immunology* 200(2): 371–372. doi:10.4049/jimmunol.1790024.

Flutter, B. and B. Gao. 2004. "MHC class I antigen presentation-recently trimmed and well presented." *Cellular & Molecular Immunology* 1(1): 22–30.

Goldberg, M.S. 2015. "Immunoengineering: How nanotechnology can enhance cancer immunotherapy." *Cell* 161(2), 201–204. doi: 10.1016/j.cell.2015.03.037.

Goldberg, M.S. 2019. "Improving cancer immunotherapy through nanotechnology." *Nature Reviews Cancer* 19(10): 587–602. doi:10.1038/s41568-019-0186-9. Epub 2019 Sep 6.

Graczyk, A., R. Pawlowska, D. Jedrzejczyk, and A. Chworos. 2020. "Gold nanoparticles in conjunction with nucleic acids as a modern molecular system for cellular delivery." *Molecules* 25(1): 204. doi:10.3390/molecules25010204.

Gregoriadis, G. 1994. "The immunological adjuvant and vaccine carrier properties of liposomes." *Journal of Drug Targeting* 2(5): 351–356. doi:10.3109/10611869408996809.

Gu, L., and D.J. Mooney. 2016. "Biomaterials and emerging anticancer therapeutics: Engineering the microenvironment." *Nature Reviews Cancer* 16(1): 56–66. doi: 10.1038/nrc.2015.3.

Hagan, C.T., Y.B. Medik, and A.Z. Wang. 2018. "Nanotechnology approaches to improving cancer immunotherapy." *Advances in Cancer Research* 139: 35–56. doi:10.1016/bs.acr.2018.05.003.

Hajighahramani, N., N. Nezafat, M. Eslami, M. Negahdaripour, S.S. Rahmatabadi, and Y. Ghasemi. 2017. "Immunoinformatics analysis and in silico designing of a novel multi-epitope peptide vaccine against staphylococcus aureus." *Infection, Genetics and Evolution* 48: 83–94. doi:10.1016/j.meegid.2016.12.010.

Hansen, J., T. Lindenstrom, J. Lindberg-Levin, C. Aagaard, P. Andersen, and E.M. Agger. 2012. "CAF05: Cationic liposomes that incorporate synthetic cord factor and poly(I:C) Induce CTL Immunity and reduce tumor Burden in mice." *Cancer Immunology, Immunotherapy* 61(6): 893–903. doi:10.1007/s00262-011-1156-6.

He, Z., A. Schulz, X. Wan, J. Seitz, H. Bludau, D.Y. Alakhova, and D.B. Darr. 2015.(2-Oxazoline) based micelles with high capacity for 3rd generation taxoids: Preparation, in vitro and in vivo evaluation." *Journal of Controlled Release* 208: 67–75. doi:10.1016/j.jconrel.2015.02.024.

Heurtault, B., P. Gentine, J.S. Thomann, C. Baehr, B. Frisch, and F. Pons. 2009. "Design of a liposomal candidate vaccine against pseudomonas aeruginosa and its evaluation in triggering systemic and lung mucosal immunity." *Pharmaceutical Research* 26(2): 276–285. doi:10.1007/s11095-008-9724-y.

Hu, J., S. Miura, K. Na, and Y.H. Bae. 2013. "pH-Responsive and charge shielded cationic micelle of poly(L-Histidine)-block-short branched PEI for acidic cancer treatment." *Journal of Controlled Release* 172(1): 69–76. doi:10.1016/j.jconrel.2013.08.007.

Hu, Y., G.R. Petroni, W.C. Olson, A. Czarkowski, M.E. Smolkin, W.W. Grosh, K.A. Chianese-Bullock, and C.L. Slingluff Jr. 2014. "Immunologic hierarchy, class II MHC promiscuity, and epitope spreading of a melanoma helper peptide vaccine." *Cancer Immunology, Immunotherapy* 63(8): 779–786. doi:10.1007/s00262-014-1551-x.

Hueman, M.T., A. Stojadinovic, C.E. Storrer, Z.A. Dehqanzada, J.M. Gurney, C.D. Shriver, S. Ponniah, and G.E. Peoples. 2007. "Analysis of naive and memory CD4 and CD8 T cell populations in breast cancer patients receiving a HER2/neu peptide (E75) and GM-CSF vaccine." *Cancer Immunology, Immunotherapy* 56(2): 135–146. doi:10.1007/s00262-006-0188-9.

Hui E.P., G.S. Taylor, H. Jia, B.Y.M. Brigette, S.L. Chan, R. Ho, W.L. Wong. 2013. "Phase I trial of recombinant modified vaccinia ankara encoding epsteinBARR viral tumor antigens in nasopharyngeal carcinoma patients." *Cancer Research* 73(6): 1676–1688. doi:10.1158/0008-5472.CAN-12-2448.

Iwama, T., T. Uchida, Y. Sawada, N. Tsuchiya, S. Sugai, N. Fujinami, and T. Nakatsura. 2016. "Vaccination with liposome-coupled glypican-3-derived epitope peptide stimulates cytotoxic T lymphocytes and inhibits GPC3-expressing tumor growth in mice." *Biochemical and Biophysical Research Communications* 469(1): 138–143. doi:10.1016/j.bbrc.2015.11.084.

Jacoberger-Foissac, C., H. Saliba, C. Seguin, A. Brion, Z. Kakhi, B.S. Frisch, S. Fournel, and B. Heurtault. 2019. "Optimization of peptide-based cancer vaccine compositions, by sequential screening, using versatile liposomal platform." *International Journal of Pharmaceutics* doi:10.1016/j.ijpharm.2019.03.002.

Jahan, S.T., S.M. Sadat, and A. Haddadi. 2018. "Design and immunological evaluation of anti-CD205-tailored PLGA-based nanoparticulate cancer vaccine." *International Journal of Nanomedicine* 13: 367–386. doi:10.2147/IJN.S144266. eCollection 2018.

Jeanbart, L. and M.A. Swartz. 2015. "Engineering opportunities in cancer immunotherapy." *Proceedings of the National Academy of Sciences* 112(47): 14467–14472. doi: 10.1073/pnas.1508516112.

Jiang, W., R.K. Gupta, M.C. Deshpande, and S.P. Schwendeman. 2005. "Biodegradable poly(lactic-coglycolic acid) microparticles for injectable delivery of vaccine antigens." *Advanced Drug Delivery Reviews* 57(3): 391–410. doi:10.1016/j.addr.2004.09.003.

Johnson, B.E., T. Mazor, C. Hong, M. Barnes, K. Aihara, C.Y. McLean, S.D. Fouse et al. 2014. "Mutational analysis reveals the origin and therapy-driven evolution of recurrent glioma." *Science* 343(6167): 189–193. doi. org/10.1126/science.1239947.

Josefowicz, S.Z. and A. Rudensky. 2009. "Control of regulatory T cell lineage commitment and maintenance." *Immunity* 30(5): 616–625. doi:10.1016/j.immuni.2009.04.009.

Van Kaer, L. 2002. "Major histocompatibility complex class IRestricted antigen processing and presentation." *Tissue Antigens* 60(1): 1–9. doi:10.1034/j.1399-0039.2002.600101.x.

Kakhi, Z., B. Frisch, B. Heurtault, and F. Pons. 2016. "Liposomal constructs for antitumoral vaccination by the nasal route." *Biochimie* 130: 14–22. doi:10.1016/j.biochi.2016.07.003.

Kakimi, K., T. Karasaki, H. Matsushita, and T. Sugie. 2016. "Advances in personalized cancer immunotherapy." *Breast Cancer* 24(1): 16–24. doi:10.1007/s12282-016-0688-1.

Kennedy, R. and E. Celis. 2008. "Multiple roles for CD4+ T cells in antitumor immune responses." *Immunological Rewievs* 222(1): 129–144. doi:10.1111/j.1600-065X.2008.00616.x.

Khazaie, K., A. Bonertz, and P. Beckhove. 2009. "Current developments with peptide-based human tumor vaccines." *Current Opinion Oncology* 21(6): 524–530. doi:10.1097/CCO.0b013e328331a78e.

Kim, H., T. Griffith, and J. Panyam. 2019. "Poly(D,L-lactide-co-glycolide) nanoparticles as a vaccine delivery platform for TLR7/8 agonist-based cancer vaccine." *Journal of Pharmacology and Experimental Therapeutics* 370(3): 715–724. doi:10.1124/jpet.118.254953.

Klug F., M. Miller, H.H. Schmidt, and S. Stevanović. 2009. "Characterization of MHC ligands for peptide-based tumor vaccination." *Current Pharmaceutical Design* 15(28): 3221–3236. doi:10.2174/138161209789105180.

Konsoulova, A. 2015. "Immunopathology and Immunomodulation: Principles of Cancer Immunobiology and Immunotherapy of Solid Tumors." *Intechopen* doi:10.5772/61211.

Krishnamachari, Y., S.M. Geary, C.D. Lemke, and A.K. Salem. 2011. "Nanoparticle delivery systems in cancer vaccines." *Pharmaceutical Research* 28(2): 215–236. doi:10.1007/s11095-010-0241-4.

Kruger, C., T.F. Greten, and F. Korangy. 2007. "Immune based therapies in cancer." *Histology & Histopathology* 22(6): 687–696. doi:10.14670/HH-22.687.

Kumar, C.S. and F. Mohammad. 2011. "Magnetic nanomaterials for hyperthermia-based therapy and controlled drug delivery." *Advanced Drug Delivery Reviews* 63(9): 789–808. doi:10.1016/j.addr.2011.03.008.

Kumar, V., V. Bhardwaj, L. Soares, J. Alexander, A. Sette, and E. Sercarz. 1995. "Major histocompatibility complex binding affinity of an antigenic determinant is crucial for the differential secretion of interleukin 4/5 or interferon gamma by T cells." *Proceedings of the National Academy of Sciences* 92(21):9510–4. doi:10.1073/pnas.92.21.9510.

Kwon, S., D. Kim, B.K. Park, G. Wu, M.C. Park, Y.W. Ha, Y. Lee. 2012. "Induction of immunological memory response by vaccination with TM4SF5 epitope-CpG-DNA-liposome complex in a mouse hepatocellular carcinoma model." *Oncology Reports* 29(2): 735–740. doi:10.3892/or.2012.2130.

Lage, A. 2014. "Immunotherapy and complexity: Overcoming barriers to control of advanced cancer." *MEDICC Review* 16(3–4): 65–72.

Lai, C., S. Duan, F. Ye, X. Hou, X. Li, J. Zhao, and X. Lu. 2018. "The enhanced antitumor-specific immune response with mannose- and CpG-ODN-coated liposomes delivering TRP2 peptide." *Theranostics* 8(6): 1723–1739. doi:10.7150/thno.22056. eCollection 2018.

Lakshmi, B.N., K.E. Reddy, S. Shantikumar, and S. Ramakrishna. 2013. "Immune system: A double-edged sword in Cancer." *Inflammation Research: Official Journal of The European Histamine Research Society* 62(9): 823–834. doi:10.1007/s00011-013-0645-9.

Lazoura, E., and V. Apostolopoulos. 2005. "Rational peptide-based vaccine design for cancer immunotherapeutic applications." *Current Medicinal Chemistry* 12(6): 629–639. doi:10.2174/0929867053202188.

Lee, Y., Y.S. Lee, S.Y. Cho, and H.J. Kwon. 2015. "Perspective of peptide vaccine composed of epitope peptide, CpG-DNA, and liposome complex without carriers." *Advances in Protein Chemistry and Structural Biology* 75–97. doi:10.1016/bs.apcsb.2015.03.004. Epub 2015 Apr 8.

Lehner, M., P. Morhart, A. Stilper, D. Petermann, P. Weller, D. Stachel, and W. Holter. 2007. "Efficient Chemokine-dependent migration and primary and secondary IL-12 secretion by human dendritic cells stimulated through toll-like receptors." *Journal of Immunotherapy* 30(3): 312–322. doi:10.1097/01.cji.0000211345.11707.46.

Lesterhuis, W.J., J.B.A.G. Haanen, and C.J.A. Punt. 2011. "Cancer immunotherapy–revisited." *Nature Reviews Drug Discovery* 10(8): 591–600. doi:10.1038/nrd3500.

Liao, S.J., D.R. Deng, D. Zeng, L. Zhang, X.J. Hu, WN Zhang, L. Li et al. 2013. "HPV16 E5 peptide vaccine in treatment of cervical cancer in vitro and in vivo." *Journal of Huazhong University of Science and Technology Medical Sciences* 33(5): 735–742. doi:10.1007/s11596-013-1189-5.

Löffler, M.W., P.A. Chandran, K. Laske, C. Schroeder, I. Bonzheim, M. Walzer, and F.J. Hilke. 2016. "Personalized peptide vaccine-induced immune response associated with long-term survival of a metastatic cholangiocarcinoma patient." *Journal of Hepatology*, 65(4): 849–855. doi:10.1016/j.jhep.2016.06.027.

Lu, F., A. Mencia, L. Bi, A. Taylor, Y. Yao, and H. HogenEsch. 2015. "Dendrimer-like alpha-dglucan nanoparticles activate dendritic cells and are effective vaccine adjuvants." *Journal of Controlled Release* 204: 51–59. doi:10.1016/j.jconrel.2015.03.002. Epub 2015 Mar 3.

Mai, J., Y. Huang, C. Mu, G. Zhang, R. Xu, X. Guo, and X. Xia. 2014. "Bone marrow endothelium-targeted therapeutics for metastatic breast cancer." *Journal of Controlled Release* 187: 22–29. doi:10.1016/j.jconrel.2014.04.057.

Manolova, V., A. Flace, M. Bauer, K. Schwarz, P. Saudan, and M.F. Bachmann. 2008. "Nanoparticles target distinct dendritic cell populations according to their size." *The European Journal of Immunology* 38(5): 1404–1413. doi:10.1002/eji.200737984.

Marin-Acevedo, J.A., B. Dholaria, A.E. Soyano, K.L. Knutson, S. Chumsri, and Y. Lou. 2018. "Next generation of immune checkpoint therapy in cancer: New developments and challenges." *Journal of Hematology & Oncology* 11(1): 39. doi:10.1186/s13045-018-0582-8.

Medzhitov, R. 2001. "Toll-like receptors and innate immunity." *Nature Reviews Immunology* 1(2): 135–145. doi:10.1038/35100529.

Melero, I., G. Gaudernack, W. Gerritsen, C. Huber, G. Parmiani, S. Scholl, N. Thatcher, J. Wagstaff, C. Zielinski, I. Faulkner. 2014. "Therapeutic vaccines for cancer: An overview of clinical trials." *Nature Reviews Clinical Oncology* 11: 509–524.

Melief, C.J.M., S.H. van der Burg. 2008. "Immunotherapy of established (pre)malignant disease by synthetic long peptide vaccines." *Nature Reviews Cancer* 8(5): 351–360. doi:10.1038/nrc2373.

Melief, C.J.M., V.H. Thorbald, A. Ramon, O. Ferry, and H. van der Burg Sjoerd. 2015. "Therapeutic cancer vaccines." *Journal of Clinical Investigation* 125(9): 3401–3412. doi:10.1172/JCI80009.

Mellman, I., G. Coukos, and G. Dranoff. 2011. "Cancer immunotherapy comes of age." *Nature* 480(7378): 480–489. doi:10.1038/nature10673.

Micha, J.P., B.H. Goldstein, C.L. Birk, M.A. Rettenmaier, J.V. Brown. 2006. "Abraxane in the treatment of Ovarian cancer: The absence of hypersensitivity reactions." *Gynecologic Oncology* 100(2): 437–438. doi:10.1016/j.ygyno.2005.09.012. Epub 2005 Oct 14.

Milani, A, D. Sangiolo, F. Montemurro, M. Aglietta, and G. Valabrega. 2013. "Active immunotherapy in HER2 overexpressing breast cancer: Current status and future perspectives." *Annals of Oncology* 24(7): 1740–1748. doi:10.1093/annonc/mdt133.

Mittendorf, E.A., J.P. Holmes, S. Ponniah, and G.E. Peoples. 2008. "The E75 HER2/ neu peptide vaccine." *Cancer Immunology, Immunotherapy* 57(10): 1511–1521. doi:10.1007/s00262-008-0540-3.

Molino, N.M., A.K.L. Anderson, E.L. Nelson, and S.W. Wang. 2013. "Biomimetic protein nanoparticles facilitate enhanced dendritic cell activation and cross-presentation." *ACS Nano* 7(11): 9743–9752 doi:10.1021/nn403085w.

Molino, N.M., M. Neek, J.A. Tucker, E.L. Nelson, and S.W. Wang. 2017. "Display of DNA on nanoparticles for targeting antigen presenting cells." *ACS Biomaterials Science & Engineering* 3(4): 496–501. doi:10.1021/acsbiomaterials.7b00148.

Moretta, A., C. Bottino, M. Vitale, D. Pende, C. Cantoni, M.C. Mingari, R. Biassoni, and L. Moretta. 2001. "Activating receptors and coreceptors involved in human natural killer cell-mediated cytolysis." *The Annual Review of Immunology* 19(1): 197–223. doi:10.1146/annurev.immunol.19.1.197.

Morgan, R.A., M.E. Dudley, J.R. Wunderlich, M.S. Hughes, J.C. Yang, R.M. Sherry, R.E. Royal et al. "Cancer regression in patients after transfer of genetically engineered lymphocytes." *Science* 314(5796): 126–129. doi:10.1126/science.1129003.

Motzer, R.J., N.M. Tannir, D.F. McDermott, O.A. Frontera, B Melichar, T.K. Choueiri, E.R. Plimack et al. 2018. "Nivolumab plus ipilimumab versus sunitinib in advanced renal-cell carcinoma." *The New England Journal of Medicine* 3788(14): 1277–1290. doi:10.1056/NEJMoa1712126.

Mu, C., X. Wu, H. Ma, W. Tao, G. Zhang, X. Xia, J. Shen. 2016. "Effective concentration of a multikinase inhibitor within bone marrow correlates with in vitro cell killing in therapy-resistant chronic myeloid leukemia." *Molecular Cancer Therapeutics* 15(5): 899–910. doi:10.1158/1535-7163.MCT-15-0577-T.

Murray, J.L., M.E. Gillogly, D. Przepiorka, H. Brewer, N.K. Ibrahim, D.J. Booser, G.N. Hortobagyi et al. 2002. "Toxicity, immunogenicity, and induction of E75-specific tumor-lytic CTLs by HER-2 peptide E75 (369–377) combined with granulocyte macrophage colony-stimulating factor in HLA-A2 patients with metastatic breast and Ovarian cancer." *Clinical Cancer Research* 8(11): 3407–3418.

Napolitani, G., A. Rinaldi, F. Bertoni, F. Sallusto, and A. Lanzavecchia. 2005. "Selected toll-like receptor agonist combinations synergistically trigger a t helper type 1–polarizing program in dendritic cells." *Nature Immunology* 6(8): 769–776. doi:10.1038/ni1223.

Neefjes, J., M.L.M. Jongsma, P. Paul, and O. Bakke. 2011. "Towards a systems understanding of MHC class I and MHC class II antigen presentation." *Nature Reviews Immunology* 11(12): 823–836. doi:10.1038/nri3084.

Nezafat, N., Y. Ghasemi, G. Javadi, M.J. Khoshnoud, and E. Omidinia. 2014. "A novel multi-epitope peptide vaccine against cancer: An in silico approach." *Journal of Theoretical Biology* 349: 121–134. doi:10.1016/j.jtbi.2014.01.018.

Nordly, P., F. Rose, D. Christensen, H.M. Nielsen, P. Andersen, E.M. Agger, and C. Foged. 2011. "Immunity by formulation design: Induction of high CD8þ T-cell responses by poly(I:C) incorporated into the CAF01 adjuvant via a double emulsion method." *Journal of Controlled Release* 150(3): 307–317. doi:10.1016/j.jconrel.2010.11.021. Epub 2010 Nov 25.

Novellino, L., C. Castelli, and G.A. Parmiani. 2005. " A listing of human tumor antigens recognized by T cells: March 2004 update." *Cancer Immunology, Immunotherapy* 54(3): 187–207. doi:10.1007/s00262-004-0560-6.

Obara, W., M. Kanehira, T. Katagiri, R. Kato, Y. Kato, and R. Takata. 2018. "Present status and future perspective of peptide-based vaccine therapy for urological cancer." *Cancer Science* 109(3): 550–559. doi:10.1111/cas.13506

Özcan, Ö.Ö., and M. Karahan. 2019. "Cancer diagnostic, imaging and treatment by nanoscale structures targetting." *Biotechnologia Acta* 12(6): 12–24. doi:10.15407/biotech12.06.012.

Palucka, K., J. Banchereau, and I. Mellman. 2010. "Designing vaccines based on biology of human dendritic cell subsets." *Immunity* 33(4): 464–478. doi:10.1016/j.immuni.2010.10.007.

Park, Y.M., S.J. Lee, Y.S. Kim, M.H. Lee, G.S. Cha, I.D. Jung, T.H. Kang, H.D. Han. 2013. "Nanoparticle-based vaccine delivery for cancer immunotherapy." *Immune Network* 13(5): 177–183. doi:10.4110/in.2013.13.5.177. Epub 2013 Oct 26.

Parmiani, G., C. Castelli, P. Dalerba, R. Mortarini, L. Rivoltini, F.M. Marincola, and A. Anichini. 2002. "Cancer immunotherapy with peptide-based vaccines: What have we achieved? Where are we going?" *Journal of the National Cancer Institute* 94(11): 805–818. doi:10.1093/jnci/94.11.805.

Parmiani, G., V. Russo, A. Marrari,G. Cutolo, C. Casati, L. Pilla, C. Maccalli, L. Rivoltini, and
C. Castelli. 2007. "Universal and stemness-related tumor antigens: Potential use in cancer immunotherapy." *Clinical Cancer Research* 13(19): 5675–5679. doi:10.1158/1078-
0432.CCR-07-0879.

Parmiani, G., V. Russo, C. Maccalli, D. Parolini, N. Rizzo, M. Maio. 2014. "Peptide-based
vaccines for cancer therapy." *Human Vaccines & Immunotherapeutics* 10(11): 3175–
3178. doi:10.4161/hv.29418.

Pol, J., N. Bloy, A. Buque, A. Eggermont, I. Cremer, C. Sautes-Fridman, J. Galon. 2015. "Trial
watch: Peptide-based anticancer vaccines." *Oncoimmunology* 4(4):e974411. doi:10.416
1/2162402X.2014.974411. eCollection 2015 Apr.

Ponte, J.F., P. Ponath, R. Gulati, M. Slavonic, M. Paglia, A. O'Shea, M. Tone, H. Waldmann,
L. Vaickus, M. Rosenzweig. 2010. "Enhancement of humoral and cellular immunity
with an anti-glucocorticoid-induced tumour necrosis factor receptor monoclonal antibody." *Immunology* 130(2): 231–242. doi:10.1111/j.1365-2567.2009.03228.x.

Porter, D.L., B.L. Levine, M. Kalos, A. Bagg, and C.H. June. 2011. "Chimeric antigen receptor-modified T cells in chronic lymphoid leukemia." *The New England Journal of
Medicine* 365(8): 725–733. doi:10.1056/NEJMoa1103849.

Ramakrishnan, R., S. Antonia, and D.I. Gabrilovich. 2008. "Combined modality immunotherapy and chemotherapy: A new perspective." *Cancer Immunol Immunother* 57(10):
1523–1529. doi:10.1007/s00262-008-0531-4.

Reck, M., D. Rodríguez-Abreu, A.G. Robinson, R. Hui, T. Csőszi, A. Fülöp, M. Gottfried
et al. 2016. "Pembrolizumab versus chemotherapy for PD-L1-positive non-small-cell
lung cancer." *The New England Journal of Medicine* 375(19):1823–1833. doi:10.1056/
NEJMoa1606774. Epub 2016 Oct 8.

Reed, S.G., M.T. Orr, and C.B. Fox. 2013. "Key roles of adjuvants in modern vaccines."
Nature Medicine 19(12): 1597–1608. doi:10.1038/nm.3409.

Reinherz, E.L., D.B. Keskin, and B. Reinhol. 2014. "Forward vaccinology: CTL targeting
based upon physical detection of HLA-bound peptides." *Frontiers in Immunology* 5:
418. doi: 10.3389/fimmu.2014.00418.

Reyes, D., L. Salazar, E. Espinoza, C. Pereda, E. Castellón, R. Valdevenito,C. Huidobro,
M.I. Becker,A. Lladser, M.N. López, and F. Salazar-Onfray. 2013. "Tumour cell lysate-
loaded dendritic cell vaccine induces biochemical and memory immune response in
castration-resistant prostate cancer patients." *British Journal of Cancer* 109(6):1488–
1497. doi:10.1038/bjc.2013.494.

Riley, R.S., C.H. June, R. Langer, and M.J. Mitchell. 2019. "Delivery technologies for cancer immunotherapy." *Nature Reviews. Drug Discovery* 18(3): 175–196. doi:10.1038/
s41573-018-0006-z.

Rosenberg, S.A., J.C. Yang, and N.P. Restifo. 2004. "Cancer immunotherapy: Moving beyond
current vaccines." *Nature Medicine* 10(9): 909–915. doi:10.1038/nm1100.

Roth, A., F. Rohrbach, R. Weth, B. Frisch, F. Schuber, and W.S. Wels. 2005. "Induction
of effective and antigen-specific antitumour immunity by a liposomal ErbB2/HER2
peptide-based vaccination construct." *British Journal of Cancer* 92(8): 1421–1429.
doi:10.1038/sj.bjc.6602526.

Schneble, E.J., J.S. Berry, F.A. Trappey, G.T. Clifton, S. Ponniah, E. Mittendorf, and G.E.
Peoples. 2014. "The HER2 peptide Nelipepimut-S (E75) Vaccine (NeuVaxTM) in
breast cancer patients at risk for recurrence: Correlation of immunologic data with
clinical response." *Immunotherapy* 6(5): 519–531. doi:10.2217/imt.14.22.

Schreiber, R.D., L.J. Old, and M.J. Smyth. 2011. "Cancer immunoediting: Integrating immunity's roles in cancer suppression and promotion." *Science Magzine* 331(624): 1565–
1570. doi:10.1126/science.1203486

Schwartzentruber, D.J., D.H. Lawson, J.M. Richards, R.M Conry, D.M. Miller, J. Treisman, F.
Gailani, et al. 2011. "gp100 peptide vaccine and interleukin-2 in patients with advanced

melanoma." *The New England Journal of Medicine* 364(22): 2119–2127. doi:10.1056/NEJMoa1012863.

Schwendener, R.A. 2014. "Liposomes as vaccine delivery systems: A review of the recent advances." *Therapeutic Advances in Vaccines* 2(6): 159–182. doi:10.1177/2051013614541440

Shen, H., C. Rodriguez-Aguayo, R. Xu, V. Gonzalez-Villasana, J. Mai, Y. Huang, and G. Zhang, 2013. "Enhancing chemotherapy response with sustained EphA2 silencing using multistage vector delivery." *Clinical Cancer Research* 19(7):1806–1815. doi:10.1158/1078-0432.CCR-12-2764.

Shevach, E.M. 2002. "CD4+ CD25+ suppressor T Cells: More questions than answers. *Nature Reviews Immunology* 2(6): 389–400. doi:10.1038/nri821.

Silva, J.M., M. Videira, R. Gaspar, V. Préat, and H.F. Florindo. 2013. "Immune system targeting by biodegradable nanoparticles for cancer vaccines." *Journal of Controlled Release* 168(2): 179–199. doi:10.1016/j.jconrel.2013.03.010.

Simons, J.W., and N. Sacks. 2006. "Granulocyte-Macrophage colony-stimulating factor-transduced allogeneic cancer cellular immunotherapy: The GVAX vaccine for prostate cancer." *Urologic Oncology: Seminars and Original Investigations* 24(5): 419–424. doi: 10.1016/j.urolonc.2005.08.021

Steinman, R.M. 2012. "Decisions about dendritic cells: Past, present, and future." *The Annual Review of Immunology* 30(1): 1–22. doi:10.1146/annurev-immunol-100311-102839.

Sunshine, J.C., K. Perica, J.P. Schneck, and J.J. Green. 2014. "Particle shape dependence of CD8+ T cell activation by artificial antigen presenting cells." *Biomaterials* 35(1): 269–277. doi:10.1016/j.biomaterials.2013.09.050.

Tan, T.T. and L.M. Coussens. 2007. "Humoral immunity, inflammation and cancer." *Current Opinion In Immunology* 19(2): 209–216. doi:10.1016/j.coi.2007.01.001.

Tanaka, T., H. Kitamura, R. Inoue, S. Nishida, A.T. Takaya, S. Kawami, and T. Torigoe. 2013. "Potential survival benefit of anti-apoptosis protein: Survivin-derived peptide vaccine with and without interferon alpha therapy for patients with advanced or recurrent urothelial cancer - results from phase I clinical trials." *Clinical and Developmental Immunology* 1–9. doi:10.1155/2013/262967.

Thomann, J.S., B. Heurtault, S. Weidner, M. Brayé, J. Beyrath, S. Fournel, F. Schuber, and B. Frisch. 2011. "Antitumor activity of liposomal ErbB2/HER2 epitope peptide-based vaccine constructs incorporating TLR agonists and mannose receptor targeting." *Biomaterials* 32(20): 4574–4583. doi:10.1016/j.biomaterials.2011.03.015.

Thundimadathil, J. 2012. "Cancer treatment using peptides: Current therapies and future prospects." *Journal of Amino Acids* 2012: 967347. doi:10.1155/2012/967347

Tsiantoulas, D, C.J. Diehl, J.L. Witztum, and C.J. Binder. 2014. "B cells and humoral immunity in atherosclerosis." *Circulation Research* 114(11): 1743–1756. doi:10.1161/CIRCRESAHA.113.301145.

Van der Burg, S.H., M.S. Bijker, M.J. Welters, R. Offringa, and C.J.M. Melief. 2006. "Improved peptide vaccine strategies, creating synthetic artificial infections to maximize immune efficacy." *Advanced Drug Delivery Reviews* 58(8): 916–930. doi:10.1016/j.addr.2005.11.003.

Van der Meel, R., L.J. Vehmeijer, R.J. Kok,G. Storm, and E.V. van Gaal. 2013. "Ligand-Targeted particulate nanomedicines undergoing clinical evaluation: Current status." *Advanced Drug Delivery Reviews* 65(10):1284–1298. doi:10.1016/j.addr.2013.08.012. Epub 2013 Sep 6.

Varypataki, E.M., N. Benne, J. Bouwstra, W. Jiskoot, and F. Ossendorp. 2017. "Efficient eradication of established tumors in mice with cationic liposome-based synthetic long-peptide vaccines." *Cancer Immunology Research* 5(3): 222–233. doi:10.1158/2326-6066. CIR-16-0283. Epub 2017 Jan 31.

Vasievich, E.A., S. Ramishetti, Y. Zhang, and L. Huang. 2012. "Trp2 peptide vaccine adjuvanted with (R)-DOTAP inhibits tumor growth in an advanced melanoma model." *Molecular Pharmaceutics* 9(2): 261–268. doi:10.1021/mp200350n. Epub 2011 Dec 28.

Wada, S., E. Yada, J. Ohtake, Y. Fujimoto, H. Uchiyama, S. Yoshida, and T. Sasada. 2016. "Current status and future prospects of peptide-based cancer vaccines." *Immunotherapy* 8(11): 1321–1333. doi:10.2217/imt-2016-0063

Walker, E.B., W. Miller, D. Haley, K. Floyd, B. Curti, and W.J. Urba. 2009. "Characterization of the class I-restricted gp100 melanoma peptide-stimulated primary immune response in tumor-free vaccine-draining lymph nodes and peripheral blood." *Clinical Cancer Research* 15(7): 2541–2551. doi:10.1158/1078-0432.CCR-08-2806.

Wang, X., S. Li, Y. Shi, X. Chuan, J. Li, T. Zhong, H. Zhang, W. Dai, B. He, and Q. Zhang. 2014. "The development of site-specific drug delivery nanocarriers based on receptor mediation." *Journal of Controlled Release* 193: 139–153. doi:10.1016/j.jconrel.2014.05.028.

Weber, J.S., and J.J. Mulé. 2015. "Cancer immunotherapy meets biomaterials." *Nature Biotechnology* 33(1): 44–45. doi:10.1038/nbt.3119

Wei, H.J., A.T. Wu, C.H. Hsu, Y.P. Lin, W.F. Cheng, C.H. Su, W.T. Chiu et al. 2011. "The development of a novel cancer immunotherapeutic platform using tumor-targeting mesenchymal stem cells and a protein vaccine." *Molecular Therapy* 19(12): 2249–2257. doi:10.1038/mt.2011.152.

Welters, M.J., G.G. Kenter, S.J. Piersma, A.P. Vloon, M.J. Löwik, D.M. Berends-van der Meer, J.W. Drijfhout, et al. 2008. "Induction of tumor-specific CD4+ and CD8+ T-cell immunity in cervical cancer patients by a human papillomavirus type 16 E6 and E7 long peptides vaccine." *Clinical Cancer Research: An Official Journal of the American Association for Cancer Research* 14(1): 178–187. doi:10.1158/1078-0432. CCR-07-1880.

Woo S.R., L. Corrales, and T.F. Gajewski. 2015. Innate immune recognition of cancer." *Annual Review of Immunology* 33(1): 445–474. doi:10.1146/annurev-immunol-032414-112043.

Wölfel, T., M. Hauer, J. Schneider, M. Serrano, C. W€olfel, E. Klehmann-Hieb, E. De Plaen, T. Hankeln, K.H. Meyer zum Buschenfelde, and D. Beach. 1995. "A p16INK4a-insensitive € CDK4 mutant targeted by cytolytic T lymphocytes in a human melanoma." *Science* 269(5228): 1281–1284. doi:10.1126/science.7652577.

Xiang, J., L. Xu, H. Gong, W. Zhu, C. Wang, J. Xu, L. Feng, L. Cheng, R. Peng, Z. Liu. 2015. "Antigen-loaded upconversion nanoparticles for dendritic cell stimulation, tracking, and vaccination in dendritic cell-based immunotherapy." *ACS Nano* 9(6): 6401–6411. doi:10.1021/acsnano.5b02014.

Xu, R., G. Zhang, J., X. Deng, V. Segura-Ibarra, S. Wu, and J. Shen. 2016. "An injectable nanoparticle generator enhances delivery of cancer therapeutics." *Nature Biotechnology* 34(4): 414–418. doi:10.1038/nbt.3506.

Yang, J., Q Zhang, K. Li, H. Yin, and J.N. Zheng. 2015. " Composite peptide-based vaccines for cancer immunotherapy (Review)." *International Journal of Molecular Medicine* 35(1): 17–23. doi:10.3892/ijmm.2014.2000.

Yong, X., Y.F. Xiao, G. Luo, B. He, M.H. Lü, C.J. Hu, H. Guo, and S.M. Yang.2012. "Strategies for enhancing vaccine-induced CTL antitumor immune responses." *Journal of Biomedicine and Biotechnology* doi:10.1155/2012/605045.

Zaks, K., M. Jordan, A. Guth, K. Sellins, R. Kedl, A. Izzo, C. Bosio, and S. Dow. 2006. "Efficient immunization and cross-priming by vaccine adjuvants containing TLR3 or TLR9 agonists complexed to cationic liposomes." *Journal of Immunology* 176(12):7335–7345. doi:10.4049/jimmunol.176.12.7335.

Zhang, H., and J. Chen. 2018. "Current status and future directions of cancer immunotherapy." *Journal of Cancer* 9(10): 1773–1781. doi:10.7150/jca.24577.

Zhao, L., A. Seth, N. Wibowo, C. Zhao, N. Mitter, C. Yu, and A.P.J. Middelberg. 2014. "Nanoparticle vaccines." *Vaccines* 32(3): 327–337. doi:10.1016/j.vaccine.2013.11.069.

Zitvogel, L., L. Apetoh, F. Ghiringhelli, and G. Kroemer. 2008. "Immunological aspects of cancer chemotherapy." *Nature Reviews Immunology* 8(1): 59–73. doi:10.1038/nri2216.

Zong, H., S. Sen, G. Zhang, C. Mu, Z.F. Albayati, D.G. Gorenstein, and X. Liu. 2016. "In vivo targeting of leukemia stem cells by directing parthenolide-loaded nanoparticles to the Bone Marrow Niche" *Leukemia* 30(7): 1582–1586. doi:10.1038/leu.2015.343.

12 Antioxidant Effects of Peptides

Rümeysa Rabia Kocatürk, Fatmanur Zehra Zelka,
Öznur Özge Özcan, and Fadime Canbolat

CONTENTS

12.1 INTRODUCTION

There are thousands of peptides identified and synthesized so far. Peptides have been found to have a large capacity for biological activity and therapeutic potential. Therefore, many studies have been conducted with peptides and their effects are explained more and more day by day. Peptides are used as medicines for chronic diseases in many areas, especially in the pharmaceutical industry (Wetzler and Hamilton 2018). Peptides are structures that are made up of naturally occurring amino acids. Peptides are described in detail in other chapters. Bioactive peptides have different biological functions and these biological functions contribute to human health. Biologically active peptides have many uses in medicine and new ones are added to this situation daily. The therapeutic use of peptides has received great scientific and industrial interest in the past decade. There was also an increase in the use of natural products such as plants in the production of these products, counting industrial proteins, vaccines, antibodies, and different medications in the fight against diseases (Daniell et al. 2009; Marauyama et al. 2014). The therapeutic use of peptides comes with many advantages such as high specificity, high biological activity, low cost, and high membrane penetration ability, but issues such as stability, toxicity, and immunogenicity remain the main concerns of those who develop peptides (Baig et al. 2018). Some peptides have been found to have extensive therapeutic effects such as antioxidant, antimicrobial, antihypertensive, anticancer/antiproliferative, anti-inflammatory, and immunomodulatory and antithrombotic effects. Peptides are of great importance in vaccine production and the properties mentioned are promising for effective vaccines. Also, the antioxidant feature of the peptides is interesting. Studies show that antioxidants can assume an essential role in fighting diseases (Carvalho-Queiroz et al. 2015; Baig et al. 2018; Girija 2018; Huang 2018; Özcan et al. 2020).

Based on this information, we will focus on the antioxidant characteristics of peptides. Topics discussed in this chapter include antioxidants, antioxidant roles, the molecular mechanism of peptides as antioxidants, other effects of peptides and therapeutic use of bioactive peptide properties, and whether bioactive peptides can be used as vaccine agents. In conclusion, we want to emphasize the importance of bioactive peptides, their antioxidant roles, and their potentially successful role in vaccine production.

12.2 ANTIOXIDANTS

12.2.1 DEFINITION OF ANTIOXIDANTS

The first definition for antioxidants was made by Halliwell et al. They described the antioxidants as agents that fundamentally delay or hinder the oxidation of the oxidizable substrates (carbohydrates, lipids, proteins, or nucleic acids) and are at a much lower frequency compared to the substrates mentioned (Halliwell, Gutteridge and Cross 1992). In time, Halliwell and Gutteridge (1990) described antioxidants as substances that target a molecule and act by preventing, delaying, or eliminating oxidative damage. Khlebnikov et al. (2017) made another definition of antioxidants as substances that can inhibit the production of reactive oxygen species or dispose of them directly or indirectly and act as a regulator of the antioxidant defense mechanisms (Khlebnikov et al. 2017). As seen, many definitions were made and antioxidants were understood as neutralizing the free radicals and by this action, they have a high capability for protecting cells getting damaged and in the end protect the organs and prevent them from getting damaged (Nasri 2016).

12.2.2 Reactive Oxygen Species and Antioxidants

Under normal physiological conditions, human beings have a balance between the produced reactive oxygen species (ROS) and antioxidants. Cells and tissues in our bodies constantly create free radicals due to metabolism, as well as environmental factors, for example, radiation, pollution, microorganisms, allergens, cigarettes, smoke, and pesticides increase the number of exposed free radicals that the body encounters (Hekimi, Lapointe, and Wen 2011). With increasing ROS types, the body's balance is disturbed, superoxide radicals begin to accumulate in the cells and an endogenous defense system is inadequate. This accumulation and increase of free radicals, for example superoxide radicals, is defined as oxidative stress and oxidative stress is not beneficial for body composition. It causes many distortions at the molecular level. The increase in ROS is toxic to the cell, damaging the proteins, lipids, and nucleic acids (DNA and RNA) inside the cell, disrupting the intracellular signaling pathways and adversely affecting health (Aslankoç et al. 2019).

12.2.3 Oxidative Stress

Oxidative stress is a general name for the imbalance resulting from the accumulation of free radicals and reactive metabolites that need to be neutralized by antioxidant mechanisms. This imbalance causes damage and disruption to important biomolecules and cells, which have a potential impact on the whole organism (Durackova 2010). There are numerous ailments related to oxidative stress, for example: Alzheimer's, Parkinson's, Huntington's, immune system disorders, amyotrophic lateral sclerosis, diabetes, cancer, cardiovascular disorders, inflammation, chronic kidney disease, acute respiratory distress syndrome, atherosclerosis, neurological disease, obesity, pulmonary fibrosis, and rheumatoid arthritis (Reuter et al. 2010; Pisoschi and Pop 2015; Yan et al. 2020). Moreover, the progressive oxidative damage that is developed by free radicals can expose aging and the development of degenerative diseases. Also, cataracts and atherosclerosis can be exposed due to aging (Percival 1998).

12.2.4 Reactive Oxygen Types

Free radicals are molecular structures. They are molecules that have one or more unpaired electrons in their outermost orbitals and because of this they are unstable and are likely to bond. To the intensive request for consolidating with other radicals or non-radical agents, free oxygen radicals can cause many biological impacts in the body. The formation of exogenous and endogenous free radicals cannot be prevented, but free radicals can be neutralized (Poljsak et al. 2011). Free radicals are constantly produced and being exposed in humans at metabolic processes (cellular respiration) and with the effect of environmental oxidants (drug toxicity, cigarette smoke, ultraviolet radiation, air pollution, intense physical activity, and alcohol) (Sen et al. 2010; Poljsak et al. 2011). Molecular oxygen is the source of ROS and has two unpaired electrons. The oxidation reactions that create ROS occur by enzymes that are known to contain transition metals (such as Fe, Cu). The formation of free radicals is exposed when molecular oxygen gets a single electron transfer. Besides, unpaired electrons containing molecular oxygen play an important role in the formation of ROS (Gutteridge 1994). Hydrogen peroxide (H_2O_2), superoxide anion (O^{2-}), single oxygen ($1/2\ O_2$), and hydroxyl radical (OH). Oxygen molecules such as H_2O_2, O^{2-}, $1/2\ O_2$, hypochlorous acid (HOCl), and hydroxyl radical (OH) that make up ROS are extremely reactive and have toxic effects on cells. Due to this toxic effect, it also damages the organs (Kohen and Nyska 2002; Pham-Huy et al. 2008; Fransen et al. 2012). In addition, reactive nitrogen types (RNS) and reactive sulfur types (RSS) are free radicals that can also cause damage. Respectively they are derived with the reaction of ROS and nitric oxides, derived with the reaction of ROS and thiols. If these reactive molecules are not neutralized, they can accumulate and cause various complications in the body (Durackova 2010; Corpas and Barroso 2015). On the other hand, free radicals can damage any protein. Therefore, since free radicals damage proteins, they affect the activity of enzymes and the function of the structural protein. In addition, the production of protein hydroperoxides is with the oxidation of proteins. The oxidation of proteins is caused by ROS/RNS that interact with transition metal ions. If proteins are oxidized, they become inactive and are eventually removed. However, these inactive proteins can accumulate in some cases and cause damage due to various diseases as well as aging (Sen and Chakraborty 2011).

12.2.5 Antioxidant Mechanism

We have mentioned that free radicals are continuously produced during normal metabolism in the body and are neutralized by antioxidant defense systems. Antioxidants can remove or delay cell damage by removing free radicals in the cell. The body can produce antioxidants naturally in the body, or they can be obtained from external foods that we eat (Lobo et al. 2010). Antioxidant agents can show their effects against oxidant molecules in four ways; the first of them is the scavenging effect and works by preventing the formation of radicals, they make the formed radicals less harmful. Enzymes such as superoxide dismutase (SOD), glutathione peroxidase (GPx), and some metal-binding proteins are examples of antioxidants that have a scavenging effect. The second one is the removal/quenching effects, these effects are formed by compounds that interact with oxidants and inactivate their activities by extinguishing them as a result of transferring hydrogen to them. Examples of these removal/quenching molecules are vitamins (vitamins A, C, and E), flavonoids, mannitol, anthocyanins, etc. The third effect is the chain-breaking effect, it breaks the reactions that continue as a chain and disable them as a result of binding oxidant molecules to them. Examples of chain-breaking molecules are uric acid, bilirubin, albumin, etc. The fourth effect is the repair effect; it removes the harmful effects of oxidant molecules as a result of repairing the damaged biomolecule. Examples of repairing molecules are DNA repair enzymes, methionine sulfoxide reductase, etc. (Arkan 2011).

12.2.6 CLASSIFICATION OF ANTIOXIDANTS

Antioxidant defense systems are complex. Antioxidants can be classified as endogenous and exogenous (Aydemir and Karadağ-Sarı 2009; Sen and Chakraborty 2011; Karabulut and Gülay 2016). With the presence of endogenous and exogenous antioxidants, the oxidant/antioxidant balance is protecting, and free radicals are neutralizing (Sen and Chakraborty 2011). The antioxidant mechanisms in the cell are the first, second, and third antioxidant defense mechanisms. The first line of defense is superoxide dismutase (SOD), catalase (CAT), and glutathione peroxidase (GPx) antioxidant defense systems (Niki 2014; Karabulut and Gülay 2016). The second group of antioxidants is responsible for interference with free radicals and is therefore responsible for preventing oxidative chain reactions. This group consists of flavonoids, glutathione enzyme, vitamins C and E, albumin, and carotenoids. The third group of defense consists of antioxidant enzymes that have the ability to repair the damage that is caused by free radicals to biomolecules, for example methionine-sulfoxide reductases, lipases, DNA repair enzymes, proteases, and transferases (Mut-Salud et al. 2016).

12.2.6.1 Enzymatic Antioxidants

Endogenous antioxidants can be classified in two subgroups as enzymatic (superoxide dismutase, catalase, glutathione peroxidase, glutathione reductase) and nonenzymatic antioxidants (Sen et al. 2010; Pham-Huy, He, and Pham-Huy 2008; Aydemir and Karadağ-Sarı 2009).

12.2.6.1.1 Superoxide Dismutase

The production of a relatively reactive superoxide anion that is proxidative (pH < 4.8) occurs by adding an electron to molecular oxygen. The reason for being peroxidative is due to the reduction in transition metals, the release of protein-bound metals, and the ability to form perhydroxyl radicals that can directly catalyze lipid oxidation under acidic conditions (Decker 2002). Also, the superoxide anion, the only enzyme in the organism that uses free radicals as a substrate, is found in cytoplasm and manganese-type mitochondria containing copper and zinc and creates a heavy burden on the cell in increasing glycolysis (Karabulut and Gülay 2016). SOD isoforms contain copper plus zinc or manganese in their active sites. In addition, these SOD isoforms are known to catalyze the transformation of superoxide anion to hydrogen peroxide by the conversion reaction:

$$2O^-_2 + 2H^+ \rightarrow O_2 + H_2O_2 \text{ Peroxidases}$$

A superoxide radical is converted to H_2O_2 through SOD. Although H_2O_2 is not a radical and does not react with most of the biologically important molecules, it plays a role in the catalysis of Cu and Fe ions as a precursor in the formation of the most reactive oxygen type hydroxyl radical (OH^-) by Fenton reaction (Karabulut and Gülay 2016). Fenton reactions are responsible for hydroxyl radical development and thus trigger radical chain responses and support this free radical formation. In this regard, it is a source of ROS. In addition, transient metal ion chelation contributes to natural antioxidants and antioxidants inhibit Fenton reactions (Huang, Ou, and Prior 2005).

Fenton Reaction (Bazinet and Doyen 2015; Aslankoç et al. 2019)

$$Fe^{2+} + HO^- \rightarrow Fe^{3+} + HO^-$$

Haber-Weiss Reaction (transient metal ion chelation) (Aslankoç et al. 2019)

$$\bullet O_2^- + H_2O_2 \rightarrow \bullet OH + OH^- + O_2$$

12.2.6.1.2   Catalase

Catalase (CAT) is an enzyme containing both groups found in many biological systems (Elias, Kellerby, and Decker 2008). CAT is largely found in intracellular organelles such as peroxisomes, and to a lesser extent in the mitochondria and endoplasmic reticulum and enzymatically acts as an antioxidant and molecularly catalyzes the conversion of hydrogen peroxide to water and oxygen, as in the following reaction (Limon-Pacheco and Gonsebatt 2009):

$$2H_2O_2 \rightarrow 2H_2O + O_2$$

12.2.6.1.3   Glutathione Peroxidase

Glutathione peroxidase (GSH-Px) is an enzymatic antioxidant and is found in many biological tissues (Elias, Kellerby and Decker 2008). GSH-Px is formed by the combination of four protein subunits. Each protein subunit contained in this antioxidant enzyme contains a selenium atom. GSH-Px is an enzyme that metabolizes H_2O_2 and organic hydroperoxides (lipid hydroperoxides, DNA hydroperoxides) using glutathione (GSH) as the electron source. GSH-Px is responsible for protecting cells against oxidative damage caused by $2H_2O$ in the cytoplasm of cells and preventing the formation of OH from H_2O_2 (Karabulut and Gülay 2016).

$$H_2O_2 + 2GSH \rightarrow 2H_2O + GSSG \quad \text{(Elias, Kellerby, and Decker 2008)}$$

12.2.6.1.4   Glutathione Reductase

Glutathione reductase is an flavoprotein enzyme containing flavin adenine dinucleotide (FAD). Glutathione reductase is converted back into GSH by transferring one electron of NADPH to disulfide bonds of oxidized glutathione. Therefore, NADPH is necessary to prevent free radical damage and its most important source is the hexose monophosphate (pentose phosphate) pathway (Sen et al. 2010; Karabulut and Gülay

2016). To convert hydrogen or lipid (LOOH) hydroperoxide to water glutathione reductase uses selenium in its active site and also reduced GSH (see the reaction below).

$$LOOH + 2GSH \rightarrow LOH + 2H_2O + GSSG \quad \text{(Elias, Kellerby, and Decker 2008)}$$

12.2.6.2 Nonenzymatic Antioxidants

Nonenzymatic antioxidants include glutathione, melatonin, uric acid, bilurubine-coenzyme Q10, selenium, α-lipoic acid (Karabulut and Gülay 2016) and there are natural nonenzymatic antioxidants like phenolic compounds, vitamin (ascorbic acid, tocopherols, folic acid), carotenoids, minerals, and other antioxidant proteins (ceruloplasmin, haptoglobin, lactoferrin, albumin, transferrin (Mamta et al. 2014), hemopexin, haptoglobin, ferritin) (Elias, Kellerby, and Decker 2008). There are also amino acids that have antioxidant effects (cysteine, methionine, tryptophan, phenylalanine, histidine (Elias, Kellerby, and Decker 2008), tyrosine, alanine, valine, leucine, isoleucine, proline) (Matsui et al. 2018).

12.2.7 How to Evaluate Antioxidant Capacity

There are known to be two main chemical mechanisms to evaluate the capacity of antioxidants. The main role of antioxidants is free radical scavenging and in the main time the radical scavenging is liable for electron migration (EM) and hydrogen atom transfer (HAT) (Zou et al. 2016). Many methods are used to evaluate the antioxidant capacity, for example the oxygen radical absorbance capacity (ORAC) assay, the total radical trapping antioxidant parameter (TRAP) assay, the crocin bleaching assay (CBA), and the lipid peroxidation assay (LPA) (Frankel and Finley 2008; Samaranayaka and Li-Chan 2011; Zhang, Mu, and Sun 2014), inhibited oxygen uptake (IOC), hydroxyl radical scavenging activity by p-NDA (p-butrisidunethyl aniline), 2,2′-azino-bis(3-ethylbenzothiazoline-6-sulfonic acid) (ABTS) radical scavenging, scavenging of superoxide radical formation by alkaline (SASA), and scavenging of H_2O_2 radicals are the antioxidant evaluation methods in HAT-based reactions (Frankel and Finley 2008). On the other hand, at EM-based redox reactions, trolox equivalent antioxidant capacity (TEAC), copper II reduction capacity (CRC) assay, 2,2-diphenyl-1-picrylhydrazyl (DPPH) radical scavenging activity (Frankel and Finley 2008; Zou et al. 2016), ferric reducing antioxidant power (FRAP), N-dimethyl-p-phenylenediamine (DMPD), and DPPH free radical scavenging total phenols by Folin-Ciocalteu N assays are also used (Frankel and Finley 2008).

12.3 PEPTIDES AS ANTIOXIDANTS

12.3.1 Antioxidant Activity of Proteins

Proteins have antioxidant functions, for instance, they hinder the oxidation of lipids by specific mechanisms, for example iron-binding proteins and antioxidant enzymes or other non-specific mechanisms. The activity of antioxidant proteins transpires through multiple interactions between the capability to remove free radicals, neutralize reactive oxygen species, reduce hydroperoxides, enzymatically eliminate specific oxidizers, chelate prooxidative transition metals, and alter the physical properties of food systems. Peptides are molecules made of proteins and are most promising as protein antioxidants, some peptides are molecules made of proteins and are most promising as protein antioxidants, some studies demonstrate that peptides compared with intact proteins have essentially higher activity (Elias, Kellerby, and Decker 2008).

12.3.2 Bioactive Peptides

Bioactive peptides are specific protein fragments that are made up of amino acids and they have many benefits on human body function. These amino acids are linked by peptide bonds and may exhibit hormone or drug-like activities. According to metabolic studies made in living organisms, peptides are known to have an essential role in human health (Singh, Vij, and Hati 2014). Peptides usually contain 2–20 amino acids and they are generally rich in hydrophobic amino acids (Chakrabarti, Guha, and Majumder 2018). Also, the positive effects of peptides on health (antimicrobial, antithrombotic, antihypertensive, opioid, immunomodulator, cholesterol-lowering, mineral binding, and antioxidative) have attracted the attention of scientists and many scientific studies have been conducted on the mechanisms and their implied roles of bioactive peptides in the interception and treatment of different diseases (Cicero, Fogacci, and Colletti 2017). For a bioactive peptide to be considered bioactive, it must provide a physiologically measurable biological effect. This also applies to other dietary components. Besides, these bioactive peptides have the potential to positively affect human health they can also have some side effects, for instance toxicity, allergenicity, and mutagenicity (Moller et al. 2008). When the relationship between the chemical structure and activity of a peptide is compared it cannot always be predicted, on the other hand, it is known that the activity of the peptide depends on its characteristics such as amino acid sequence, the chain length of the peptide, the amino acid type in N- and C-terminal and the polarity of the amino acids that make up the peptides (Li and Yu 2015).

12.3.3 Production of Bioactive Peptides

The use of bioactive peptides is of great interest, and many studies are being conducted in the food industry to find and utilize new peptide sequences from protein-rich products (Chakrabarti, Guha, and Majumder 2018). Vegetable, animal, and marine foods contain a large number of bioactive peptides, and they occur in enzymatic hydrolysis, fermentation, chemical hydrolysis, or the processes of gastrointestinal digestion, and consequently, bioactive peptides are formed by the extraction of these foods (Rutherfurd-Markwick 2012; Cicero, Fogacci, and Colletti 2017). Animal-based peptides can be obtained from meat proteins, eggs, and milk (whey and casein), and plant-based peptide plant sources are obtained from oats, soy, canola, wheat, legumes (chickpeas, beans, peas, and lentils), flaxseeds, and hemp seeds. Additionally, proteins from marine sources such as fish, salmon, squid, sea urchin, sea horse, oyster, and snow crab were also used (Kitts and Weiler 2003; Moller et al. 2008). The truth that peptides are inactive in the protein sequence is found as a result of enzymatic hydrolysis in vivo/in vitro (Hartmann and Meisel 2007; Sila and Bougatef 2016). The most common method of obtaining bioactive peptides is the hydrolysis of protein molecules with proteolytic. The efficiency and activity of bioactive peptides have an important place. The biological activity of the entire hydrolysate is evaluated and purified by enzymes and identified to find the most amino acid-strong sequence of bioactive peptides (Chakrabarti, Guha, and Majumder 2018). There are methods to estimate the yield of bioactive peptides from food protein sources. Frequently used methods are the quantitative structure-activity relationship (QSAR) and bioinformatics based in-silico methods (Gu, Majumder, and Wu 2011; Chakrabarti, Guha, and Majumder 2018). Proteins can be hydrolyzed to peptides with the proteolytic enzymes of bacteria. Hydrolysates can be developed into functional foods after in vivo testing for a biologically active hydrolysate (Daliri, Oh, and Lee 2017). The chemical synthesis of bioactive peptides is constantly increasing because the quantity of these peptides in nature is very low (Perez et al. 2012).

12.3.3.1 Enzyme Hydrolysis

In enzyme hydrolysis proteins are treated with different enzymes at a given pH and temperature, and enzyme hydrology has some advantages. These advantages are that is has shorter reaction times compared to microbial fermentation and this method is easy to scale (Daliri, Oh, and Lee 2017).

12.3.3.2 Fermentation

The fermentation method is related to the cultivation of yeasts, fungi, bacterial microorganisms, and uses enzymes of microorganisms and hydrolyzes the protein to shorter peptides. Microorganisms must grow to use the fermentation method. Therefore, this method is much more temporary than enzyme hydrolysis. Extending hydrolysis relies on the microbial strain, protein source, and fermentation time (Chakrabarti, Guha, and Majumder 2018).

12.3.4 ANTIOXIDANT PEPTIDE SEQUENCES

In different studies, antioxidant peptides have been identified and the antioxidant properties of these antioxidant peptides have been found to vary. For example, in the study with combinatorial tripeptides, six different antioxidant assays of the libraries demonstrated that various aspects related to structure-activity correlations affected antioxidant activity (Ohashi et al. 2015). The antioxidant effect of the N-terminal and central amino acid of the tripeptides was found to be greater than the amino acid at the C-terminal (Li et al. 2011). Two tripeptides containing Tyr have been shown to exhibit higher clearance activity than two tripeptides containing His against the hydrophilic ABTS and AAPH radical. This result shows that the amino acid residues of Tyr and Trp play an important role in expressing radical scavenging activities, namely antioxidant activity. In addition, tripeptides other than Cys amino acid-containing peptides have been shown to show poor cleansing activity against the hydrophobic DPPH radical, i.e. their antioxidant properties have been found to be lower than those peptides.

Tripeptides generally show high correlations with linoleic acid peroxidation, ORAC and ABTS assays, antioxidant activity, and low correlations with peroxynitrite (PN) scavenge activity, FRAP, and DPPH assays. The highest correlations were observed during ORAC and ABTS analyses. Concomitant use of these assays has the potential as an effective approach to screening and determining antioxidant activity after antioxidant peptide isolation (Ohashi et al. 2015). Tripeptides related to Pro-His-His and tripeptides contains either two Tyr or two His amino acid residues has been identified as an active core of the antioxidant peptides. On the other hand, when two Tyr amino acid-containing tripeptides and two His amino acid-containing tripeptides were compared it is found that Tyr amino acid-containing tripeptides showed higher activities in the peroxidation of the linoleic acid. Also, Cys amino acid-containing tripeptides showed a high PN scavenging activity. On the other hand, powerful synergistic effects with phenolic (e.g. butylated hydroxyanisole and δ-tocopherol) antioxidants were showed on Tyr-His-Tyr tripeptides. Replacing Pro-Her-Her's N-terminal or C-terminal to other amino acid residues has been shown to affect the activity for being antioxidant. These results showed that the amino acid sequence, composition, and size is important for the antioxidant feature of the peptides. In addition, this study is proof that it is responsible for the inhibition of oxidative reactions initiated by different amino acid residues and peptide sequences, different types of free radicals, or pro-oxidants. This inhibition can occur in different situations. For instance, metal ions, as well as in different molecular conditions, e.g. lipid, aqueous, or emulsion systems, different pH situations, and the presence of different compounds in food matrices or biological systems, etc. (Huang, Ou, and Prior 2005; Ohashi et al. 2015). In a study, it is demonstrated that amino active hydrogen atoms and carboxyl active hydrogen do have an important antioxidant activity. In consideration of these peptides, Gly-Tyr-Gly, Gly-Tyr-Tyr-Gly, and Tyr-Gly-Gly-Tyr antioxidant activity, the peptides that contain tyrosine amino acid showed different activities according to the activity on the properties of ROS and/or RNS. However, the antioxidant activity against the hypochlorite ion was not affected by the number and the position of tyrosine amino acid residues. Also, the number of tyrosine amino acid residues altered antioxidant activity against the peroxyl radical, but when it comes to the position of tyrosine amino acid it did not have any great importance. Besides the position of the tyrosine, amino acid was only an important factor in radical scavenging activity on PN. This suggests that antioxidant peptides may be designed according to the properties of amino acid side chain groups (Matsui et al. 2018). Antioxidant peptides are critical for future destiny of health science because oxidative stress arises because of the imbalance between the production of ROS and antioxidative defenses and this can cause problems as is explained earlier. It has been shown that there is a link between oxidative stress and many diseases such as, diabetes, cancer, Alzheimer's, joint inflammation, and schizophrenia. Oxidative stress is known to be reduced by consuming antioxidant-rich foods. Antioxidants can slow down or prevent this oxidation (Chakrabarti, Jahandideh, and Wu 2014; Li and Yu 2015; Pisoschi and Pop 2015; Liu et al. 2016). Peptides in foodstuffs show antioxidant activity (Nongonierma et al. 2015; Safitri, Herawati, and Hsu 2017). This antioxidant activity depends on the correct positioning of amino acids in the protein sequence (Ünal, Şener, and Cemek 2018). For example, in studies conducted to prevent linoleic acid oxidation, peptides containing proline at the N-terminal have a more effective antioxidant role than those containing proline at the C-terminal and a higher metal-chelate level at the C-terminal than those containing histidine residues at the C-terminal (Ünal, Şener, and Cemek 2018). Also, antioxidant peptides can be produced either with animal or plant sources.

12.3.4.1 Animal-Based Bioactive Peptides

First of all, milk is found as a good source of peptides. The Glu-Leu amino acid sequence is derived from casein protein hydrolysate, which has been shown to have a really good antioxidant activity (Suetsuna, Ukeda, and Ochi 2000). Milk includes many kinds of peptides. It has been found that peptides from milk proteins inhibit the enzymatic or chemical peroxidation of essential fatty acids and most of the peptides are encoded by the U-casein sequence. The addition of leucine or proline to the N-terminal of the His-His dipeptide has been reported to increase antioxidant activity and also has synergistic effects with antioxidants (Hartmann and Meisel 2007). With biological activities of *Lactobacillus plantarum* it has been found as an antioxidant peptide strain from fermented milk and evaluated that *L. plantarum* 55 strain was the most active as an antioxidant (Aguilar-Toalá et al. 2017). Whey protein concentrates are a protein complex derived from milk and compared to other protein sources, whey protein concentrates contain branched-chain amino acids leucine, isoleucine, and valine in high concentrations. In a study on the production of potential antioxidant hydrolysates by using alcalase enzyme at certain time intervals from whey, it was determined that the smallest molecular weight of the four fractions obtained had the highest antioxidant activity (Peng, Xiong, and Kong 2009). In enzymatic hydrolysis of α-lactoglobulin and α-lactalbumin have found that tryptophan, one of the amino acids, had the highest antioxidant capacity, followed by methionine and cysteine, respectively. The antioxidant feature is due to its behavior as a hydrogen donor. In addition, the peptide with the Trp-Tyr-Ser-Leu-Ala-Met-Ala-Ala-Ser-Asp-Ile series was compared to butylated hydroxyanisole (BHA) and the peptide had higher antioxidant potential (Hernandez-Ledesma et al. 2005). With applying QSAR modeling on β-lactoglobulin there are found tripeptides containing Cys and Trp generally have higher antioxidant activity (Tian, Fang, and Jiang 2015). Another study was conducted with casein and whey protein in camel milk. Peptides that are digested with pepsin from camel milk significantly increased the tolerance of yeast cells to peroxide-induced oxidative stress. The results showed that camel milk casein and whey proteins have bioactive peptides, which are important radical-relieving activities, so they can be potential peptides for the prevention and treatment of oxidative stress-related diseases (Ibrahim, Isono, and Miyata 2018).

In a study various activities of the egg yolk protein were evaluated, and hydrolysis was performed with pepsin to obtain the peptides. Among the peptides obtained,

Tyr-Ile-Asn-Gln-Met-Pro-Gln-Lys-Ser-Arg-Glu peptide sequence has been reported to have very strong antioxidant activity (Zambrowicz et al. 2015). In another study with egg yolk, phosphite was digested with phosphite trypsin enzyme and the soluble peptides were exchanged through a 1 kDa membrane. The retentate (that is, the unfiltered portion) was then called the oligophosphospeptides of the phosphite, which were divided into three fractions on the anion exchange column. (Xu, Katayama, and Mine 2007). Trypsin enzyme digested egg white peptides showed a huge increasing effect in plasma radical scavenging (antioxidant activity) in spontaneously hypertensive rats (Manso et al. 2008). The antioxidant activity of the properties of the peptides belonging to traditionally produced goat and cow cheeses were examined and found that the ripening period increased the antioxidant activity of the produced peptides (Öztürk 2015). Ovotransferrin, an egg yolk protein, 278P, and thermolysis enzymes have been used, and ovotransferrin peptides have been found to have high antioxidant activity. In addition, ovotransferrin peptides have more powerful anticancer activity than single-step enzyme hydrolysates as well as natural ovotransferrin. (Lee et al. 2017)

A peptide isolated from the hydrolysate of dark chicken meat (Tyr-Ala-Ser-Gly-Arg) was seen to be matched with the amino acid at the 143–147 amino acid residues in chicken β-actin and also revealed a high potential of antioxidant activities on peroxyl radicals (Fukada et al. 2016).

In a study, the maturation period increases the antioxidant activity of peptides in goat and cow cheeses produced in traditional methods (Öztürk 2015).

The protein isolates of sea jellyfish (*R. Esculentum*) have many benefits, especially as they have important radical savage activities that make them a good antioxidant (Yu et al. 2006).

Proteases have been used to evaluate different proteins of sardine (*Sardinella aurita*). Hydrolysate, which has the highest antioxidant activity, was obtained by extracting sardine with a natural enzyme. The peptide with the Leu-His-Tyr amino acid sequence from these has found as a strong antioxidant (Bougatef et al. 2010).

Some antioxidant peptides have been found from *Mactra veneriformis* protein hydrolysate (Leu-Tyr-Glu-Gly-Tyr, Trp-Asp-Asp-Met-Glu-Lys, Thr-Asp-Tyr, Leu-Asp-Tyr and Trp-Gly-Asn-Val-Ser- Gly-Ser-Pro) (R. Liu et al. 2015). Peptides with high antioxidant properties from their proteins in horse mackerel (*Magalaspis cordyla*) were also isolated. The sequence of the isolated amino acid was determined as Ala-Cys-Phe-Leu (518.5 Da) and it was determined that this fraction was more effective in preventing the oxidation of polyunsaturated fatty acids compared to a-tocopherol (Sampath Kumar, Nazeer, and Jaiganesh 2011).

In another study protein hydrolysate provided from Flavourzyme-treated round scad (*Decapterus maruadsi*) had high contents of antioxidant-potentiating amino acids, for example leucine, arginine, histidine, and lysine (Thiansilakul, Benjakul, and Shahidi 2007). Also was it determined that oxidative reactions were prevented by adding pigs' blood plasma (2.5%) containing proteins that have antioxidant activity, for example serum transferrin and albumin to the pork meat products by drying in a spray dryer (Faraji, Decker, and Aaron 1991).

In another study porcine plasma using alcalase enzyme was digested; then the hydrolysate was appraised for antioxidant activity using metal chelation assays, scavenging of DPPH radical, liposome oxidation, and reduction of iron (Liu et al. 2010). On the other hand, in a different study, peptides (Asp-Ala-Gln-Glu-Lys-Leu-Glu, Glu-Glu-Leu-Asp-Asn-Ala-Leu-Asn, Ile-Glu-Ala-Glu-Gly-Glu, and Val-Pro-Ser-Ile-Asp-Asp-Gln-Glu-Glu-Leu-Met) defined as active antioxidant components of pig myo brillar protein hydrolysate (papain hydrolysis derived), the hydrolysate were appraised in oleic acid test system for their antioxidant potency (Saiga and Nishimura 2013). When these four active peptides are compared, the hindrance of linoleic acid peroxidation is deemed to depend on the length of the chain, while the Ile-Glu-Ala-Glu-Gly-Glu peptide had the least activity in comparison to longer peptides (Aluko 2015).

12.3.4.2 Plant-Based Bioactive Peptides

Antioxidant peptides isolated from a soybean protein hydrolysate is very important and there have been many studies about it. Triple peptides from soy bean containing two Tyr residues are stronger because they contain two His residues to inhibit linoleic acid peroxidation, and also a high antioxidant activity if the peptide contains Trp amino acid or Tyr tripeptide at the C-terminal (Saito et al. 2003). In 1996 a peptide from soybean has found. After it was designed new peptides were based on this peptide: Leu-Leu-Pro-His-His. In comparison, among the tested peptides Pro-Her-Her was found the most antioxidative sequence. So, the antioxidant activity is associated with histidine and proline amino acids (Chen et al. 1996). In a study, three separate enzymes were used and have demonstrated that the ultrafiltration membrane fractions and the soy protein hydrolysates may hinder lipid peroxidation in cooked minced meat (Zhang, Li, and Zhou 2010). Additionally, in this study results with peptide (Leu-Met-Trp) revealed that the low molecular weight of peptide sequences had a greater effect on peroxyl radical scavengers compared with higher molecular weight peptide sequences (Aluko 2015). In another study, it is found that a peptide (Ala-Arg-Glu-Glu-Thr-Val-Val-Pro-Gly) isolated from whole-wheat products operates a protective role in vascular smooth muscle cells on high glucose-induced oxidative stress and may have a potential therapeutic effect (Chen et al. 2017).

Antioxidant peptides are also isolated from hydrolysates of walnut flour proteins. There are six important peptides (Val-Arg-Asn, Asn-Pro-Ala-Asn, Ala-His-Ser-Val-Gly-Pro, Ser-Ser-Glu, Thr-Tyr, and Ser-Gly-Gly-Tyr) and those were found to have more antioxidant potential than other peptides of walnut (Feng et al. 2018).

In another study, peptides that have antioxidative properties from millet were identified using trypsin enzyme and the peptides showed a high antioxidant activity when tested by different free radicals (Agrawal, Joshi, and Gupta 2016).

Cereal peptides and hydrolyzed proteins have been found to exhibit oxidative stress-reducing effects in experiments (in chemical-based analyzes, cell cultures, animal models, and food systems) (Esfandi, Walters, and Tsopmo 2019).

The antioxidant peptide is reported to be purified from chickpea (*Cicer arietium* L.) protein hydrolysates. The amino acid sequence of this purified peptide is Asn-Arg-Tyr-His-Glu and amino acids to each other were determined as 1: 1: 1: 1: 1. This antioxidant peptide has been found to effectively inhibit DPPH, superoxide, and hydroxyl free radicals from free radical sources. It is reported that the peptide has Cu + 2 and Fe + 2 chelate activities also inhibit autooxidation of linoleic acid (Zhang et al. 2011).

A protein isolate from beans (*Phaseolus vulgaris* L.) and a purified protein product (phaseolin) were digested with pepsin and pancreatin and then the antioxidant activities were tested from the protein hydrolysates (Carrasco-Castilla et al. 2012).

The yellow eld pea seed hydrolysate has been shown to be an important contributor to the peptide antioxidant properties of hydrophobicity and net charge with a series of trials on thermolysin digestion including column chromatography divisions. First, pea protein hydrolysate was divided into fractions (F1–F5), which differ in the total content of hydrophobic amino acids by reverse phase high performance liquid chromatography (HPLC). (Pownall, Udenigwe, and Aluko 2010).

Spirulina platensis is the only blue-green algae that is grown commercially for food use. It is a good food source due to its high protein content and natural biochelate vitamins. A mixture of these two sources rich in protein was found to be successful in preventing liver damage induced by CCl4 (hepatotoxin) in an in vivo study (Gad et al. 2011).

Also, with assay on shrimp peptides Gln and Lys amino acids are also known as potential antioxidants (Wu et al. 2019).

Alfalfa leaf: Due to its high protein content and nutritional value, the Food and Agriculture Organization recommended it for human consumption as a potential protein source. Mice fed with the proteins of these leaves were observed to have an increase in GSH-Px and SOD enzymes and a decrease in the concentration of malonaldehyde (MDA). In the leaf, albumin was found to be the dominant protein, while gluten and globin were found in small amounts. It is found that cloverleaf peptides have a good antioxidant activity (Xie et al. 2008). In a study they have found that hydrolysate of carrot seed protein peptides can be promising oxidative damage scavengers in food. They have antioxidant activities (Ye et al. 2018). In another study, peptides were produced from pollen using the alcalase enzyme. It was purified by size-exclusion chromatography after enzymatic hydrolysis. Reverse-phase HPLC was used to lyse the antioxidant activity of pollen-derived peptides. The DPPH radical removal activity of the fractions obtained was 66.61% (Maqsoudlou et al. 2018) (Table 12.1).

12.4   OTHER EFFECTS OF PEPTIDES

12.4.1   ANTIHYPERTENSIVE ACTIVITY

Hypertension is a cardiovascular disease that affects approximately one quarter of the world's population and is a controllable risk factor that plays a role in related complications. The angiotensin I-converting enzyme (ACE) is a dipeptityl carboxypeptitase and has the role of turning angiotensin I to angiotensin II. Angiotensin II has a general vasoconstriction impact. It plays an important physiological role in controlling blood pressure, liquid, and salt balance in vertebrates (Hartmann and Meisel 2007; Hayes et al. 2016) and peptides that repress the ACE enzyme are

TABLE 12.1

Antioxidant Peptides

Peptide Source	Hydrolysis Method	Amino Acid Sequence	Activity	References
Rice	Alcalase Neutrase	Thr-Gln-Val-Tyr Arg-Pro-Asn-Tyr-Thr-Asp-Ala, Thr-Ser-Gln-Leu-Leu-Ser-Asp-Gln, Thr-Arg-Thr-Gly-Asp-Pro-Phe-Phe, Asn-Phe-His-Pro-Gln Phe-Arg-Asp-Glu-His-Lys- Lys and Lys-His-Asp- Arg-Gly-Asp-Glu-Phe	Antioxidant activity with DPPH and ABTS radical scavenging activity and in FRAP-Fe^{3+} reducing assay.	Li et al. 2007J. Zhang et al. 2010Yan et al. 2015
Chickpea protein	Alcalase	Asp-His-Gly and Val-Gly-Asp-Ile	Scavenging effect of DPPH radicals was seen.	Ghribi et al. 2015
Wheat germ	Bacillus licheniformis	Ile-Val-Tyr	This sequence found as contributor to the ACE inhibition.	Matsui, Li, and Osajima 1999
Hemp seed protein (Cannabis sativa L.) hydrolysate	Pepsin Pancreatin	Trp-Val-Tyr-Tyr and Pro- Ser-Leu-Pro-Ala	DPPH radical scavenging assay and metal chelation activity for measurement of antioxidant activity.	Girgih et al 2014
Soy(b-conglycinin)	Alkaline protease	Leu-Leu-Pro-His-His	Measurement of antioxidant activity by ferric thiocyanate assay.	Chen, Muramoto, and Yamauchi 1995
Egg (egg white)	Protease Pepsin-	Tyr-Ala-Glu-Glu-Arg-Tyr-Pro-Ile-Leu Asp-His-Thr-Lys-Glu, Met-Pro-Asp-Ala-His-Leu, Phe-Phe-Gly-Phe-Asn Ala-Glu-Glu-Arg-Tyr-Pro, Asp-Glu-Asp-Thr-Gln-Ala-Met-Pro	Antioxidant activity has been measured by DPPH, ORAC radical scavenging assays.	Davalos et al. 2004J.B. Liu et al. 2015 Nimalaratne, Bandara, and Wu 2015
Wheat germ protein	Alcalase	Gly-Asn-Pro-Ile-Pro-Arg-Glu-Pr o-Gly-Gln-Val-Pro-Ala-Tyr	Antioxidant activity conducted by ABTS assay.	Karami et al. 2019

(Continued)

TABLE 12.1 (CONTINUED)
Antioxidant Peptides

Peptide Source	Hydrolysis Method	Amino Acid Sequence	Activity	References
Corn gluten meal	Alkaline protease and Flavourzyme	Leu-Pro-Phe, Leu-Leu-Pro-Phe, Phe-Leu-Pro-Phe and Asp-Pro-His	Antioxidant activities by ABTS, and hydroxyl radicals scavenging assay.	Zhuang, Tang, and Yuan 2013
Sorghum	Alcalase	Leu-Asp-Ser-Cys-Lys-Asp-Tyr-Val-Met-Glu	DPPH, ABTS, Fe^{2+} chelating activity and reducing power assay have been conducted to evaluate the antioxidant activity.	Agrawal, Joshi, and Gupta 2017
Sweet potato	Alcalase	Tyr-Tyr-Ile-Val-Ser	OH radical scavenging.	Zhang, Mu, and Sun 2014
Palmaria palmata protein	Corolase PP	Ser-Asp-Ile-Thr-Arg-Pro-Gly-Gly-Asn-Met	Showed high antioxidant activity at ORAC and FRAP radicals.	Harnedy, O'Keeffe, and FitzGerald 2017
Grass carp (Ctenopharyngodon idella) skin	Alcalase	Pro-Tyr-Ser-Phe-Lys, Gly-Phe-Gly-Pro-Glu-Leu and Val-Gly-Gly-Arg-Pro	Antioxidant activities by DPPH, OH radical, ABTS radical scavenging and inhibit lipid peroxidation.	Cai et al. 2015
Sea squirt (Halocynthia roretzi) protein	Pepsin	Leu-Glu-Trp, Tyr-Tyr-Pro-Tyr-Gln-Leu and Met-Thr-Thr-Leu	Measured antioxidant activity by DPPH and ORAC radicals.	Kim et al. 2018
Bluefin leatherjacket skin (Navodon septentrionalis) and heads	Trypsin, Flavourzyme, neutrase, papain, alcalase, and pepsin, and protein hydrolysate (BSH) Papain	Gly-Ser-Gly-Gly-Leu, Gly-Pro-GlyGly-Phe-Ile and Phe-Ile-Gly-Pro Trp-Glu-Gly-ProLys, Gly-Pro-Pro and Gly-Val-Pro-Leu-Thr	Antioxidant activity by DPPH, HO, O_2^- radical scavenging assays. DPPH, hydroxyl, ABTS, superoxide radicals scavenging and inhibit the peroxidation of linoleic acid to measure the antioxidant activity.	Chi, Wang, Hu et al. 2015 Chi, Wang, Wang et al. 2015

(Continued)

TABLE 12.1 (CONTINUED)

Antioxidant Peptides

Peptide Source	Hydrolysis Method	Amino Acid Sequence	Activity	References
Oyster (*Saccostrea cucullata*)	Protease	Leu-Ala-Asn-Ala-Lys, Pro-Ser-Leu-Val-Gly-Arg-Pro-Pro-Val-Gly-Lys-Leu-Thr-Leu and Val-Lys-Val-Leu-Leu-Glu-His-Pro-Val-Leu	DPPH radicals scavenging assay and human colon carcinoma (HT-29) cell lines inhibition.	Umayaparvathi et al. 2014
Oyster (*Crassostrea talienwhanensis*)	Subtilisin	Pro-Val-Met-Gly-Ala and Glu-His-Gly-Val	High antioxidant activity with DPPH radical scavenging assay.	Q. Wang et al. 2014
Palm Kernel Cake Proteins	Papain	Gly-Ile-Phe-Glu-Leu-Pro-Trp-Arg-Pro-Ala-Thr-Asn-Val-Phe	Antioxidant activity measurement by metal chelating ability and DPPH radical scavenging.	Zarei et al. 2014
Royal jelly protein	Protease N	Ala-Leu, Phe-Arg, Phe-Lys, Ile-Arg, Lys-Leu, Lys-Phe, Lys-Tyr, Tyr-Asp, Arg-Tyr, Tyr-Tyr, Lys-Asn-Tyr-Pro, Leu-Asp-Arg	High antioxidant activity at peroxidation of linoleic acid	Guo, Kouzuma, and Yonekura 2009
Blood clam (*Tegillarca granosa*) muscle	Trinitrobenzene sulfonic acid Neutrase	Trp-Pro-Pro	Inhibit lipid peroxidation, DPPH, HO, O_2 and ABTS radical scavenging activity have been observed to prove the antioxidant activity.	Chi, Hu, Wang, Lee et al. 2015
Croicine croaker (*Pseudosciaena crocea*) muscle	Pepsin and Alcalase	Tyr-Leu-Met-Ser-Arg, Val-Leu-Tyr-Glu-Glu and Met-Ile-Leu-Met-Arg	DPPH, superoxide, ABTS and hydroxyl radical scavenging. Lipid peroxidation activities have also been observed.	Chi, Hu, Wang, Ren et al. 2015

(Continued)

TABLE 12.1 (CONTINUED)
Antioxidant Peptides

Peptide Source	Hydrolysis Method	Amino Acid Sequence	Activity	References
Spotless smoothhound (*Mustelusgriseus*) muscle	Papain	Gly-Ala-Ala, GlyPhe-Val-Gly, Gly-Ile-Ile-Ser-His-Arg, Glu-Leu-Leu-Ile and Lys-Phe-Pro-Glu	Hydroxyl, ABTS, superoxide radical scavenging activities have been detected to prove antioxidant activity.	Wang et al 2014
Marine *Sepia brevimana* mantle *Sphyrna lewini* muscle	Trypsin, α-chymotrypsin, and pepsin	Ile/Leu-Asn-Ile/Leu-Cys-Cys-Asn,	Antioxidant activities have measured by DPPH radicals scavenging and lipid peroxidation.	Sudhakar and Nazeer 2015
Sphyrna lewini muscle	Papain	Trp-Asp-Arg and Pro-Tyr-Phe-Asn-Lys	ABTS, DPPH radicals scavenging assays and peroxyl free radical scavenging in β-carotene linoleic acid assay was made to measure the antioxidant capacity.	Wang et al. 2012
Tilapia (*Oreochromis niloticus*) gelatin	Properase E.	Leu-Ser-Gly-Tyr-Gly-Pro	Found to have scavenging of hydroxyl radicals to prove the antioxidant activity.	Sun, Zhang, and Zhuang 2013Zhuang and Sun 2011
Whey	Kasein	High levels of glutamylcysteine	They found high glutathione-promoting activity	Bounous and Gold 1991
Otolithes ruber muscle protein hydrolysate	Pepsin	Lys-Thr-Phe-Cys-Gly-Arg- His	Hydroxyl and DPPH radical scavenging activities. Inhibition of the lipid peroxidation and DNA damage. Also improved the endogenous cellular antioxidant enzymes (catalase (CAT), superoxide dismutase (SOD) and glutathione-S-transferase (GST)) in study with Wistar rats.	Nazeer, Kumar, and JaiGanesh 2012

potentially assumed as agents that lower the blood pressure (Kannan, Hettiarachchy, and Marshall 2012). Therefore, a great deal of research has been carried out on the peptide production which shows antihypertensive activity from milk, cheese, meat, fish, and a wide variety of plants and algae. However, no correlation has been reported between in vivo antihypertensive effects and the results of in vitro studies investigating the inhibition of ACE enzymes. In this manner, there is no assurance that the result that is taken from ACE inhibition acquired in vitro will have a similar impact in vivo (Majumder and Wu 2013; Miralles, Amigo, and Recio 2018; Ünal, Şener, and Cemek 2018; Girija 2018).

12.4.2 Anticancerogenic Activity

The yellow eld pea seed hydrolysate has been shown to be an important contributor to the peptide antioxidant properties of net charge and hydrophobicity through a progression of investigations including column chromatography divisions of a thermolysin digestion. Due to its low toxicity and small size, it has advantages of high tissue penetration, permeability, and cell dissemination. Peptides can also influence at least one specific molecular pathway that is associated with cancer development and the pathways are not commonly genotoxic. The main mechanisms that can be described are via inhibiting the cell migration, antioxidant activity, gene transcription/inhibition of proliferation on cells, inhibition of the formation of blood vessels on tumor cells, induction of apoptosis, cytotoxicity, and induction of disorganization of tubulin structure. Peptides that have a potent cytotoxic activity can be derived from various substances such as eggs, milk, marine organisms, and plants (Su et al. 2014; Cicero, Fogacci, and Colletti 2017; Ünal, Şener, and Cemek 2018).

12.4.3 Cholesterol Reducer and Anti-Obesity Activity

Obesity is a fact that shortens the life span and also impairs the quality of life with its side effects (Yardım et al. 2017). There are findings that obesity increases the risk of heart disease, hypertension, and diabetes. When the cholesterol and/or triglyceride levels increase in plasma, development of atherosclerosis can increase. In addition, a decrease in high-density lipoprotein levels can also contribute to atherosclerosis development. Protein hydrolysates containing a specific peptide sequence have been found to exhibit lipid-lowering (hypolipidemic) effects in in vitro, ex vivo, and in vivo studies. Hypolipidemic peptides display their activities by binding bile acids, breaking down cholesterol micelles, and hindering their absorption. It also has the ability to influence the activities of liver and fat cell enzymes and to activate gene expression of lipogenic proteins, receptors and the 3-hydroxy-3-methylglutaryl CoA reductase enzyme. The 3-hydroxy-3-methylglutaryl CoA reductase enzyme acts as the main regulatory enzyme in cholesterol biosynthesis. Inhibition of this enzyme is the main target to reduce the cholesterol biosynthesis rate (Mazorra-Manzano, Ramírez-Suarez, and Yada 2017; White, Garber and Anantharamaiah 2014) and finally, it has role in the lipid metabolism and obesity (Ünal, Şener, and Cemek 2018).

12.4.4 ANTI-DIABETIC ACTIVITY

Diabetes is the most common non-communicable disease that occurs, due to the insufficiency of the pancreas in insulin production or the ineffectiveness of secreted insulin. This type of deficiency causes increased blood sugar and damages the body tissue, especially blood vessels and nerves. While the number of patients with diabetes in the world was 108 million in 1980, it increased to 422 million in 2014 (Agarwa and Gupta 2016; Arrutia et al. 2016; Brown et al. 2017; Ünal, Şener, and Cemek 2018). Another alluring alternative in pharmacological mediations is the utilization of bioactive peptides capable of inhibiting key enzymes that regulate blood sugar levels (e.g. dipeptitil peptitase IV (DPP-IV) and alpha-glycosidase). DPP-IV (EC 3.4.14.5) is the key compound in intestinal processing and ingestion systems. The movement of this enzyme is associated with the breakdown of incretions such as glucagon-like peptide (GLP-1) and gastric inhibitory peptide (GIP). GLP-1 and GIP peptides are known to assume a significant job in the regulation of glucose homeostasis in the digestive tract in the first minutes of food intake. The existence of bioactive peptides in protein hydrolysates, which can hinder DPP-IV in vitro and lower blood sugar levels in vivo, increased the production of hydrolysates with antidiabetic activity. Milk protein hydrolysates containing bioactive peptide capable of inhibiting DPP-IV were produced using microbial, animal, and vegetable proteases (Mazorra-Manzano et al. 2017; Yan et al. 2019).

12.5 ROLES OF ANTIOXIDANTS IN IMMUNITY

The immune system is the name given to the sum of the processes that protect the whole organism against diseases, recognize and destroy pathogens and tumor cells. The immune system scans every foreign substance that enters or comes into contact with the body and distinguishes it from the healthy body cells and tissues of the living body and provides defense against harmful substances. The immune system should also detect and respond to abnormal cells and molecules that occur at certain intervals in the body and prevent the development of diseases (Calder 2013). The immune system is very delicate to the balance of oxidants and antioxidants, and a small change in this balance can cause undesirable results. Many of the immune cell functions are dependent on ROS, and ROS increase cytotoxic activity in phagocytes and exhibit microbicidal action, so it is important for immunity. Also, they provide, for example, control of signal transmission of gene expression, the integrity and function of membrane lipids, cellular proteins, and nucleic acids in immune system cells. Without sufficient amounts of antioxidants, ROS produced by phagocytic immune cells can be harmful to the cells themselves and cause oxidative stress. Compared to other somatic cells, there are higher amounts of antioxidants in the immune system and antioxidants have an important place in the continuation of the immune system (Amir Aslani and Ghobadi 2016).

12.6 THERAPEUTIC USE OF BIOACTIVE PEPTIDE PROPERTIES

Food-borne bioactive peptides have health benefits. They are successfully used in the treatment of many diseases. Bioactive peptides are more bioavailable and less allergic than other peptides. Peptides have multiple benefits on biological functions such as blood pressure, metabolic risk factors (coagulation, obesity, lipoprotein metabolism, peroxidation), immunity, hypertension, cardiovascular disease, inflammation, intestinal and neurological functions, diabetes, microbial infections, dental health, and mineral metabolism (Bouglé and Bouhallab 2015; Priya 2019). The use of bioactive peptides using nanoparticle and microparticle systems can increase the advantages of bioactive peptides and achieve therapeutic effects (McClements 2018). As a therapeutic agent, bioactive proteins have many uses. Lately, there has been an expansion in the utilization of natural products, such as plants for enormous scope in the production of significant proteins, including, vaccines, antibodies, industrial proteins, and different pharmaceuticals (Daniell et al. 2009; Marauyama et al. 2014). The use of plants offers numerous points of advantages compared to mammalian culture cells due to their low prices and safety. In addition, past examinations have exhibited the potential ability of plants to create complex proteins, for example, secretory antibodies, comprising of four distinctive polypeptide chains that are covalently connected by disulfide bonds (Orzáez, Granell, and Blázquez 2009; Marauyama et al. 2014). This increases the availability of the proteins obtained from these creatures as vaccine production and gives us ideas about producing vaccines against diseases. The peptides produced from the natural compounds described in this chapter have high antioxidant properties and the use of these antioxidant peptides will provide many benefits. In addition, these bioactive peptides do not only have antioxidant roles, they have many benefits, and it is believed that these bioactive peptide sequences may be used in the vaccine production and steps described in other chapters of this book.

12.7   CAN BIOACTIVE PEPTIDES BE USED AS VACCINE AGENTS?

Peptide vaccines, which are also mentioned in other chapters of this book, are known to have great advantages. These vaccines are generally shown in many preclinical and clinical studies, with low cost, easy stability, relative safety, and easy synthesis. In addition, these peptide vaccines can be used in almost all diseases ranging from many virus infections to Alzheimer's disease and even allergies (Yang and Kim 2015). It is a matter of curiosity whether antioxidant natural peptides can be used in vaccine structure in addition to the vaccines created with synthetic peptides; studies on this subject argue that successful results can be obtained. First of all, the peptides that will be used to deal with conditions such as nanoparticle production in vaccine studies using peptides must have certain properties. In this respect, the choice of peptides with antioxidant properties can change the effectiveness of the molecule to be produced (Spicer et al. 2018). For example, in a study with GV1001 16-amino-acid peptide, the antioxidant properties of the peptide draw attention (Park et al. 2016). In addition, there are opinions that oral vaccines exist in the studies and that these vaccines obtained with plants can be both cheap and healthy. In particular, studies with amyloid peptide and studies in Alzheimer's disease are indicative of this condition (Yoshida et al. 2011; Yoshida et al. 2019). It can be said that plant-induced antigens used in the production of oral vaccines have fewer side effects than vaccines administered by injection. In addition to the immunological safety benefit, plant-derived vaccines are thought to be safer and cheaper than those produced from animal cells or microbes. In addition, a vaccine produced from animal cells or microbes requires improvement during the manufacturing process because animal cells may contain viruses, microbes, endotoxins, and prions that can infect humans. However, plant-based vaccines can be administered directly to humans, and therefore a plant-borne vaccine can be produced relatively cheaply. This study highlights the oral vaccines but suggests that plant-based vaccines can be advantageous (Yoshida et al. 2019). In addition, it is thought that nanoparticle vaccines with plant protein can be used for viruses and can be a good protector (Santoni, Zampieri, and Avesani 2020). The use of VP6-ferritin nanoparticle vaccine obtained from milk protein in rotavirus is thought to induce mucosal and humoral immunogenicity and may reduce the side effects of this disease in infants, and is a promising candidate for this disease (Li et al. 2019). In another study, it is thought that an extremely potent seafood antimicrobial peptide epinecidin-1 with anticancer and immunomodulatory activities is promising for treatment (Neshani et al. 2019).

It raises the question that peptides with antioxidant properties can be obtained naturally and may be more safe and effective for humans. However, there are not many studies in this area, so experimental studies are required to test its accuracy. In the study, oral vaccination was made using plants and its advantages were explained.

12.8 CONCLUSION

Peptides are very broad therapeutics that have many uses in many areas. The antioxidant peptides are described in this chapter and the vaccine use of peptides with many features are promising for the future. In particular, the utilization of the characteristic bioactive peptides on vaccine production can have many promising advantages; it can reduce sensitivity for individuals and can increase immunity. Studies in this area are very few, so experimental studies are needed. The aim of this chapter is to create an awareness of natural bioactive peptides and give a comprehensive introduction to antioxidant peptides.

REFERENCES

Agarwa, P. and R. Gupta. 2016. "Alpha-amylase inhibition can treat diabetes mellitus." *Research and Reviews Journal of Medical and Health Sciences* 5(4): 1–8. doi:10.1155/2017/3592491.

Agrawal, H., R. Joshi, and M. Gupta. 2017. "Isolation and characterisation of enzymatic hydrolysed peptides with antioxidant activities from green tender sorghum." *Lebensmittel-Wissenschaft & Technologie- Food Science and Technology* 84: 608–616. doi:10.1016/j.fbio.2018.08.010.

Agrawal, H., R. Joshi, and M. Gupta. 2016. "Isolation, purification and characterization of antioxidative peptide of pearl millet (Pennisetum glaucum) protein hydrolysate." *Food Chemistry* 204: 365–372. doi:10.1016/j.foodchem.2016.02.127.

Aguilar-Toalá, J.E., L. Santiago-López, C.M. Peres, C. Peres, H.S. Garcia, B. Vallejo-Cordoba, and A. Hernández-Mendoza. 2017. "Assessment of multifunctional activity of bioactive peptides derived from fermented milk by specific Lactobacillus plantarum strains." *Journal of Dairy Science* 100(1), 65–75. doi:10.3168/jds.2016-11846.

Aluko, R.E. 2015. "Amino acids, peptides, and proteins as antioxidants for food preservation." *Handbook of Antioxidants for Food Preservation* 105–140. doi:10.1016/b978-1-78242-089-7.00005-1.

Amir Aslani, B. and S. Ghobadi. 2016. "Studies on oxidants and antioxidants with a brief glance at their relevance to the immune system." *Life Sciences* 146: 163–173. doi:10.1016/j.lfs.2016.01.014.

Arkan, T. 2011. *Antioxidant properties of various solvent extracts from Daphne oleoides subp. oleoides and Daphne sericea.* Konya: Master Thesis, Selçuk University, Institute of Science.

Arrutia, F., Á. Puente, F.A. Riera, C. Menéndez, and U.A. González. 2016. "Influence of heat pre- treatment on BSA tryptic hydrolysis and peptide release." *Food Chemitry* 202: 40–48. doi:10.1016/j.foodchem.2016.01.107.

Aslankoç, R, D. Demirci, Ü. Inan, M. Yildiz, A. Özturk, M. Çetin, E. Savran, and B. Yilmaz. 2019. "The Role of Antioxidant Enzymes in Oxidative Stress - Superoxide Dismutase (SOD), Catalase (CAT) and Glutathione Peroxidase (GPX)." *Süleyman Demirel University, Faculty of Medicine Journal* 26(3): 362–369. doi: 10.17343/sdutfd.566969.

Aydemir, B. and E. Karadağ Sarı. 2009. "Antioxidants and Their Relationship with Growth Factors." *Kocatepe Veterinary Journal* 2(2): 56–60.

Baig, M.H., K. Ahmad, M. Saeed, A.M. Alharbi, G.E. Barreto, G.M. Ashraf, and I. Choi. 2018. "Peptide based therapeutics and their use for the treatment of neurodegenerative and other diseases." *Biomedicine & Pharmacotherapy* 103: 574–581. doi:10.1016/j.biopha.2018.04.025.

Bazinet, L. and A. Doyen. 2015. "Antioxidants, mechanisms, and recovery by membrane processes." *Critical Reviews in Food Science and Nutrition* 57(21): 10245–10251. doi: 10.1080/10408398.2014.912609.

Bougatef, A., N. Nedjar-Arroume, L. Manni, R. Ravallec, A. Barkia, D. Guillochon, and M. Nasri. 2010. "Purification and identification of novel antioxidant peptides from enzymatic hydrolysates of sardinelle (Sardinella aurita) by-products proteins." *Food Chemistry* 118, 559–565. doi:10.1016/j.foodchem.2009.05.021.

Bouglé, D. and S. Bouhallab. 2015. "Dietary bioactive peptides: Human studies." *Critical Reviews in Food Science and Nutrition* 57(2): 335–343. doi:10.1080/10408398.2013.8 73766.

Bounous, G. and P. Gold. 1991. "The biological activity of undenatured dietary whey proteins: Role of glutathione." *Clinical and Investigative Medicine. Medecine Clinique et Experimentale* 14(4): 296–309.

Brown, A., D. Anderson, K. Racicot, S.J. Pilkenton, and E. Apostolidis. 2017. "Evaluation of chemical phytochemical enriched commercial plant extracts on the *in vitro* inhibition of α- glucosidase." *Frontiers Nutrition* 4(56): 1–8. doi:10.3389/fnut.2017.00056.

Cai, L.Y., X.S. Wu, Y.H. Zhang, X.X. Li, S. Ma, and J.R. Li. "Purification and characterization of three antioxidant peptides from protein hydrolysate of grass carp (*Ctenopharyngodon idella*) skin." *Journal of Functional Foods* 16: 234–242. doi:10.1016/j.jff.2015.04.042.

Calder, P.C. 2013. "Feeding the immune system." *Proceedings of the Nutrition Society* 72(03): 299–309. doi:10.1017/s0029665113001286

Carrasco-Castilla, J., A.J. Hernandez-Alvarez, C. Jimenez-Martinez, C. Jacintho-Hernandez, M. Alaiz, J. Giron-Calle, J. Vioquei, and G. Dávila-Ortiz. 2012. "Antioxidant and metal chelating activities of peptide fractions from phaseolin and bean protein hydrolysates. *Food Chemistry* 135(3): 1789–1795. doi:10.1016/j.foodchem.2012.06.016.

Carvalho-Queiroz, C., R. Nyakundi, P. Ogongo, H. Rikoi, N.K. Egilmez, I.O. Farah, T.M. Kariuki, and P.T. LoVerde. 2015. "Protective potential of antioxidant enzymes as vaccines for schistosomiasis in a non-human primate model." *Frontiers in Immunology* 6: 273. doi:10.3389/fimmu.2015.00273.

Chakrabarti, S., F. Jahandideh, and J. Wu. 2014. "Food-Derived bioactive peptides on inflammation and oxidative stress." *BioMed Research International* 2014: 608979. doi:10.1155/2014/608979.

Chakrabarti, S., S. Guha, and K. Majumder. 2018. "Food-Derived bioactive peptides in human health: Challenges and opportunities." *Nutrients* 10(11): 1738. doi:10.3390/nu10111738.

Chen, H.M., K. Muramoto, and F. Yamauchi. 1995. "Structural analysis of antioxidative peptides from soybean." *Journal of Agricultural and Food Chemistry* 43:574–578. doi:10.1021/jf00051a004.

Chen, H.M., K. Muramoto, F. Yamauchi, and K. Nokihara. 1996. "Antioxidant activity of designed peptides based on the antioxidative peptide isolated from digests of a soybean protein." Agricultural and *Environmental Chemistry* 44: 2619–2623.

Chen, S., D. Lin, Y. Gao, X. Cao, and X. Shen. 2017. "A novel antioxidant peptide derived from wheat germ prevents high glucose-induced oxidative stress in vascular smooth muscle cells in vitro." *Food & Function* 8: 142–150. doi:10.1039/c6fo01139j.

Chi, C.F., B. Wang, Y.M. Wang, B. Zhang, and S.J. Deng. 2015. "Isolation and characterization of three antioxidant peptides from protein hydrolysate of bluefin leatherjacket (*Navodon septentrionalis*) heads." *Journal of Functional Foods* 12: 1–10. doi:10.1016/j.jff.2014.10.027.

Chi, C.F., F.Y. Hu, B. Wang, T. Li, and G.F. Ding. 2015. "Antioxidant and anticancer peptides from the protein hydrolysate of blood clam (*Tegillarca granosa*) muscle." *Journal of Functional Foods* 15: 301–313. doi:10.1016/j.jff.2015.03.045.

Chi, C.F., B. Wang, F.Y. Hu, Y.M. Wang, B. Zhang, S.J. Deng, and C.W. Wu. 2015. "Purification and identification of three novel antioxidant peptides from protein hydrolysate of bluefin leatherjacket (*Navodon septentrionalis*) skin." *Food Research International* 73: 124–129. doi:10.1016/j.foodres.2014.08.038.

Chi, C.F., F.Y. Hu, B. Wang, X.J. Ren, S.J. Deng, and C.W. Wu. 2015. "Purification and characterization of three antioxidant peptides from protein hydrolyzate of croceine croaker (*Pseudosciaena crocea*) muscle." *Food Chemistry* 168: 662–667. doi:10.1016/j.foodchem.2014.07.117.

Cicero, A.F.G., F. Fogacci, and A. Colletti. 2017. "Potential role of bioactive peptides in prevention and treatment of chronic diseases: A narrative review. *British Journal of Pharmacology* 174(11), 1378–1394. doi:doi: 10.1111/bph.13608.

Corpas, F.J. and J.B. Barroso. 2015. "Reactive sulfur species (RSS): Possible new players in the oxidative metabolism of plant peroxisomes." *Frontiers in Plant Science* 6:116. doi:10.3389/fpls.2015.00116

Daliri, E.B.M., D.H. Oh, and B.H. Lee. 2017. "Bioactive peptides." *Foods* 6: 32. doi:10.3390/foods6050032.

Daniell, H., Singh, N.D., Mason, H., and S.J. Streatfield. 2009. "Plant-made vaccine antigens and biopharmaceuticals." *Trends in Plant Science* 14: 669–679. doi:10.1016/j.tplants.2009.09.009.

Davalos, A.M.Miguel, B. Bartolome, and R. Lopez-Fandino. 2004. "Antioxidant activity of peptides derived from egg white proteins by enzymatic hydrolysis." *Journal of Food Protection* 67(9): 1939–1944. doi:10.4315/0362-028x-67.9.1939.

Decker, E.A. 2002. "Antioxidant mechanisms." In Akoh, C.C., and Min, D.B. (Eds), *Lipid chemistry*, 2nd ed. New York: Marcel Dekker Inc.

Durackova, Z. 2010. "Some current insights into oxidative stress." *Physiological Reseaech* 59: 459–469.

Elias, R.J., S.S. Kellerby, and E.A. Decker. 2008. "Antioxidant activity of proteins and peptides. *Critical Reviews in Food Science and Nutrition* 48(5): 430–441. doi:10.1080/10408390701425615.

Esfandi, R., M.E. Walters, and A. Tsopmo. "Antioxidant properties and potential mechanisms of hydrolyzed proteins and peptides from cereals." *Heliyon* 5: e01538. doi:10.1016/j.heliyon.2019.e01538.

Faraji, H., E.A. Decker, and D.K. Aaron. 1991. "Suppression of lipid oxidation in phosphatidylcholine liposomes and ground pork by spray dried porcine plasma." *Agricultural and Food Chemistry* 39: 1288–1290.

Feng, L., F. Peng, X. Wang, M. Li, H. Lei, and H. Xu. 2018. "Identification and characterization of antioxidative peptides derived from simulated in vitro gastrointestinal digestion of walnut meal proteins." *Food Research International (Ottawa, Ont.)* 116: 518–526. doi:10.1016/j.foodres.2018.08.068.

Frankel, E.N. and J.W. Finley. 2008. "How to standardized the multiplicity of methods to evalu- ate natural antioxidants." *Journal of Agricultural and Food Chemistry* 56: 4901–4908. doi:10.1021/jf800336p.

Fransen, M., M. Nordgren, B. Wang, and O. Apanasets. 2012. "Role of peroxisomes in ROS/RNS-metabolism: Implications for human disease." *Biochimica et Biophysica Acta— Molecular Basis of Disease* 1822(9): 1363–1373. doi:10.1016/j.bbadis.2011.12.001

Fukada, Y., S. Mizutani, S. Nomura, W. Hara, R. Matsui, K. Nagai, Y. Murakami, N. Washio, N. Ikemoto, and M. Terashima. 2016. Antioxidant activities of a peptide derived from chicken dark meat. *Journal of Food Science and Technology* 53(5): 2476–2481. doi:10.1007/s13197-016-2233-9.

Gad, A.S., Y.A. Khadrawy, A.A. El-Nekeety, S.R. Mohamed, N.S. Hassan, and A.M. Abdel-Wahhab. 2011. "Antioxidant activity and hepatoprotective effects of whey protein and Spirulina in rats." *Nutrition* 27(5): 582–589. doi:10.1016/j.nut.2010.04.002.

Ghribi, A.M., A. Sila, R. Przybylski, N. Nedjar-Arroume, I. Makhlouf, C. Blecker, H. Attia, P. Dhulster, A. Bougatef, and S. Besbes. 2015. "Purification and identification of novel antioxidant peptides from enzymatic hydrolysate of chickpea (*Cicer arietinum* L.) protein concentrate." *Journal of Functional Foods* 12: 516–525. doi:10.1016/j.jff.2014.12.011.

Girgih, A.T., R. He, S. Malomo, M. Offengenden, J. Wu, and R.E. Aluko. 2014. "Structural and functional characterization of hemp seed (Cannabis sativa L.) protein-derived antioxidant and antihyper- tensive peptides." *Journal of Functional Foods* 6: 384–394. doi:10.1016/j.jff.2013.11.005.

Girija, A.R. 2018. "Peptide nutraceuticals. Peptide applications in biomedicine." *Biotechnology and Bioengineering* 157–181. doi:10.1016/b978-0-08-100736-5.00006-5.

Gu, Y., K. Majumder, and J. Wu. 2011. "QSAR-aided in silico approach in evaluation of food proteins as precursors of ACE inhibitory peptides." *Food Research International* 44: 2465–2474. doi:10.1016/j.foodres.2011.01.051.

Guo, H., Y. Kouzuma, and M. Yonekura. 2009. "Structures and properties of antioxidative peptides derived from royal jelly protein." *Food Chemistry* 113: 238–245 doi:10.1016/j.foodchem.2008.06.081.

Gutteridge, J.M.C. 1994. "Biological origin of free radicals, and mechanisms of antioxidant protection" *Chemico-Biological Interactions* 91: 133–140.

Halliwell, B. and J.M.C. Gutteridge. 1990. "The antioxidants of human extracellular uids." *Archives of Biochemistry and Biophysics* 280(1): 1–8.

Halliwell, B., J.M.C. Gutteridge, and C.E. Cross. 1992. "Free radicals, antioxidants, and human disease: Where are we now?" e Journal of Laboratory and Clinical Medicine 119(6): 598–620.

Harnedy, P.A., M.B. O'Keeffe, and R.J. FitzGerald. 2017. "Fractionation and identification of antioxidant peptides from an enzymatically hydrolysed Palmaria palmata protein isolate." *Food Research International (Ottawa, Ont.)* 100(Pt 1): 416–422. doi:10.1016/j.foodres.2017.07.037.

Hartmann, R., and H. Meisel. 2007. "Food-derived peptides with biological activity: From research to food applications." *Current Opinion in Biotechnology* 18: 163– 169. doi:10.1016/j.copbio.2007.01.013.

Hayes, M., L. Mora, K. Hussey, and R.E. Aluko. 2016. "Boarfish protein recovery using the pH- shift process and generation of protein." *Innovative Food Science and Emerging Technologies* 37(5): 253–260.

Hekimi, S., J. Lapointe, and Y. Wen. 2011. "Takinga'good'lookatfree radicals in the aging process." *Trends in Cell Biology* 21(10): 569–576. doi:10.1016/j.tcb.2011.06.008.

Hernandez-Ledesma, B., A., Davalos, B. Bartolome, and L. Amigo. 2005. "Preparation of antioxidant enzymatic hydrolysates from á-lactalbumin and â-lactoglobulin. Identification of active peptides by HPLC-MS/MS." *Journal of Agricultural and Food Chemistry* 53(3): 588–593. doi:10.1021/jf048626m.

Huang D., B. Ou, and R.L. Prior. 2005. "The chemistry behind antioxidant capacity assays." *Journal of Agricultural Food Chemistry* 53(6): 1841–1856. doi:10.1021/jf030723c.

Huang, D. 2018. "Dietary antioxidants and health promotion." *Antioxidants (Basel, Switzerland)* 7(1): 9. https://doi.org/10.3390/antiox7010009.

Ibrahim, H.R., H. Isono, and T. Miyata. 2018. "Potential antioxidant bioactive peptides from camel milk proteins." *Animal Nutrition* 273–280. doi:10.1016/j.aninu.2018.05.004

Kannan, A., N. Hettiarachchy, and M. Marshall. 2012. *Food proteins and peptides as bioactive agents. Bioactive food proteins and peptides: Applications in human health.* N.S. Hettiarachchy, eds. Boca Raton: CRC Press, Taylor And Francis Group, 1–27.

Karabulut H., and M.Ş. Gülay. 2016. "Antioxidants." *Veterinary Journal of Mehmet Akif Ersoy University* 1(1): 65–76. doi:10.24880/maeuvfd.260790.

Karami, Z., S.H. Peighambardoust, J. Hesari, and B. Akbari-Adergani. 2019. "Response surface methodology to optimize hydrolysis parameters in production of antioxidant peptides from wheat germ protein by alcalase digestion and identification of antioxidant peptides by LC-MS/MS." *Journal of Agricultural Science and Technology* 21(4): 829–844. http://journals.modares.ac.ir/article-23-17648-en.html.

Khlebnikov, I.A. Schepetkin, N.G. Domina, L.N. Kirpotina, and M.T. Quinn. 2017. "Improved quantitative structure- activity relationship models to predict antioxidant activity of avonoids in chemical, enzymatic, and cellular systems." *Bioorganic and Medicinal Chemistry* 15(4): 1749– 1770. doi:10.1016/j.bmc.2006.11.037.

Kim, S.S., C.B. Ahn S.W. Moon, and J.Y. Je. 2018. "Purification and antioxidant activities of peptides from sea squirt (Halocynthia roretzi) protein hydrolysates using pepsin hydrolysis. *Food Bioscience* 25: 128–133. doi:10.1016/j.fbio.2018.08.010.

Kitts, D.D., and K. Weiler. 2003. "Bioactive proteins and peptides from food sources. Applications of bioprocesses used in isolation and recovery." *Current. Pharmaceutical Design* 9: 1309–1323.

Kohen, R. and A. Nyska. 2002. "Oxidation of biological systems: Oxidative stress phenomena, antioxidants, redox reactions, and methods for their quantification." *Toxicologic Pathology* 30(6): 620–650.

Lee, J.H., S.H. Moon, H.S. Kim, E. Park, D.U. Ahn, and H.D. Paik, 2017. "Antioxidant and anticancer effects of functional peptides from ovotransferrin hydrolysates." *Journal of the Science of Food and Agriculture* 97(14): 4857–4864. doi:10.1002/jsfa.8356.

Li, G.H., M.R. Qu, J.Z. Wan, and J.M. You. 2007. "Antihypertensive effect of rice protein hydrolysate with in vitro angiotensin I-converting enzyme inhibitory activity in spontaneously hypertensive rats." *Asia Pacific Journal of Clinical Nutrition* 16(Suppl. 1): 275–280.

Li, Y. and J. Yu. 2015. "Research progress in structure-activity relationship of bioactive peptides." *Journal of Medicinal Food* 18(2): 147–156. doi:10.1089/jmf.2014.0028.

Li, Y.W., B. Li, J.G. He, and P. Qian. 2011. "Quantitative structure- activity relationship study of antioxidative peptide by using different sets of amino acids descriptors." *Journal of Molecular Structure* 998: 53–61. doi:10.1016/j.molstruc.2011.05.011.

Li, Z., K. Cui, H. Wang, F. Liu, K. Huang, Z. Duan, F. Wang, D. Shi, and Q. Liu. 2019. "A milk-based self-assemble rotavirus VP6-ferritin nanoparticle vaccine elicited protection against the viral infection. *Journal of Nanobiotechnology.* 17(1): 13. doi:10.1186/s12951-019-0446-6.

Limon-Pacheco, J. and M.E. Gonsebatt. 2009. "The role of antioxidants and antioxidant-related enzymes in protective responses to environmentally induced oxidative stress." *Mutation Research* 674(1–2): 137–147. doi:10.1016/j.mrgentox.2008.09.015

Liu, J.B., Y. Jin, S.Y. Lin, G.S. Jones, and F. Chen. 2015. "Purification and identification of novel antioxidant peptides from egg white protein and their antioxidant activities." *Food Chemistry* 175: 258–266. doi:10.1016/j.foodchem.2014.11.142.

Liu, Q., B. Kong, Y.L. Xiong, and X. Xia. 2010. "Antioxidant activity and functional properties of porcine plasma protein hydrolysate as influenced by the degree of hydrolysis." *Food Chemistry* 118(2): 403–410. doi:10.1016/j.foodchem.2009.05.013.

Liu, R., W. Zheng, J. Li, L. Wang, H. Wu, X. Wang, and L. Shi. 2015. "Rapid identification of bioactive peptides with antioxidant activity from the enzymatic hydrolysate of Mactra veneriformis by UHPLC-Q-TOF mass spectrometry." *Food Chemistry* 167: 484–489. doi:10.1016/j.foodchem.2014.06.113.

Liu, R., Xing, L., Fu, Q., Zhou, G.H., and W.G. Zhang. 2016. "A review of antioxidant peptides derived from meat muscle and by-products." *Antioxidants (Basel)* 5(3): 32. doi:10.3390/antiox5030032.

Lobo, V., A. Patil, A. Phatak, and N. Chandra. 2010. "Free radicals, antioxidants and functional foods: Impact on human health." *Pharmacognosy Reviews* 4(8): 118–126. doi:10.4103/0973-7847.70902

Majumder, K., and J. Wu. 2014. "Molecular targets of antihypertensive peptides: understanding the mechanisms of action based on the pathophysiology of hypertension." *International Journal of Molecular Sciences* 16(1): 256–283. doi:10.3390/ijms16010256.

Mamta, K., G.S. Misra, S.K. Dhillon, and M.V. Brar. 2014. "Antioxidants." In Brar, S., Dhillon, G., and Soccol, C. (Eds), *Biotransformation of waste biomass into high value biochemicals.* New York: Springer.

Manso, M.A., M. Miguel, J. Even, R. Hernández, A. Aleixandre, and R. López-fandino. 2008. "Effect of the long-term intake of an egg white hydrolysate on the oxidative status and blood lipid profile of spontaneously hypertensive rats." *Food Chemistry* 109: 361–367. doi:10.1016/j.foodchem.2007.12.049.

Maqsoudlou, A., A.S. Mahoonak, L. Mora, H. Mohebodini, F. Toldrá, and M. Ghorbani. 2018. "Peptide identification in alcalase hydrolysated pollen and comparison of its bioactivity with royal jelly." *Food Research International* doi:10.1016/j.foodres.2018.09.027.

Maruyama, N., K. Fujiwara, K. Yokoyama, C. Cabanos, H. Hasegawa, K. Takagi, and K. Nishizawa. 2014. "Stable accumulation of seed storage proteins containing vaccine peptides in transgenic soybean seeds." *Journal of Bioscience and Bioengineering* 118(4): 441–447. doi:10.1016/j.jbiosc.2014.04.004

Matsui, R., R. Honda, M. Kanome, A. Hagiwara, Y. Matsuda, T. Togitani, N. Ikemoto, and M. Terashima. 2018. "Designing antioxidant peptides based on the antioxidant properties of the amino acid side-chains." *Food Chemistry* 15(245): 750–755. doi:10.1016/j.foodchem.2017.11.119.

Matsui, T., C.H. Li, and Y. Osajima. 1999. "Preparation and characterization of novel bioactive peptides responsible for angiotensin I-converting enzyme inhibition from wheat germ." *Journal of Peptide Science: An Official Publication of the European Peptide Society* 5(7): 289–297. doi:10.1002/(SICI)1099-1387(199907)5:7<289::AID-PSC196 >3.0.CO;2-6.

Mazorra-Manzano, M.A., J.C. Ramírez-Suarez, and R.Y. Yada. 2017. "Plant proteases for bioactive peptides release: A review." *Critical Reviews in Food Science and Nutrition* 58(13), 2147–2163. doi:10.1080/10408398.2017.1308312.

McClements, D.J. 2018. Encapsulation, protection, and delivery of bioactive proteins and peptides using nanoparticle and microparticle systems: A review." *Advances in Colloid and Interface Science* 253: 1–22. doi:10.1016/j.cis.2018.02.002.

Miralles, B., L. Amigo, and I. Recio, 2018. "Critical review and perspectives on food-derived antihypertensive peptides." *Journal Agricultural Food Chemistry* 66(36): 9384–9390. doi:10.1021/acs.jafc.8b02603.

Moller, N.P., K.E. Scholz-Ahrens, N. Roos, and J. Schrezenmeir. "Bioactive peptides and proteins from foods: Indication for health effects." *European Journal of Nutrition* 47(4):171–182. doi:10.1007/s00394-008-0710-2.

Mut-Salud, N., P.J. Álvarez, J.M. Garrido, E. Carrasco, A. Aránega, and F. Rodríguez-Serrano. 2016. "Antioxidant intake and antitumor therapy: Toward nutritional recommendations for optimal results." *Oxidative Medicine and Cellular Longevity* 1–19. doi:10.1155/2016/6719534

Nasri, H. 2016. "Antioxidants; from laboratory investigations to clinical studies." *Annals of Research in Antioxidants* 1(2): e19.

Nazeer, R.A., N.S. Kumar, and R.Jai Ganesh. 2012. "In vitro and in vivo studies on the antioxidant activity of fish peptide isolated from the croaker (Otolithes ruber) muscle protein hydrolysate." *Peptides* 35(2): 261–268. doi:10.1016/j.peptides.2012.03.028.

Neshani, A., H. Zare, M.R. Akbari Eidgahi, A. Khaledi, and K. Ghazvini. 2019. "Epinecidin-1, a highly potent marine antimicrobial peptide with anticancer and immunomodulatory activities." *BMC Pharmacology and Toxicology* 20(1): 33. doi:10.1186/s40360-019-0309-7.

Niki, E. 2014. "Antioxidant defenses in eukaryotic cells." In Poli, G., Albano, E., and Dianzani, M.U. (Eds), *Free radicals: From basic science to medicine*. Basel: Birkhauser Verlag, 365–373.

Nimalaratne, C., N. Bandara, and J.P. Wu. 2015. "Purification and characterization of antioxidant peptides from enzymatically hydrolyzed chicken egg white." *Food Chemistry* 188: 467–472. doi:10.1016/j.foodchem.2015.05.014.

Nongonierma, A.B., S.L. Maux, C. Dubrulle, C. Barre, and R.J. FitzGerald. 2015. "Quinoa (Chenopodium quinoa Willd.) protein hydrolysates with *in vitro* dipeptidyl peptidase IV (DPP-IV) inhibitory and antioxidant properties." *Journal of Cereal Science* 65: 112–118. doi:10.1016/j.jcs.2015.07.004.

Ohashi, Y., R. Onuma, T. Naganuma, T. Ogawa, R. Naude, K. Nokihara, and K. Muramoto. 2015. "Antioxidant properties of tripeptides revealed by a comparison of six different assays." *Food Science and Technology Research* 21(5): 695–704. doi:10.3136/fstr.21.695.

Ohata, M., S. Uchida, L. Zhou, and K. Arihara. 2016. "Antioxidant activity of fermented meat sauce and isolation of an associated antioxidant peptide." *Food Chemistry* 194: 1034–1039. doi:10.1016/j.foodchem.2015.08.089.

Orzáez, D., A. Granell, and M.A. Blázquez. 2009. "Manufacturing antibodies in the plant cell." *Biotechnology Journal* 4(12): 1712–1724. doi:10.1002/biot.200900223.

Özge Özcan, Ö., M. Karahan, P. Vijayaraj Kumar, S.L. Tan, and Y. Na Tee. 2020. "New generation peptide-based vaccine prototype." *Current and Future Aspects of Nanomedicine*. IntechOpen. doi:10.5772/intechopen.89115.

Öztürk, H.I. 2015. *Determination of Some Quality Properties, Bioactive Peptide Content and Functional Properties of Tulum Cheeses Produced by Traditional Method.* Doctoral dissertation, Selçuk University Institute of Science.

Park, H.H., H.J. Yu, S. Kim, G. Kim, N.Y. Choi, E.H. Lee, Y.J. Lee, M.Y. Yoon, K.Y. Lee, and S.H. Koh. 2016. "Neural stem cells injured by oxidative stress can be rejuvenated by GV1001, a novel peptide, through scavenging free radicals and enhancing survival signals." *Neurotoxicology* 55: 131–141. doi:10.1016/j.neuro.2016.05.022.

Peng, X., Y.L. Xiong, and B. Kong, 2009. "Antioxidant activity of peptide fractions from whey protein hydrolysates as measured by electron spin resonance. *Food Chemistry* 113, 196–201. doi:10.1016/j.foodchem.2008.07.068.

Percival, M. 1998. "Antioxidants." *Clinical Nutrition Insights* 10: 1–4.

Perez Espitia, P.J., N. de Fátima F. Soares, J.S. dos Reis Coimbra, N.J. de Andrade, R. Souza Cruz, and E.A.A. Medeiros. 2012. "Bioactive peptides: Synthesis, properties, and applications in the packaging and preservation of food." *Comprehensive Reviews in Food Science and Food Safety* 11(2): 187–204. doi:10.1111/j.1541-4337.2011.00179.x.

Pham-Huy, L.A., H. He, and C. Pham-Huy. 2008. "Free radicals, antioxidants in disease and health." *International Journal of Biomedical Science* 4(2): 89–96.

Pisoschi, A.M. and A. Pop. 2015. "The role of antioxidantss in the chemistry of oxidative stress: A review." *European Journal of Medicinal Chemistry* 97: 55–74. doi:10.1016/j. ejmech.2015.04.040.

Poljsak, B., P. Jamnik, P. Raspor, and M. Pesti. 2011. "Oxidation-antioxida- tion-reduc-tion processes in the cell: Impacts of environmental pollution." In Jerome, N. (Ed.), *Encyclopedia of environmental health.* Elsevier, 300–306.

Pownall, T.L., Udenigwe, C.C., and Aluko, R.E. 2010. "Amino acid composition and antioxidant properties of pea seed (*Pisum sativum* L.) enzymatic protein hydrolysate fractions." *Journal of Agricultural and Food Chemistry* 58(8): 4712–4718. doi:10.1021/ jf904456r.

Priya, S. 2019. "Therapeutic perspectives of food bioactive peptides: A mini review." *Protein & Peptide Letters* 26: 664. doi:10.2174/0929866526666190617092140.

Reuter, S., S.C. Gupta, M.M. Chaturvedi, and B.B. Aggarwal. 2010. "Oxidative stress, inflammation, and cancer: How are they linked?" *Free Radical Biology and Medicine* 49(11): 1603–1616. doi:10.1016/j.freeradbiomed.2010.09.006.

Rutherfurd-Markwick, K.J. 2012. "Food proteins as a source of bioactive peptides with diverse functions." *British Journal of Nutrition* 108(Suppl. 2): S149–S157. doi:10.1017/ S000711451200253X.

Safitri, N.M., E.Y. Herawati, and J.L. Hsu. 2017. "Antioxidant activity of purified active peptide derived from spirulina platensis enzymatic hydrolysates." *Research Journal of Life Science* 4(2): 119–128. doi:10.21776/ub.rjls.2017.004.02.5.

Saiga, A.E., and T. Nishimura. 2013. "Antioxidative properties of peptides obtained from por- cine myo brillar proteins by a protease treatment in an Fe (II)-induced aqueous lipid per- oxidation system." *Bioscience Biotechnology and Biochemistry* 77(11): 2201–2204. doi:10.1271/bbb.130369.

Saito K, Jin DH, Ogawa T, Muramoto K, Hatakeyama E, Yasuhara T, and Nokihara K. 2003. "Antioxidative properties of tripeptide libraries prepared by the combinatorial chemistry." *Journal of Agricultural and Food Chemistry* 51(12): 3668–3674. doi:10.1021/ jf021191n.

Samaranayaka, A.G.P. and E.C.Y. Li-Chan. "Food-derived peptidic antioxidants: A review of their production, assessment, and potential applications." *Journal of Functional Foods* 3: 229–254. doi:10.1016/j.jff.2011.05.006

Sampath Kumar, N.S., R.A. Nazeer, and R. Jaiganesh, 2011. "Purification and biochemical characterization of antioxidant peptide from horse mackerel (Magalaspis cordyla) viscera protein." *Peptides* 32(7): 1496–1501. doi:10.1016/j.peptides.2011.05.020.

Santoni, M., R. Zampieri, and L. Avesani. 2020. "Plant virus nanoparticles for vaccine applications." *Current Protein & Peptide Science* 21(4): 344–356. doi:10.2174/1389203721
666200212100255.

Sen, S., and R. Chakraborty. 2011. "The role of antioxidants in human health." *American
Chemical Society, Oxidative Stress: Diagnostics, Prevention and Therapy* Chapter 1:
1–37. doi:10.1021/bk-2011-1083.ch001

Sen, S., R. Chakraborty, C. Sridhar, Y.S.R. Reddy, and B. De. 2010. "Free radicals, antioxidants, diseases and phytomedicines: Current status and future prospect." *International
Journal of Pharmaceutical Sciences Review and Research* 3(1): 91–100.

Sila, A. and A. Bougatef. 2016. "Antioxidant peptides from marine by-products: Isolation,
identification and application in food systems." A review. *Journal of Functional Foods*
21: 10–26. doi:10.1016/j.jff.2015.11.007.

Singh, B.P., S. Vij, and S. Hati. 2014. "Functional significance of bioactive peptides derived
from soybean." *Peptides* 54: 171–179. doi:10.1016/j.peptides.2014.01.022.

Spicer, C.D., C. Jumeaux, B. Gupta, and M.M. Stevens. 2018."Peptide and protein nanoparticle conjugates: Versatile platforms for biomedical applications." *Chemical Society
Review* 47(10): 3574–3620. doi:10.1039 / c7cs00877e.

Su, X., C. Dong, J. Zhang, L. Su, X. Wang, H. Cui, and Z. Chen, 2014. "Combination therapy
of anti-cancer bioactive peptide with Cisplatin decreases chemotherapy dosing and
toxicity to improve the quality of life in xenograft nude mice bearing human gastric
cancer." *Cell & Bioscience* 4(1): 7. doi:10.1186/2045-3701-4-7.

Sudhakar, S. and R.A. Nazeer. "Preparation of potent antioxidant peptide from edible part
of shortclub cuttlefish against radical mediated lipid and DNA damage." *Lebensmittel-
Wissenschaft & Technologie- Food Science and Technology* 64: 593–601. doi:10.1016/j.
lwt.2015.06.031.

Suetsuna, K., H. Ukeda, and H. Ochi. 2000. "Isolation and characterization of free radical scavenging activities peptides derived from casein." *The Journal of Nutritional
Biochemistry* 11: 128–131.

Sun, L.P., Y.F. Zhang, and Y.L. Zhuang. 2013. "Antiphotoaging effect and purification of an
antioxidant peptide from tilapia (*Oreochromis niloticus*) gelatin peptides." *Journal of
Functional Foods* 5: 154–162. doi:10.1016/j.jff.2012.09.006.

Thiansilakul, Y., S. Benjakul, and F. Shahidi. 2007. "Compositions, functional properties and
antioxidative activity of protein hydrolysates prepared from round scad (*Decapterus
mayuadsi*)." *Food Chemistry* 103: 1385–1394. doi:10.1016/j.foodchem.2006.
10.055.

Tian, M., B. Fang, L. Jiang, H. Guo, J. Cui, and F. Ren. 2015. "Structure-activity relationship
of a series of antioxidant tripeptides derived from β-Lactoglobulin using QSAR modeling." *Dairy Science & Technology* 95: 451–463. doi:10.1007/s13594-015-0226-5

Umayaparvathi, S., S. Meenakshi, V. Vimalraj, M. Arumugam, G. Sivagami, and T.
Balasubramanian. 2014."Antioxidant activity and anticancer effect of bioactive pep-
tide from enzymatic hydrolysate of oyster (*Saccostrea cucullata*). *Biomedicine &
Preventive Nutrition* 4: 343–353. doi: 10.1016/j.bionut.2014.04.006.

Ünal, M., A. Şener, and K. Cemek. 2018. "Effects of Bioactive Peptides on Health." *The
Journal of Food* 43(6): 930–942. doi:10.15237/gida.GD18048.

Wang, B., Y.D. Gong, Z.R. Li, D. Yu, C.F. Chi, and J.Y. Ma. 2014. "Isolation and characterisation of five novel antioxidant peptides from ethanol-soluble proteins hydrolysate
of spotless smoothhound (*Mustelus griseus*) muscle." *Journal of Functional Foods* 6:
176–185. doi:10.1016/j.jff.2013.10.004.

Wang, B., Z.R. Li, C.F. Chi, Q.H. Zhang, and H.Y. Luo. 2012. "Preparation and evaluation
of antioxidant peptides from ethanol-soluble proteins hydrolysate of *Sphyrna lewini*
muscle." *Peptides* 36: 240–250. doi:10.1016/j.peptides.2012.05.013.

Wang, Q., W. Li, Y. He, D. Ren, F. Kow, L. Song, and X. Yu. 2014. "Novel antioxidative peptides from the protein hydrolysate of oysters (Crassostrea talienwhanensis)." *Food Chemistry* 145: 991–996. doi:10.1016/j.foodchem.2013.08.099.

Wetzler, M. and P. Hamilton. 2018. "Peptides as therapeutics: Peptide applications in biomedicine." *Biotechnology and Bioengineering* Elsevier: 215–230.

White, C. R., D. W. Garber, and G. M. Anantharamaiah. 2014. "Anti-inflammatory and cholesterol-reducing properties of apolipoprotein mimetics: a review." *Journal of Lipid Research* 55(10): 2007–2021. doi:10.1194/jlr.R051367.

Wu, D., N. Sun, J. Ding, B.W. Zhu, and S.Y. Lin. 2019. "Evaluation and structure-activity relationship analysis of antioxidant shrimp peptides." *Food Function* 10: 5605–5615. doi:10.1039/C9FO01280J

Xie, Z., J. Huang, X. Xu, and Z. Jin. 2008. "Antioxidant activity of peptides isolated from alfalfa leaf protein hydrolysate." *Food Chemistry* 111(2): 370–376. doi:10.1016/j.foodchem.2008.03.078.

Xu, X., S. Katayama, and Y. Mine. 2007. "Antioxidant activity of tryptic digests of hen egg yolk phosvitin. *Journal of the Science of Food and Agriculture* 87(14): 2604–2608. doi:10.1002/jsfa.3015.

Yan, J., J. Zhao, R. Yang, and W. Zhao. 2019. "Bioactive peptides with antidiabetic properties: a review." *International Journal of Food Science and Technology* 54: 1909–1919. doi:10.1111/ijfs.14090.

Yan W., G. Lin, R. Zhang, Z. Liang, and W. Wu. 2020. "Studies on the bioactivities and molecular mechanism of antioxidant peptides by 3D-QSAR, in vitro evaluation and MD simulations." *Food & Function* 1(4):3043–3052. doi:10.1039/C9FO03018B.

Yan, Q.J., L.H. Huang, Q. Sun, Z.Q. Jiang, and X. Wu. 2015. "Isolation, identification and synthesis of four novel antioxidant peptides from rice residue protein hydrolyzed by multiple proteases." *Food Chemistry* 179: 290–295. doi:10.1016/j.foodchem.2015.01.137.

Yang, H. and D.S. Kim. 2015. "Peptide immunotherapy in vaccine development." *Advances in Protein Chemistry and Structural Biology* 1–14. doi:10.1016/bs.apcsb.2015.03.001.

Yardıma, N., S. Kocadağ, E.Z. Kelat, Ö.S. Adıgüzel, M. Atabey, and M. Saygı. 2017. Obesity and Diabetes Clinical Guidelines for Primary Health Care Institutions. Public Health Agency of Turkey, Publication No: 1070.

Yaya, H., X. Lujuan, Z. Guanghong, and Z. Wangang. 2015. Effect of extraction methods on the antioxidant activity of crude peptides from Jinhua ham." *Journal of Food and Nutrition Research* 36: 115–118. doi:10.12691/jfnr-4-6-6.

Ye, N., P. Hu, S. Xu, M. Chen, S. Wang, J. Hong, T. Chen, and T. Cai. 2018. "Preparation and Characterization of Antioxidant Peptides from Carrot Seed Protein." *Journal of Food Quality* 1–9. doi:10.1155/2018/8579094.

Yoshida, T., E. Kimura, S. Koike, J. Nojima, E. Futai, N. Sasagawa, Y. Watanabe, and S. Ishiura. 2011. "Transgenic rice expressing amyloid β-peptide for oral immunization." *International Journal of Biological Sciences* 7(3): 301–7. doi:10.7150 / ijbs.7.301.

Yoshida, T., Y. Watanabe, and S. Ishiura. 2019. "Production of the herb Ruta chalepensis L. expressing amyloid β-GFP fusion protein." *Proceedings of the Japan Academy Serie B Physical Biological Sciences* 95(6): 295–302. doi:10.2183 / pjab.95.021.

Yu, H.H., X.G. Liu, R.E. Xing, S. Liu, Z.Y. Guo, P.B. Wang, C. Li, and P. Li. 2006. "In vitro determination of antioxidant activity of proteins from jellyfish Rhopilema esculentum." *Food Chemistry* 95: 123–130. doi:10.1016/j.foodchem.2004.12.025.

Zambrowicz, A., M. Pokora, B. Setner, A. Dąbrowska, M.Szołtysik, K. Babij, and J. Chrzanowska. 2015. "Multifunctional peptides derived from an egg yolk protein hydrolysate: Isolation and characterization." *Amino Acids* 47(2): 369–380. doi:10.1007/s00726-014-1869-x.

Zarei, M., A. Ebrahimpour, A. Abdul-Hamid, F. Anwar, F.A. Bakar, R. Philip, and N. Saari. 2014. "Identification and characterization of papain-generated antioxidant peptides from palm kernel cake proteins." *Food Research International* 62: 726–734. doi:10.1016/j.foodres.2014.04.041.

Zhang, J., H. Zhang, L. Wang, X. Guo, X. Wang, and H. Yao. 2010. "Isolation and identification of antioxidative peptide from rice endosperm protein enzymatic hydrolysate by consecutive chromatography and MALDI-TOF/TOF MS/MS." *Food Chemistry* 119(1): 226–234. doi:10.1016/j.foodchem.2009.06.015.

Zhang, L., J. Li, and K. Zhou. 2010. "Chelating and radical scavenging activities of soy protein hydrolysates prepared from microbial proteases and their effect on meat lipid peroxidation." *Bioresource Technology* 101(7): 2084–2089. doi:10.1016/j.biortech.2009.11.078.

Zhang, M, T.H. Mu, and M.J. Sun. 2014. "Purification and identification of antioxidant peptides from sweet potato protein hydrolysates by Alcalase." *Journal of Functional Foods* 7: 191–200. doi:10.1016/j.jff.2014.02.012.

Zhang, T., Y. Li, M. Miao, and B. Jiang 2011. "Purification and characterizations of a new antioxidant peptide from chickpea (Cicer arietium L.) protein hydrolysates." *Food Chemistry* 128(1): 28–33. doi:10.1016/j.foodchem.2011.02.072.

Zhuang, H., N. Tang, and Y. Yuan. 2013. "Purification and identification of antioxidant peptides from corn gluten meal." *Journal of Functional Food* 5(4): 1810–1821. doi:10.1016/j.jff.2013.08.013.

Zhuang, Y. and L. Sun. 2011. "Preparation of reactive oxygen scavenging peptides from tilapia (Oreochromis niloticus) skin gelatin: Optimization using response surface methodology." *Journal of Food Science* 76(3): C483–C489. doi:10.1111/j.1750-3841.2011.02108.x.

Zou, T.B., T.P. He, H.B. Li, H.W. Tang, and E.Q. Xia. 2016. "The structure-activity relationship of the antioxidant peptides from natural proteins." *Molecules* 21(1): 72. doi:10.3390/molecules21010072.

13 Peptide Vaccine

Joel Lim Whye Ern, Tan Shen Leng, Tee Yi Na,
and Palaniarajan Vijayaraj Kumar

CONTENTS

13.1 INTRODUCTION

For over two centuries, vaccination has been considered as one of the most effective methods to prevent various diseases. A vaccine teaches the body's immune system to perceive and fight pathogens such as viruses or bacteria that have been the root cause of diseases. Vaccines protect and keep the human body safe against more than 25 fatal diseases, including tetanus, diphtheria, mumps, measles, whooping cough, meningitis, polio, and cervical cancer (WHO 2019). Despite conventional vaccine approaches which have been widely known to be effective at reducing mortality and morbidity caused by infectious diseases, this approach was also known to be able to cause allergenic and reactogenic responses because of the incorporation of large proteins or whole organisms that may contain unnecessary antigenic load (Li et al. 2014).

On the other hand, a peptide vaccine is an alternative approach that incorporates short peptide fragments to direct the inauguration of targeted immune responses and overcome allergenic and reactogenic responses. Typically, a conventional vaccine formulation may accommodate between tens to more than hundreds of proteins depending on the type of pathogen being utilized. Based on the high number of proteins being used, the majority of the proteins are not needed or irrelevant for the induction of body's protective immune system and may cause allergenic and reactogenic responses (Li et al. 2014). Therefore, it is important to remove the unnecessary proteins from the vaccine formulations. Based on this knowledge, a justification has been made to focus on subunit vaccines which utilize the selection of a few desirable and relevant proteins of the microbes in vaccine formulations (Thompson and Staats

2011). As shown in Figure 13.1, a peptide vaccine contains only epitopes adequate in activating positive T and B memory cell-mediated immune responses (Li et al. 2014; Apellaniz and Nieva 2015; Skwarczynski and Toth 2016). Peptide vaccines are commonly composed of 20–30 amino acids representing specific epitopes of an antigen associated with diseases (Yang and Kim 2015). The peptides are able to activate the pertinent cellular and humoral responses and remove allergenic and reactogenic responses at the same time because the epitopes are antigenic determinants which reside in bigger proteins. However, due to the peptide vaccines having relatively smaller size, they are unsteadily immunogenic on their own and dependent on carrier molecules for a better chemical stability.

In the history of vaccines, the first successful vaccination in the world was done by Edward Jenner in the West in 1796. Through his observation that a person who had previously caught cowpox did not later catch smallpox, he had successfully developed the immunity against smallpox by the inoculation of cowpox virus into a patient. Later on, the first ever smallpox vaccine was invented in 1798 (The Immunisation Advisory Centre 2020). The initial peptide vaccination investigation was done through the study of virus-derived CD8 T-cell epitopes in the late 1980s. It was reported small synthetic peptides were able to be perceived by the CD8 cytotoxic T lymphocytes (CTL) in mice that were vaccinated. It was also reported that MHC class I molecules were presented with those peptides in vivo and thereupon capable of successfully activating protective T cell responses and resist a related virus threat (Li et al. 2014). The manufactured peptides practiced were usually greater than the 9–11 amino acids of the minimal peptide-sequence perceived by CD8 T cells. The lengthy peptides require shortening to minimal MHC-I binding ligands by proteases

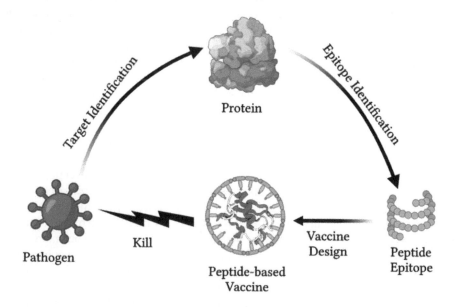

FIGURE 13.1 Mechanism of synthetic peptide-based vaccine.

and peptidases or by a professional antigen presenting cells (APC) process, then packed onto MHC-I groves.

In reality, the minimal peptides utilized in many studies activate lesser immune reactions compared to longer peptides because minimal peptides only obtain CD8 T cell response without refining by APC. In a study conducted by Bijker et al., before the minimal epitope can bind to MHC class I molecule and activate the CD8 CTL responses, long peptides require uptake and processing (Li et al. 2014; Bijker et al. 2007). Thus, the vaccinations with longer peptides are often delayed but provide a more sustainable CTL response. In the human study, the first peptide vaccination was conducted with long peptides developed from self-antigens mucin and HER-2/neu, mutated K-RAS (Hos et al. 2018). Based on the study, the safety of synthetic peptide superintendence and tumor-specific or antigen-specific T cell responses were observed. The purpose of the study was to construct better efficacious vaccines for cancer therapy by using long peptide approaches.

Peptide vaccines have great potential in the field of medicine. It is an alternative strategy to immunization that recognizes immunogens, the peptide epitopes that induce necessary response and to utilize synthetic design of peptides in the synthetization of vaccines. Since most peptide vaccines are developed synthetically, there would be no likelihood of mutation or reversion and lack of possibility of contamination by pathogenic or harmful substances (WHO 1999). Besides, synthetization of peptide vaccines means there would be a chemical manipulation of the peptide sequence which could practically improve its stability and reduce any undesirable side effects such as allergenic and reactogenic responses. This is because the composition of peptide vaccines does not include futile components such as lipopolysaccharides, lipids, and toxins (Li et al. 2014). And through the synthetization of peptides, antigens that may be difficult to synthesize in quantity from a legitimate source such as parasite antigens, would now give the possibility of eliciting immunity to antigens that are not usually recognized. For example, "self" antigens such as tumor-specific antigens in cancers which are not usually recognized, are now able to elicit a response to epitopes that remain vague during natural infection (WHO 1999). Furthermore, the cost of producing peptide vaccines are low because the process of synthetization of peptide vaccines is easy and stability is enhanced. Besides, peptide vaccines are capable of targeting diseases such as virus infections, Alzheimer's disease, and allergies without any limitation (Yang and Kim 2015).

Another advantage of peptide vaccines when compared to conventional vaccines is that peptide vaccines can be modified with self-antigen or non-self-antigen and appropriately balance the immune responses (Yang and Kim 2015). Despite many advantages of peptide vaccines, there are several disadvantages of peptide vaccines that must be acknowledged. For instance, peptide vaccines have a low immunogenicity if used alone and thus, a new generation of adjuvants is required to counter this issue (Li et al. 2014). Besides, it was published that none of the peptide vaccines were approved by the Food and Drug Administration (FDA), albeit more than 500 peptides had advanced to clinical trials (Yang and Kim 2015). There are several important problems that resulted in the failure of peptide vaccines in the clinical trials. Those problems are lack of single peptide epitopes as vaccine applicants, occurrence of immune dodging, failing to produce a controlled and prolonged immune reaction, substandard efficacy, and the improper model of clinical trials.

13.2 PEPTIDE VACCINE DELIVERY METHOD

13.2.1 LIPOSOMES

The application of liposomes had led to rapid advancement in drug delivery technology as an effective carrier for drugs, vaccines, and other bioactive agents because of their biocompatibility, biodegradability, and low toxicity. Liposomes were first discovered in the 1960s by Alec Bangham and his coworkers using the electron microscope by the negative staining technique (Bangham and Horne 1964). Since then, liposomes have gained far more attention due to their versatility. Phospholipids are amphiphilic molecules, consisting of both hydrophobic and hydrophilic components. When in contact with water, they form a self-enclosed lipid bilayered vesicle with an aqueous core. This structure allows the liposome to deliver antigens which would normally be unable to pass through the cell membrane, hydrophobic compound within the lipid bilayers, and hydrophilic compound within the aqueous core shown in Figure 13.2.

Typical vaccines are either attenuated, inactive, or purified antigen vaccines. Although attenuated vaccines produce strong and long-lasting protection, the safety concerns towards immunocompromised patients cannot be eliminated. On the other hand, inactive and purified antigen vaccines are usually safer and more stable, but they do not provide long-lasting immune response and required several doses over time. The utilization of liposomes as vaccine delivery vehicles is able overcome these

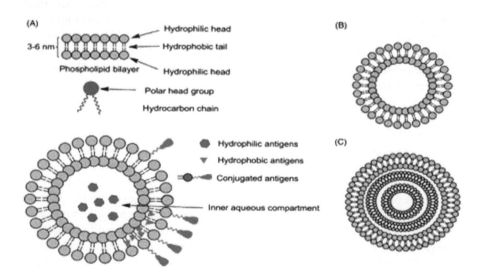

FIGURE 13.2 (A) Structure of liposome; (B) unilamellar liposome; (C) multilamellar liposome (Marasini et al. 2017).

limitations as it is ideal in combining antigen and adjuvant into an effective vaccine. The first utilization of liposome as vaccine delivery vehicles was in 1974 (Allison and Gregoriadis 1974). To date, a number of liposome-based vaccines are commercially approved for human use and several are under clinical trials (Marasini et al. 2017). Stimuvax, formerly known as BLP25 liposome vaccine, has been tested in non-small-cell lung cancer (NSCLC) patients. The formulation consists tecemotide antigen and 3-O-deacyl-4′-monophosphoryl lipid A (MPL) adjuvant anchored in the membrane of the liposome. In the early-stage trial, the results showed the survival time of patients with the vaccine increased compared to the control group (Xia et al. 2014).

The diversity of liposome is found to enhance immunogenicity as the liposome formulation can be tailored according to the antigen properties. The physicochemical properties of a liposome can be changed with different methods of preparation, lipid composition, size, and surface charge (Torchilin 2005). Alteration of the lipid composition is able to change the liposome's surface charge. However, the change in the lipid composition may also affect the membrane fluidity, rigidity, and stability. Therefore, it is difficult to assess the immune response directly on the effect of changing different liposomes' physicochemical properties (Bernasconi et al. 2016). Generally, the surface charge of a liposome can be modified by using charged lipids such as positive charge diacyl-dimethylammonium-propane (DAP), phosphatidylcholine (PC), and stearylamine. Cationic liposomes are the most studied liposome types due to their favorable interaction with the negatively charged cell membrane and their mucoadhesive properties with mucosal membrane. This leads to an increase in cellular uptake of antigen and thus a greater immune response (Christensen et al. 2011; Schmidt et al. 2016). In addition, the fluidity, stability, and permeability of liposome bilayers may influence antigen uptake which in turn affects the immune responses. The fluidity is referring to the viscosity of the lipid bilayers, which depends on the nature and temperature of the lipid. A lipid with main phase transition temperature (T_m) below 37°C is said to be in fluid-disordered phase in the body in which the molecules diffuse freely within the bilayers. A lipid with T_m above 37°C exists as a solid-ordered phase at physiological temperatures due to the individual lipids being closely packed and forming a rigid ordered gel phase (Schmidt et al. 2016). Joseph et al. have shown that liposomes composed of fluid phase lipids at physiological temperatures are more immunogenic through intranasal administration (Even-Or et al. 2011). However, many studies have shown that immunogenicity increases as fluidity decreases (Christensen et al. 2012). Besides, the degree of lipid saturation and incorporation of cholesterol may influence the T_m of the liposome bilayers which in turn affects the fluidity and stability (Bernasconi et al. 2016; Christensen et al. 2012; Perez-Cullell et al. 2000).

13.2.2 EMULSION

The use of purified antigen to evoke an immune response is often insufficient. In order to amplify the immune response, adjuvants such as aluminum hydroxide, oil, and water emulsions are used. In this topic, we will be focusing on the use of emulsion as vaccine adjuvants. Emulsion is a disperse system consisting of two or more immiscible liquids forming a dispersed phase and a continuous phase.

In vaccine formulation, emulsions protect antigens from rapid degradation, increase antigen uptake by antigen-presenting cells, and activate the non-specific immune response locally, thereby improving the immune response (Koh et al. 2006). MF59 is the first oil-in-water emulsion of squalene adjuvant, developed in the 1990s by Novartis, composed of squalene oil and surfactant (Span 85 and Tween 80) (O'Hagan et al. 2013). Squalene is a naturally occurring organic compound which offers the potential of a low level of toxicity. Studies showed that MF-59 adjuvanted influenza vaccines had superior immunogenicity than alum adjuvant for influenza vaccine with a high safety profile (O'Hagan 2007; Black 2015). However, MF59 may induce inflammatory arthritis and reactogenicity (Sanina 2019). Another squalene-based adjuvant is GSK's AS03 used in Pandemrix, a A/H1N1 pandemic flu vaccine.

13.2.3 Virus-Like Particle

There are several types of virus-like particle (VLP)-based vaccines available on the market: GlaxoSmithKline's Engerix® (HBV) and Cervarix® (HPV), and Merck and Co., Inc.'s Recombivax HB® (HBV) and Gardasil® (HPV) (Roldao et al. 2010). The US FDA approved the first vaccination of recombinant HBV in 1986 and followed by approving of the first HPV vaccine in 2006. The main characteristics of VLPs are that the virus protein can be produced either in enveloped form which is enclosed by genome of nucleic acid or envelope of lipid or exist in non-enveloped in different expression systems depending on their complexity (Dai, Wang, and Deng 2018). A rod-like or icosahedral VLP structure or viral core protein will exist at the interface of the virus and therefore known as a VLP (Hsieh 2014).

VLPs are multiprotein structures that have similar conformation to the native viruses but lack the genome of a virus where the more effective vaccination will be produced. Besides, reversion of viruses is avoided as VLPs are unable to replicate when administered to the recipients and meanwhile it could stimulate the immune system through identifying the repetitive subunits and thus produce fast and immediate cellular and humoral responses (Fuenmayor, Gòdia, and Cervera 2017). Compared to individual proteins and peptides, antibody activity or immune reactivity of VLPs is predicted to be higher due to the conformational epitopes present in VLPs looking like the native virus. Shiyu et al. mentioned the optimal size of VLPs which ranges from 20–200 nm and can be easily be identified and form the cross linkage on antigen-presenting cells (Dai, Wang, and Deng 2018). Cross-linking of the VLP's surface with the receptors on the B cells was able to stimulate the stronger B cell reactivity and was supported by Antonio et al (Roldao et al. 2010; Urakami et al. 2017) Mona et al. mentioned a VLP-based vaccine was purposely aimed to induce the antibody reactivity through the linkage of the B cell and followed by initiation of T helper cells and presentation on MHC class II molecules by APCs (Mohsen et al. 2017). Meanwhile, antigen VLP presentation not only occurred on MHC class II and facilitated the priming and presentation of CD8[+] T cell by MHC class I. Thus, this phenomenon led to additional advantages by VLP-based vaccines and it is highly encouraged to utilize it in medical therapy (Mohsen et al. 2017).

For VLP production, protein folding, and post-translational modification systems are the main criteria need to be considered in the expression system of VLP (Dai, Wang, and Deng 2018; Fuenmayor, Gòdia, and Cervera 2017).There are a few types of expression system available to produce VLPs such as bacteria and yeast, baculovirus-insect cell expression systems (BVES), mammalian cells, followed by plant expression system.

First and foremost, bacteria and yeast are used as one of the expression systems for VLPs due to their properties of cost-effective production systems. The VLPs formed through this system could exist in no-envelope form or consist of one or two protein strains. Meanwhile, a post-translational modifications (PTMs) system can be performed only by yeast instead of bacteria, which carries the main VLP immunogenicity and has been used widely in the manufacturing of recombinant proteins (Dai, Wang, and Deng 2018; Fuenmayor, Gòdia, and Cervera 2017).

Moreover, VLPs were also produced by using BVES where the insect cell lines are obtained from *Spodoptera frugiperda* (Sf9 and Sf21) and *Trichoplusia ni (Tn5)*. For instances, Cervarix® which is the HPV available in the market was produced using this system. VLPs produced through this system are able to perform complex PTMs as compared to bacteria and yeast. One of the barriers of this system is the side effects of baculovirus infection on the secretory machinery of the host insect cell and may lead to the low production of secreted and membrane-bound proteins and thus affect the amount of the extracted cell lines.

Furthermore, mammalian cells are involved in VLP production. Complex enveloped VLPs, which are made up of multiple structural proteins, can be produced by mammalian cells due to their ability to induce more complex and specific PTMs. For instance, Chinese hamster ovary (CHO) cell lines, human embryonic kidney 293(HEK293), and baby hamster kidney (BHK-21) are the mammalian cells that had been widely used in this research and development (R&D) field. The benefits of this technology are that the potential of contamination by a human virus was low as compared to other mammalian cells (Dai, Wang, and Deng 2018; Fuenmayor, Gòdia, and Cervera 2017).

Last but not least, plant expression systems are also involved in the production of VLPs. Plant species involved in VLP production are *Nicotiana tabacum, Arabidopsis thaliana*, potato, and carrot. The plant expression system was introduced as it can produce high quantities of complex VLPs at a lower price and has low potential for contamination by human pathogens (Dai, Wang, and Deng 2018; Fuenmayor, Gòdia, and Cervera 2017).

13.2.4 POLYMERS

Polymers bring benefits in various biotechnological as well as biomedical applications, especially in the delivery of drugs and antigens. Shae, Postma, and Wilson et al. mentioned polymeric delivery systems was poised to impact vaccine delivery (Shae, Postma, and Wilson 2016). By having the technology of a polymeric delivery system, multiple vaccine components can be enclosed in either synthetic or natural forms of polymers. For instance, biodegradable poly (lactide-co-glycolide) (PLGA), chitosan, alginate, and hyaluronic acid (HA) were applied in vaccine delivery applications.

Formulation of polymers with antigens can involve either physical entrapment or chemical conjugation (Shae, Postma, and Wilson 2016).

Polymeric carriers were implemented to facilitate the delivery vaccine and bio-distribution via few mechanisms. Polymers were used to stimulate the immune systems and as media for vaccine delivery (Nevagi, Skwarczynski, and Toth 2019). The desired immune response will be induced when polymers with immuno-stimulating characteristics are complexed with the antigen shown in Figure 13.3. As the polymer antigen complex is administered, it will identify by the specific receptors that surround the surface of immune cells and deliver the vaccine to the target site and thus lead to activation of the immune system. For instance, large macromolecules and nanoparticles with a size smaller than 100 nm will move from the interstitial space into the blood circulation of the lymphatic system and can be specifically targeted to the antigen-presenting cell in the lymph nodes (Shae, Postma, and Wilson 2016). This was supported by the study reported by Reddy et al. who mentioned that poly(propylene sulfide) nanoparticles with a size of 25 nm were able to be engulfed by the lymph node (Reddy et al. 2007). Besides, the immune response also will be activated by the depots effects formed by polymeric particles with a large size and injectable form of polymer as it will be engulfed by the dendritic cells and move to the lymph nodes (Shae, Postma, and Wilson 2016). To conclude, polymeric delivery systems can minimize the enzymatic degradation of antigens and facilitate the uptake of polymeric antigen complex by immune cells and thus activate the immune response. Therefore, a synergistic approach for reducing the dosage and number of immunizations can be achieved through formulating the polymers with the antigen (Nevagi, Skwarczynski, and Toth 2019).

13.2.5 DENDRIMER

For the past few years, peptide-based vaccines have gained much popularity as reassuring methods in counteracting deadly infectious diseases. Nonetheless, one

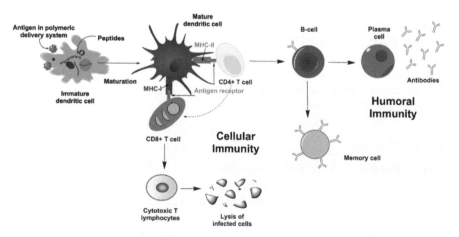

FIGURE 13.3 Vaccine-induced immune response (Nevagi Skwarczynski and Toth 2019).

of the biggest issues is to retain in vivo peptide stability and improvement of peptide immunogenicity to produce a defensive immunity which improves clearance of infections. One of the particulate systems employed to convey vaccine antigens is a dendrimer.

A dendrimer has many well-known properties such as flexibility, derivatizable, distinct branched macromolecules, sectionalized chemical polymers with sizes and physicochemical characteristics featuring biomolecules such as proteins (Boas and Heegaad 2004). The common framework of dendrimers consists of well-defined branching design from the center and out toward the border. The branching of dendrimers is very crucial in determining the "generation" of the dendrimer (Heegard, Boas, and Sorensen 2010). For example, generation 1 of a dendrimer has only a single layer of branching ends. The fundamental of branching indicates an addition of a new branching shell. For instance, an extension in the dendrimer generation will cause the number of functional groups on the surface of the dendrimer to double and be more globular in structure (Heegard, Boas, and Sorensen 2010) as shown in Figure 13.4. The dendrimer composition is incrementally more compartmentalized, with a center region progressively shielded from the surrounding area by the outermost surface shell in the higher-generation dendrimers (Heegard, Boas, and Sorensen 2010).

PAMAM dendrimers are one of the main topics of interest in peptide vaccine delivery. The PAMAM dendrimers have precise sizes, spherical shape with a positive charge at physiological pH, and are biodegradable (Zilio et al. 2017). Due to the

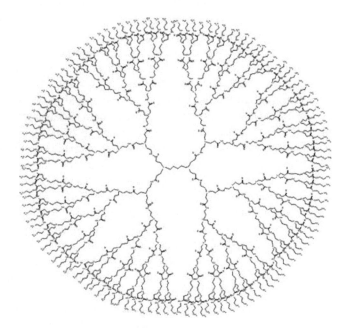

FIGURE 13.4 A fifth generation of PAMAM dendrimer used by my researchers to deliver the peptide vaccine.

aforementioned properties, functional groups such as peptides can be easily bonded to the surface of their structure. Some nucleic acids which are negatively charged can also be simply attached on the dendrimer surface because of electrostatic interaction. This enables them to be protected from extracellular nucleases due to the complex structure. Besides, the dendrimer-nucleic acid complexes have a modestly positive charge which enables adequate cell transfection consequent to internalization in the endosome (Zilio et al. 2017).

Furthermore, dendrimers have a highly multivalent exterior which permits a powerful potential for interactivity with its surroundings which can be utilized for human benefit (Boas and Heegaad, 2004). For example, using dendrimers that have several surface ligand molecules can produce a stronger binding in comparison to monomeric ligands. The preorganization of the ligands at the dendrimer exterior causes an interdependent improvement in binding affinity. This occurrence is referred as "the dendritic effect" or "the chelate effect" (Boas and Heegaad 2004).

In terms of the dendrimer's biocompatibility for the utilization of peptide vaccination, dendrimers have an interior structure that is greatly shielded from the surroundings by the outer shell and the exterior (Boas and Heegaad 2004). Thus, depending on the properties of the surface groups and of the proportion of the dendrimer, the dendrimer biological characteristics can be varied extensively. Predominately, an increasing molecular size of dendrimers gives a lower bio-permeability in any event of the nature of their surface groups (Boas and Heegaad 2004). Through renal excretion in vivo, the clearance rate of a higher-generation dendrimer from the body is typically high because the size and low flexibility which restrict the permeability of the dendrimers into the tissues. Thus, a greater number of dendrimers are deposited in the organs before the excretion happens (Boas and Heegaad 2004).

Generally, small molecular weight peptides are not immunogenic and too weak to produce immune response upon injection into the host body. In order to curb this problem, the easiest method is to increase the molecular weight of the peptides either by polymerization or through the incorporation of a multipurpose, high molecular weight carrier, a naturally derived protein. Because a dendrimer is a multivalent and has well-defined structures for carrying antigenic materials by linking antigen molecules to the surface functional groups of the dendrimer, preparation of reproducible immunogens for peptide vaccine uses is permissible (Boas and Heegaad 2004). For instance, as shown in Figure 13.5, a multiple antigenic peptide (MAP) dendrimer developed by Tam et al. in 1988 can be produced with detailed combination of B cell and T cell epitopes and is being utilized for vaccine and immunization (Tam 1988). The MAP structure is a wedge-like, asymmetrical dendrimer created by assembling successive generations of lysine residues acylating the α- and ε-primary amino groups that can be linked to a small molecular weight antigen of interest, to make the antigen more immunogenic and prevent the necessity of employing a carrier (Boas and Heegaad 2004).

For example, a MAP system for glycoimmunogens that is used in vaccination is the Tn-antigenic dendrimer which was researched by Bay et al. where the tetrameric core structure is derivatized with the Tn-antigen and with a Th-cell stimulatory peptide (Niederhafner, Šebestík, and Ježek 2003; Bay et al. 1997). This derivatization

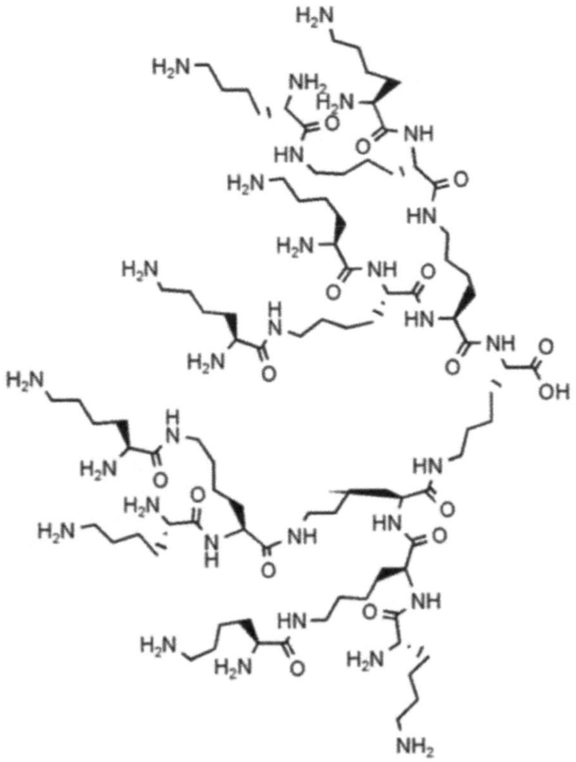

FIGURE 13.5 Multiple Antigen Peptide (MAP) dendron (Boas and Heegaad 2004).

showed a favorable result because it can react with the monoclonal antibodies against Tn. A G5-PAMAM dendrimer has been put into use which acts as a carrier of the Tn-antigen. The resulting glycoconjugates were evaluated as vaccine candidates in comparison with a carrier protein conjugated to a monomer, dimer, or trimer of the Tn-antigen. The result of the test shows that the Tn-antigen dendrimer conjugates induced zero-antibody reactions, and thus no immunogenicity, meanwhile Tn-antigen conjugated with a carrier protein showed an increase to antibody reactions (Boas and Heegaad 2004).

13.3 FUTURE DEVELOPMENT

The safety and efficacy of vaccines against infectious diseases largely depends on the quality of the vaccines. Unlike chemical drugs, vaccines are required to be stored and transported at recommended temperatures from the time of manufacture to the point of administration in order to maintain product quality. Improper vaccine storage and handling will lead to decline in a vaccine's effectiveness and the ineffectiveness of a vaccine only becomes evident when the immunized personnel acquire the disease that the vaccine was designed to prevent. Thus, vaccine formulation plays a critical role in improving vaccine stability and provides an optimal efficacy and safety as there is a need to develop vaccines which are stable in a range of conditions. In a recent study, alumina-encapsulated vaccine formulation showed improved thermostability and immunogenicity in a live-attenuated strain of human enterovirus 71 (Bay et al. 1997). Oil-in-water emulsion MF59 adjuvant required a lower dose of antigen and is more potent for both antibody and T cell responses (O'Hagan 2007). VLP technology is one of the main focuses on developing new vaccines which offer potentially safer and cheaper vaccine candidates. Engineered VLPs are able to improve the stability of a vaccine and allow a longer shelf-life, enhance thermostability, and reduce cost in maintaining cold-chain during storage and transportation (Frietze, Peabody, and Chackerian 2016).

REFERENCES

Allison, A.C., and G. Gregoriadis. 1974. "Liposome as immunological adjuvants." *Nature* 252: 252.

Apellániz, B. and J.L. Nieva. 2015. "The use of liposomes to shape epitope structure and modulate immunogenic responses of peptide vaccines against HIV MPER." *Advances in Protein Chemistry and Structural Biology* 99: 15–54. doi:10.1016/bs.apcsb.2015.03.002

Bangham, A.D. and R.W. Horne. 1964. "Negative staining of phospholipids and their structural modification by surface-active agents as observed in the electron microscope." *Journal of Molecular Biology* 8(5): 660–668.

Bay, S., R. Lo-Man, E. Osinaga, H. Nakada, C. Leclerc, and D. Cantacuzene. 1997. "Preparation of a multiple antigen glycopeptide (MAG) carrying the Tn antigen. A possible approach to a synthetic carbohydrate vaccine." *The Journal of Peptide Research* 49(6): 620–625.

Bernasconi, V., K. Norling, M. Bally, F. Höök, and N.Y. Lycke. 2016. "Mucosal vaccine development based on liposome technology." *The Journal of Immunology Research* 5482087. doi:10.1155/2016/5482087.

Bijker, M.S., C.J. Melief, R. Offringa, and S.H. Van Der Burg. 2007. "Design and development of synthetic peptide vaccines: Past, present and future." *Expert Review of Vaccines* 6(4): 591–603.

Black, S. 2015. "Safety and effectiveness of MF-59 adjuvanted influenza vaccines in children and adults." *Vaccine* 33(Suppl. 2): B3–B5.

Boas, U. and P.M. Heegaard. 2004. "Dendrimers in drug research." *Chemical Society Reviews* 33(1): 43–63.

Christensen, D., K.S. Korsholm, P. Andersen, and E.M. Agger. 2011. "Cationic liposomes as vaccine adjuvants." *Expert Review Vaccines* 10(4): 512–521.

Christensen, D., M. Henriksen-Lacey, A.T. Kamath, T. Lindenstrøm, K.S. Korsholm, J.P. Christensen, A.F. Rochat, et al. 2012. "A cationic vaccine adjuvant based on a saturated quaternary ammonium lipid have different in vivo distribution kinetics and display a distinct CD4 T cell-inducing capacity compared to its unsaturated analog." *Journal of Controlled Release: Official Journal of the Controlled Release Society* 160(3): 468–476. doi:10.1016/j.jconrel.2012.03.016.

Dai, S., H. Wang and F. Deng. 2018. "Advances and challenges in enveloped virus-like particle (VLP)-based vaccines." *The Journal of Immunology Science* 2: 36–41.

Even-Or, O., A. Joseph, N. Itskovitz-Cooper, S. Samira, E. Rochlin, H. Eliyahu, D. Simberg, I. Goldwaser, Y. Barenholz, and E. Kedar. 2011. "A new intranasal influenza vaccine based on a novel polycationic lipid-ceramide carbamoyl-spermine (CCS). II. Studies in mice and ferrets and mechanism of adjuvanticity. *Vaccine* 29(13): 2474–2486. doi:10.1016/j.vaccine.2005.12.017.

Frietze, K.M., D.S. Peabody, and B. Chackerian. 2016. "Engineering virus-like particles as vaccine platforms." *Current Opinion in Virology* 18: 44–49.

Fuenmayor, J., F. Gòdia, and L. Cervera. 2017. "Production of virus-like particles for vaccines." *New Biotechnology* 39: 174–180.

Heegaard, P.M., U. Boas, and N.S. Sorensen. 2010. "Dendrimers for vaccine and immunostimulatory uses. A review." *Bioconjugate Chemistry* 21(3): 405–418.

Hos, B.J., E. Tondini, S.I. van Kasteren, and F. Ossendorp. 2018. "Approaches to improve chemically defined synthetic peptide vaccines." *Frontiers in Immunology* 9: 884.

Hsieh, F.H. 2014. "Primer to the immune response." *Annals of Allergy, Asthma & Immunolog.*113(3): 333.

Koh, Y.T., S.A. Higgins, J.S. Weber, and W.M. Kast. 2006. "Immunological consequences of using three different clinical/laboratory techniques of emulsifying peptide-based vaccines in incomplete Freund's adjuvant." *Journal of Translational Medicine* 4(42): 1–12.

Li, W., M. Joshi, S. Singhania, K. Ramsey, and A. Murthy. 2014. "Peptide vaccine: Progress and challenges." *Vaccines* 2(3): 515–536.

Marasini, N., K.A. Ghaffar, M. Skwarczynski, and I. Toth. 2017. "Liposomes as a vaccine delivery system." *Micro and Nanotechnology in Vaccine Development* 221–239.

Mohsen, M.O., L. Zha, G. Cabral-Miranda, M.F. Bachmann (Eds) 2017. "Major findings and recent advances in virus–like particle (VLP)-based vaccines." *Seminars in Immunology*, Elsevier.

Nevagi, R.J., M. Skwarczynski, and I. Toth. 2019. "Polymers for subunit vaccine delivery." *European Polymer Journal* 114: 397–410.

Niederhafner, P., J. Šebestík, and J. Ježek. 2003. "Peptide dendrimers." *Journal of Peptide Science: An Official Publication of the European Peptide Society* 11(12): 757–788.

O'Hagan, D.T., G.S. Ott, G.V. Nest, R. Rappuoli, and G.D. Giudice. 2013. "The history of MF59® adjuvant: A phoenix that arose from the ashes." *Expert Review of Vaccines* 12(1): 13–30.

O'Hagan, D.T. 2007. "MF59 is a safe and potent vaccine adjuvant that enhances protection against influenza virus infection." *Expert Review of Vaccines* 6(5): 699–710.

Perez-Cullell, N., L. Coderch, A. Maza, J.L. Parra, and J. Estelrich. 2000. "Influence of the fluidity of liposome compositions on percutaneous absorption." *Drug Delivery* 7(1): 7–13.

Reddy, S.T., A.J. Van Der Vlies, E. Simeoni, V. Angeli, G.J. Randolph, C.P. O'Neil, L.K. Lee, M.A. Swarts, and J.A. Hubbell. 2007. "Exploiting lymphatic transport and complement activation in nanoparticle vaccines." *Nature Biotechnology* 25(10): 1159. doi:10.1038/nbt1332.

Roldao, A., M.C.M. Mellado, L.R. Castilho, M.J. Carrondo, and P.M. Alves. 2010. "Virus-like particles in vaccine development." *Expert Review of Vaccines* 9(10): 1149–1176.

Sanina, N. 2019. "Vaccine adjuvants derived from marine organisms." *Biomolecules* 9(8): 1–16.

Schmidt, S.T., C. Foged, K.S. Korsholm, T. Rades, and D. Christensen. 2016. "Liposome-based adjuvants for subunit vaccines: Formulation strategies for subunit antigens and immunostimulators." *Pharmaceutics* 8(7): 1–22.

Shae, D., A. Postma, and J.T. Wilson. 2016. "Vaccine delivery: Where polymer chemistry meets immunology." *Future Science* 7: 193–196.

Skwarczynski, M., and I. Toth. 2016. "Peptide-based synthetic vaccines." *Chemical Science* 7(2): 842–854. doi:10.1039/C5SC03892H.

Tam, J.P. 1988. "Synthetic peptide vaccine design: Synthesis and properties of a high-density multiple antigenic peptide system." *Proceedings of the National Academy of Sciences* 85(15): 5409–5413.

The Immunisation Advisory Centre. 2020. *A brief history of vaccination.* https://www.imm une.org.nz/vaccines/vaccine-development/brief-history-vaccination.

Thompson, A.L. and H.F. Staats. 2011. "Cytokines: The future of intranasal vaccine adjuvants." *Clinical and Developmental Immunology.* 2011: 1–17. doi:10.1155/2011/289597.

Torchilin, V.P. 2005. "Recent advances with liposomes as pharmaceutical carriers." *Nature Reviews Drug Discovery* 4(2): 145–160.

Urakami, A., A. Sakurai, M. Ishikawa, M.L. Yap, Y. Flores-Garcia, Y. Haseda, T. Aoshi, et al. 2017. "Development of a novel virus-like particle vaccine platform that mimics the immature form of alphavirus." *Clinical Vaccine Immunology* 24(7): e00090-17. doi:10.1128/CVI.00090-17.

World Health Organization. 1999. *Guidelines for the production and quality control of synthetic peptide vaccines. Vaccines, synthetic peptide* 889: 24–43.

World Health Organizations. 2019. *Vaccines.* https://www.who.int/topics/vaccines/en/.

Xia, W., J. Wang, Y. Xu, F. Jiang, and L. Xu. 2014. "L-BLP25 as a peptide vaccine therapy in non-small cell lung cancer: A review." *Journal of Thoracic Disease* 6(10): 1513–1520. doi:10.3978/j.iss.

Yang, H. and D.S. Kim. 2015. "Peptide immunotherapy in vaccine development: From epitope to adjuvant." *Advances in Protein Chemistry and Structural Biology* Elsevier 99: 1–14.

Zilio, S., J.L. Vella, C. Adriana, P.M. Daftarian, D.T. Weed, A. Kaifer, I. Marigo, K. Leone, V. Bronte, and P. Serafini. 2017. "4PD functionalized dendrimers: A flexible tool for in vivo gene silencing of tumor-educated myeloid cells." *The Journal of Immunology* 198(10): 4166–4177. doi:10.4049/jimmunol.1600833.

14 Vaccine Adjuvants in Immunotoxicology

Fadime Canbolat

CONTENTS

14.1 IMMUNOTOXICOLOGY

Immunotoxicology is the discipline that investigates the adverse effects of exposure to xenobiotics in an organism on the immune system (Germolec et al. 2017). Studies on animals and humans showed that the immune system was a potential target for a variety of chemicals (Luster 2013). It was suggested with the definition of regulator agents that the immune system was an important target for chemicals and drugs (Germolec et al. 2017; Luster 2013; Verdier 2002). Therefore, drug-induced toxicity and immunotoxicology is a growing sub-discipline of toxicity.

14.2   THE IMMUNE SYSTEM

The human body has the ability to recognize and deal with the substances which are foreign to its organism. It thereby defends itself against foreign substances entering the body in different ways (Elgert 2009c). Many systems, organs, and cells play a role in the immune response, which generally starts with the introduction of these foreign substances, which are defined as antigens, into the organism and continues with many interrelated biological reactions (Klostermen 2009). The organs and cells which take part in the events of immunity are called the immune system. The immune system is the whole process that protects a living being against diseases, recognizes pathogens and tumor cells, and eliminates them (Sompayrac 2019). The system scans every foreign substance that enters the body or comes into contact with it, and distinguishes them from the healthy cells and tissues of the living being (Songu and Katılmış 2012).

The immune response, which is a cellular event, includes the immune cells. The immune cells develop by differentiating from the stem cell in the bone marrow (Klostermen 2009). The stem cells of the bone marrow need to mature in the primary lymphoid organs before they become immunologically active cells (Defranco, Locksley, and Robertson 2007). Mature T and B lymphocytes are then transferred into the secondary lymphoid organs, wait for antigens, and give the immune response when they meet antigens (Sompayrac 2019).

14.2.1   The Organs and Immune Cells Included in the Immune System

The human body protects the organism against bacteria, viruses, and other foreign substances through a variety of mechanisms (Defranco, Locksley, and Robertson 2007). This protection includes physical barriers, phagocytic cells in the blood and tissues, and various blood-related molecules (Klostermen 2009; Defranco, Locksley, and Robertson 2007; Elgert 2009c; Songu and Katılmış 2012). Figure 14.1 shows the main organs taking part in the immune system.

The immune cells are produced from the stem cells in the bone marrow (Defranco, Locksley, and Robertson 2007). The immune cells, produced by differentiating from the stem cells in the bone marrow, are stored in lymphoid tissues where they mature as significant lymphocyte types (B and T lymphocytes) (Sompayrac 2019). The activation of lymphocytes depends on the processing of an antigen properly by antigen-presenting cells and the presentation of it to T lymphocytes. After the recognition of the antigen, T and B lymphocytes continue to differentiate as effector cells which eliminate antigens, and memory cells which are needed for future rechallenges from the antigen (Songu and Katılmış 2012).

14.2.1.1   B lymphocytes

B lymphocytes are responsible for the antibody-related humoral immunity (Lebien and Tedder 2008). They mature in the bone marrow in mammals (Elgert 2009c).

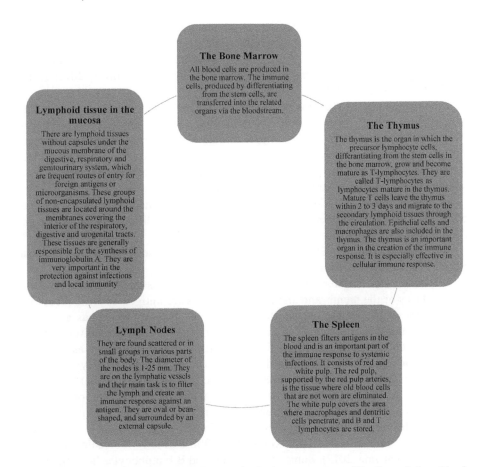

The Bone Marrow

All blood cells are produced in the bone marrow. The immune cells, produced by differentiating from the stem cells, are transferred into the related organs via the bloodstream.

Lymphoid tissue in the mucosa

There are lymphoid tissues without capsules under the mucous membrane of the digestive, respiratory and genitourinary system, which are frequent routes of entry for foreign antigens or microorganisms. These groups of non-encapsulated lymphoid tissues are located around the membranes covering the interior of the respiratory, digestive and urogenital tracts. These tissues are generally responsible for the synthesis of immunoglobulin A. They are very important in the protection against infections and local immunity

The Thymus

The thymus is the organ in which the precursor lymphocyte cells, differantiating from the stem cells in the bone marrow, grow and become mature as T-lymphocytes. They are called T-lymphocytes as lymphocytes mature in the thymus. Mature T cells leave the thymus within 2 to 3 days and migrate to the secondary lymphoid tissues through the circulation. Epithelial cells and macrophages are also included in the thymus. The thymus is an important organ in the creation of the immune response. It is especially effective in cellular immune response.

Lymph Nodes

They are found scattered or in small groups in various parts of the body. The diameter of the nodes is 1-25 mm. They are on the lymphatic vessels and their main task is to filter the lymph and create an immune response against an antigen. They are oval or bean-shaped, and surrounded by an external capsule.

The Spleen

The spleen filters antigens in the blood and is an important part of the immune response to systemic infections. It consists of red and white pulp. The red pulp, supported by the red pulp arteries, is the tissue where old blood cells that are not worn are eliminated. The white pulp covers the area where macrophages and dentritic cells penetrate, and B and T lymphocytes are stored.

FIGURE 14.1 The main organs taking part in the immune system. The figure is based basically on the following references: Klostermen 2009; Sompayrac 2019; Songu and Katılmış 2012; Defranco, Locksley, and Robertson 2007; Elgert 2009a.

Antibodies constitute a class of protein called immunoglobulin. The structure of immunoglobulins consists of antigen-binding tips and another tip which defines group characteristics (Lebien and Tedder 2008). Immunoglobulins are divided into five main groups, all of which work differently, as immunoglobulin G, immunoglobulin A, immunoglobulin M, immunoglobulin D, and immunoglobulin E (Songu and Katılmış 2012; Justiz Vaillant and Ramphul 2020).

Immunoglobulin G (IgG) is the most common group of immunoglobulins in the circulation (Engels and Wienands 2018). It is included in body fluids and can easily enter into tissues. IgG is the only immunoglobulin transferred through the placenta and it may transfer the immunity from mother to fetus (Songu and Katılmış 2012). This group of immunoglobulins provides protection against bacteria, toxins, and viruses in the body fluids (Sompayrac 2019). IgG is divided into four main subgroups (IgG1, IgG2, IgG3, and IgG4) (Justiz Vaillant and Ramphul 2020).

Immunoglobulin A (IgA) is secretory immunoglobulin available in saliva, tear-drops, colostrum, and bronchial, gastrointestinal, prostatic, and vaginal secretions. IgA prevents the involvement of bacteria in epithelial cells and works as primary defense against local infections in the mucosa (Songu and Katılmış 2012; Justiz Vaillant and Ramphul 2020).

Immunoglobulin M (IgM) is the first immunoglobulin appearing in the circulation as a response to an antigen. It is also the first antibody produced in newborns (Songu and Katılmış 2012). It cannot pass through the placenta during the intrauterine period and cannot transfer maternal immunity to the fetus. This situation has a diagnostic benefit as the presence of IgM suggests an infection ongoing due to a specific pathogen (Justiz Vaillant and Ramphul 2020).

Immunoglobulin D (IgD) is usually found on the cell membrane of B lymphocytes. It works as an antigen receptor on the onset of the differentiation of B cells (Songu and Katılmış 2012).

Immunoglobulin E (IgE) is involved in allergic responses and the fight against parasitic infections (Maurer et al. 2018).

B lymphocytes transfer the immunoglobulin molecules they synthesize on the cell surfaces of the membrane and these molecules are antigen-specific receptors (Sompayrac 2019; Songu and Katılmış 2012). One B lymphocyte has a surface immunoglobulin receptor that can bind to only one type of antigen epitope. When an antigen enters the organism, it finds and stimulates B lymphocytes carrying the receptor, specific to the antigen, on its surface. Stimulated B lymphocytes transform and turn into plasma cells (Defranco, Locksley, and Robertson 2007; Sompayrac 2019; Songu and Katılmış 2012). A plasma cell, on the other hand, synthesizes a large number of antibodies that are specific to the stimulating antigen. The plasma cell cannot proliferate, and its life cycle is short (~ 2–3 days). However, it can synthesize approximately 20,000 antibody molecules a minute (Songu and Katılmış 2012). Some of the stimulated B lymphocytes become memory cells. Memory B lymphocytes are long-lasting (sometimes for a lifetime), and when they meet the same antigen again, they rapidly proliferate and give a faster and stronger antibody response (Justiz Vaillant and Ramphul 2020; Songu and Katılmış 2012).

14.2.1.2 T lymphocytes

T lymphocytes are responsible for the immune response of cellular type (Hanna et al. 2004) precursor T cells produced in the bone marrow and mature as T lymphocytes in the thymus (Kohler and Thiel 2009). During the maturation period, many specific surface molecules and receptors fit on the surface of T lymphocyte. There is no surface immunoglobulin on the T cell surface. Instead, there is a *T cell receptor (TCR)* that specifically recognizes antigens. One T lymphocyte carries a TCR for only one type of antigen (Ozato, Tsujimura, and Tamura 2002). Moreover, *cluster of differentiation antigens (CDs)*, which are specific surface molecules on the T cell surface, are formed and the ability to respond to antigenic stimulation is obtained. One of the CDs, CD4+, shapes helper T cells and CD8+

shapes the cytotoxic T cells (Beyaz 2004). When an antigen enters the organism, it finds and stimulates T lymphocytes carrying the receptor, which is specific to the antigen, on the cell surface. Stimulated T lymphocytes transform and as a result, antigen-sensitive T lymphocytes are formed. T lymphocytes are the most important cells of the immune system and form a specific immunity which is not directly dependent on the antibody and is directed and joined by the cells (Mosmann and Coffman 1989).

T cells are mainly divided into four subgroups:

- *Cytotoxic T cells (CT):* There are CD8$^+$ cells on the surface of these cells (Seder, Darrah, and Roederer 2008). They fight cells which are capable of directly eliminating foreign cells entering the organism (Kaech, Wherry, and Ahmed 2002).
- *Helper T cells (TH):* There are CD4$^+$ cells on the surface of these cells (Seder, Darrah, and Roederer 2008). Helper T cells constitute most of the T cells. These cells activate other T cells and B cells. When the stimulation begins, these cells synthesize cytokines. The main cytokines released from helper T lymphocytes are interleukin-2 (IL-2), interleukin-3 (IL-3), interleukin-4 (IL-4), interleukin-5 (IL-5), interleukin-6 (IL-6), tumor necrosis factor (TNF-β), and interferon gamma (IFN-γ) (Kaech, Wherry, and Ahmed 2002; Seder, Darrah, and Roederer 2008; Kohler and Thiel 2009). Helper T lymphocytes have two subsets of TH1 and TH2 (Parker 1993; Mosmann and Coffman 1989). While TH1 is effective in the humoral immune response, TH2 is effective in the cellular immune response (Seder, Darrah, and Roederer 2008; Beyaz 2004; Bird et al. 1998). TH1 lymphocytes produce IFN-γ and IL-2. TH2 lymphocytes produce IL-4, IL-5, and IL-10 (Yurdakök and İnce 2008; Bird et al. 1998)
- *Suppressor T cells (TS):* These cells have CD8$^+$, similar to CT. They inhibit the activation of T and B cells (Beyaz 2004; Maughan, Preston, and Williams 2015).
- *Memory T Cells:* These cells have CD4$^+$, similar to TH (Hanna et al. 2004; Kohler and Thiel 2009; Kaech, Wherry, and Ahmed 2002). Storing the first antigenic stimulus, they detect the antigen and provide an immune response when they are stimulated with the same antigen for the second time (Beyaz 2004).
- Monocytes, Macrophages, and Dendritic Cells

Monocytes and tissue macrophages belong to the mononuclear phagocytic system, which is a part of the reticuloendothelial system. All cells of the mononuclear phagocytic system consist of common precursors in the bone marrow which produces blood monocytes (Elgert 2009b). Monocytes are transferred to various tissues and mature as macrophages in those tissues. Tissue macrophages are found in connective tissues in scattered form or in organs and other areas in the form of clusters. Macrophages have an important role in both non-specific and antigen-specific immune responses (Defranco, Locksley, and Robertson 2007). As phagocytic cells, they contribute to immunity by keeping infected agents in their body until a specific immunity develops (Hanna et al. 2004). Moreover, it functions in the early stage of the host response, during the enhancement of the inflammatory response, and the onset of a specific immunity. Dendritic cells share an important role with macrophages such as presenting antigens to T lymphocytes (Hanna et al. 2004; Vallhov et al. 2012; Maughan, Preston, and Williams 2015). Dendritic cells are found in lymphoid tissues and the areas where the body comes into contact with an antigen (Defranco, Locksley, and Robertson 2007).

14.2.1.3 Natural Killer Cells (NKs)

Among the lymphoid cells, they are the cells which eliminate infected or foreign cells, with a different structure than T and B lymphocytes (Hanna et al. 2004). Natural killer cells (NKs) are modulators of innate and adaptive immune responses (Tang et al. 2020).

NKs have the ability to directly destroy the cells they target before they recognize and become sensitive to them (Elgert 2009b).

14.2.1.4 Cytokines

Cytokines are regulatory proteins with low molecular weight which are produced during all stages of immune responses. Cytokines adjust the reactions of a host to foreign antigens or harmful agents by regulating the movement, proliferation, and differentiation of leukocytes and other cells (Beyaz 2004). Cytokines are synthesized by many types of cells, especially by activated TH and macrophages (Ozato, Tsujimura, and Tamura 2002). The production of cytokines usually occurs within a chain reaction where a cytokine affects the production of other consecutive cytokines or receptors. Excessive production of cytokines may have serious side effects. Cytokines respond by binding to specific receptors in target cells (Kishimoto, Taga, and Akira 1994). Table 14.1 shows primary cytokines in the body. A group of cytokines mediate inflammation by attracting and activating phagocytes and lead into fever and acute-phase response, while others function as maturation factors for the hematopoiesis of white and red cell (Akdoğan and Yöntem 2018). The main cytokines in our body are interleukin-1 (IL-1), IL-2, IL-3, IL-4, interleukin 5 (IL-5), IL-6, interleukin 12 (IL-12), interleukin 15 (IL-15), IFNs, TNF, and colony-stimulating factors (Beyaz 2004; Krakauer and Russo 2001).

14.2.1.5 The Immune Response Mechanism

The immune response includes a series of interactions between the components of the immune system and antigens. It consists of active and passive immunity and contains the humoral and cellular immune mechanisms (Songu and Katılmış 2012).

- Active and Passive Immunity

TABLE 14.1

Primary Cytokines in the Body

Cytokine	Properties	Reference
Interleukin 1 (IL-1)	IL-1 primarily functions as a mediator for the inflammatory response. IL-1 may stimulate the creation of an acute phase response, mobilize neutrophils, induce fever, and activate the vascular epithelium. IL-1 also works as a precursor signal in the activation of CD4+ TH cells and the development and differentiation of B cells.	(Lund 2008; Songu and Katılmış 2012)
Interleukin 2 (IL-2)	IL-2 interacts with T lymphocytes by binding to specific membrane receptors in activated T cells. This cytokine allows the immune response to peak in the presence of antigens.	(Beyaz 2004; Krakauer and Russo 2001; Songu and Katılmış 2012; Kaech, Wherry, and Ahmed 2002; Kishimoto, Taga, and Akira 1994)
Interleukin 3 (IL-3)	IL-3 is produced by activated TH. It enhances the development of many cells.	(Beyaz 2004; Krakauer and Russo 2001; Songu and Katılmış 2012)
Interleukin 4 (IL-4)	IL-4 is produced by activated TH. Together with IL-3 it provides a synergistic effect on the development of mast cells. It is involved in the development of B lymphocytes and other T lymphocytes.	(Beyaz 2004; Krakauer and Russo 2001; Kaech, Wherry, and Ahmed 2002; Songu and Katılmış 2012; Kishimoto, Taga, and Akira 1994)
Interleukin 5 (IL-5)	IL-5 is produced by activated TH. It plays a role in the production of IgA and the development of B lymphocytes.	(Beyaz 2004; Krakauer and Russo 2001; Kaech, Wherry, and Ahmed 2002)
Interleukin 6 (IL-6)	IL-6 is produced by activated TH. It is involved in the development of B lymphocytes together with IL-4 and IL-5.	(Beyaz 2004; Krakauer and Russo 2001; Kaech, Wherry, and Ahmed 2002)
Interleukin-12 (IL-12),	IL-12 is an important regulatory cytokine with a central function in initiating and regulating cellular immune responses. This cytokine belongs to a large group of cytokines that fold into a four-alpha-helix bundle. Macrophages, monocytes, dendritic cells, and B lymphocytes produce IL-12 in response to bacteria and parasites. Afterwards, IL-12 stimulates the production of IFN-gamma and TNF-alpha from both NK and TH	(Beyaz 2004; Krakauer and Russo 2001; Seder, Darrah, and Roederer 2008)

(Continued)

TABLE 14.1 (CONTINUED)
Primary Cytokines in the Body

Cytokine	Properties	Reference
Interleukin-15 (IL-15)	IL-15 is similar to IL-2 in terms of effect and uses the same receptors. It functions through stimulating the proliferation of T cells, the protection against apoptosis, enhanced memory of cytotoxic T lymphocytes, the proliferation in the intestinal epithelial cells, the chemotaxis of T lymphocytes, B cell proliferation, and conversion into IgA-producing cells in the mucosa. It was suggested that stronger cellular and humoral responses were obtained following a systemic or mucosal administration.	(Beyaz 2004; Krakauer and Russo 2001; Seder, Darrah, and Roederer 2008)
Interferons (IFNs)	IFNs are group cytokines which protect cells from the invasion of intracellular parasites. Substances which induce IFN production are not all antigenic. There are three types of IFN: interferon alpha (IFN-α) produced by leukocytes; interferon beta (INF-β) produced by fibroblasts, and interferon gamma (IFN-γ) produced by NK and T cells. The IFN produced during the immune reaction is primarily IFN-γ. IFN-γ functions to activate macrophages, produce cytotoxic lymphocytes, and increase NK activity.	(Beyaz 2004; Krakauer and Russo 2001; Kaech, Wherry, and Ahmed 2002; Seder, Darrah, and Roederer 2008)
Tumor necrosis factor (TNF)	TNF is produced by activated macrophages and other activated cells, such as T cells. Besides acting as an important mediator in the inflammatory response and indirectly affecting the fever response, TNF may function as a co-stimulator for T cells. This cytokine is a strong stimulator especially for IL-1, IL-6, and IL-8.	(Watts 2005; Seder, Darrah, and Roederer 2008; Beyaz 2004; Krakauer and Russo 2001)
Colony-stimulating factors	Colony-stimulating factors are the cytokines which stimulate cells in the bone marrow to produce a large number of thrombocytes, erythrocytes, neutrophils, monocytes, eosinophils and basophils.	(Beyaz 2004; Krakauer and Russo 2001)

Active immunity occurs when encountering a specific antigen or with the transfer of the antibodies protecting against antigens. It is acquired through immunization or getting through the disease itself (Baxter 2007). Moreover, the immune system usually responds within hours thanks to the presence of memory B and T cells when encountering the agent again later. During the active immunity process, the main mechanism providing the relation between innate immunity and adaptive immunity which is acquired weeks later is the ability to perceive the *pathogen-associated molecular patterns (PAMPs)* which innate immunity considers foreign substances (Peterfalvi et al. 2019; Maughan, Preston, and Williams 2015). This recognition occurs via the *class-specific pattern recognition receptor (PRR)* which is available in the innate immune system. The most important PRR in innate immunity is a *toll-like receptor (TLR)*. The innate immune system considers PAMPs in a microorganism as a signal of danger and stimulates the adaptive immune system (Ivanov et al. 2020). *Antigen-presenting cells (APCs),* which are present in the circulation system and are the supervisory components of the immune system, are the group of cells which encounters a pathogen first and define it as *"foreign."*

APCs migrate to the nearest lymph node along with the antigen they identify as foreign and present the antigen to T lymphocytes (Seder, Darrah, and Roederer 2008). The antigen taken into the APC through endocytosis is presented to T cells in a cavity in the molecule *major histocompatibility complex (MHC)* in the membrane after it undergoes a number of processes in the cell. Figure 14.2 illustrates the activation of T cells (Goran et al. 2020). The antigen in the MHC molecule binds to a specific TCR. This binding is necessary for the activation of T lymphocytes; however, it is not sufficient. A co-stimulatory molecule and the receptor found in T lymphocytes should also bind for the activation. Cytokines are secreted following the activation. T lymphocytes mature through these cytokines and develop into effector T lymphocytes. Effector T lymphocytes differ as TH1 and TH2 according to the cytokines they secrete (Vallhov et al. 2012). TH1 produce IFN-γ and IL-2. TH2 produce IL-4, IL-5, and IL-10 (Yurdakök and İnce 2008; Vallhov et al. 2012).

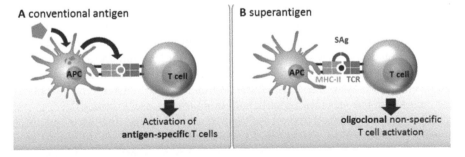

FIGURE 14.2   The Activation of T cell. In the study by Goran et al (2020) it has been illustrated that antigens are processed into antigen presented cell (APC) and presented on Major Histocompatability Complex (MHC-II) molecules to CD4+ T cells (Goran et al. 2020).

Passive immunity is the immunity transmitted from one source to another (e.g. mother to baby). Passive immunity provides short-term protection expressed in weeks or months (Baxter 2007). The humoral or antibody-mediated immunity depends on the presence of antibodies in the blood and body fluids. The combination of an antigen with an antibody creates many responses. During the development of humoral immunity, the primary immune response occurs when the antigen enters the body for the first time (Zeitlin et al. 2000). Secondary or memory response is created in the second or subsequent encounters with the antigen. During the secondary response, the increase in the level of antibodies occurs in a shorter time and reaches a higher level due to the pre-existing memory cells (Beyaz 2004). The effect of T lymphocytes and macrophages is dominant in cellular immunity (Hanna et al. 2004). There are four main stages in the creation of the immune response: firstly the recognition of an antigen and the activation of the antigen-presenting cells, secondly the presentation of the antigen to T lymphocytes produced by the antigen-presenting cells, thirdly the formation of the co-stimulatory molecules, and the secretion of cytokines is the last step (Songu and Katılmış 2012).

14.3   VACCINES AND ADJUVANTS

Vaccines are the biological substances which help living beings acquire immunity to protect themselves from diseases (Aytar and Başbülbül 2019). Vaccines consist of the agents which cause disease or are a mixture of antigen substances, which are produced synthetically, to form a defense mechanism and produce antibodies against possible diseases in the immune system (Aytar and Başbülbül 2019).

During vaccination, the most common side effects depending on the interaction with the immune system are hypersensitivity and autoimmune responses (Brennan and Dougan 2005; Verdier 2002).

The aim of vaccination is to create a strong, protective, and long-lasting immune response to infections. Live vaccines prepared with microorganisms that are weakened in a way not to cause a disease, but may proliferate and enhance in a recipient after a while (Pulendran and Ahmed 2011; Francis 2018). A strong immunity can be achieved with a single dose of live vaccine. However, inactivated vaccines produced from all dead microorganisms or fractions do not show antigenic proliferation (Pulendran and Ahmed 2011). Therefore, a vaccine formulation includes substances (adjuvants) which increase the immunogenicity of an antigen but are not antibodies or immunogenic. Inactivated vaccines, which are less immunogenic, should be repeated at several intervals and given with adjuvants to increase the potency (Singh and Hagan 1999; Maughan, Preston, and Williams 2015). Therefore, adjuvants are not included in live vaccines, as sufficient immune response is created.

Adjuvants are designed to increase the potency of vaccines that have low immunogenic properties alone (Maughan, Preston, and Williams 2015). Adjuvants alone are not immunogenic substances. In a vaccine, they are substances which create a stronger immune response and enhance their effects (potentiation) than administering the vaccine alone (Oleszycka and Lavelle 2014). Vaccine adjuvants are widely used to increase the immunogenic effect, especially in the development of weak immunogenic protein, peptide, and DNA vaccines. Adjuvants generally show their effect by increasing the stimulation and efficacy of the cells in the immune system, maintaining the balance of TH (TH1/TH2), and thereby reducing the amount of the antigens needed for the expected immunity to develop. Adjuvants may work by affecting the signals of onset in the immune response, in other words, by affecting the uptake, processing, and presentation of an antigen by antigen-presenting cells (Batista-Duharte, Lindblad, and Oviedo-Orta 2011; Fratalay and Öner 2001). For example, in vaccines containing a very well purified protein or subunit that cannot be recognized by the innate immune system, the use of PAMPs as an adjuvant stimulates the pathogen recognition mechanism, i.e. the innate immunity, to acquire the adaptive immunity (Yurdakök and İnce 2008).

General use of adjuvants are as follows:

- Increasing the immunogenicity of well purified or recombinant antigens
- Creating a stronger and long-lasting immune response
- Lowering the number of antigens or vaccination required for the primary immune response, and thereby cutting the cost of a vaccine
- Increasing the efficacy of a vaccine in those poorly responding (newborns, the elderly, those with immunodeficiency)
- Enhancing the uptake of antigens by the mucosa and increasing the efficacy of mucosal vaccines (stimulating the mucosal immunity)
- Stimulating the cellular immunity
- Helping prevent antigen competition in combined vaccines (Brennan and Dougan 2005)

An ideal adjuvant should provide the cellular or humoral response, develop immune memory, that is long-term immunity, be safe and have few side effects, have an autoimmunity-stimulating effect, be affordable and have long shelf-life, and should not be mutagenic, carcinogenic, and teratogenic.

14.3.1 Types of Adjuvants

There are several vaccine adjuvants which are licensed and in preclinical testing phase. Aluminum salts (aluminum hydroxide $Al(OH)_3$ and aluminum phosphate $Al(PO)_4$), emulsion-structured adjuvants (chiron microfluidized oil/water emulsion (MF59), adjuvant system 03 oil in water emulsion (AS03), combined adjuvants (adjuvant system 04 (AS04) = $Al(OH)_3$ + monophosphoryl lipid A (MPL)) are among the main types of adjuvants (Garçon et al. 2011; Singh and Hagan 1999; Hogen Esch, O'Hagan, and Fox 2018).

Vaccine adjuvants which are in the preclinical testing phase are as follows: Freund's adjuvant, *Mycobacterium vaccae*, cytokine/growth factors, CpG motifs, syntex adjuvant formulation/2 (SAF/2), muramyl tripeptide (MTP), liposomes, saponins (use in combination in licensed vaccines are separate) (Brennan and Dougan 2005; Batista-Duharte, Lindblad, and Oviedo-Orta 2011). Table 14.2 shows the most commonly used types of adjuvants.

TABLE 14.2
The Most Commonly Used Types of Adjuvants

Inorganic Salts				Oligonucleotides			
Adjuvant	**Component**	**Site of effect**	**References**	**Adjuvant**	**Component**	**Site of effect**	**References**
Aluminum salts	Al(OH)$_3$ and Al(PO$_4$)	Ig G TH2	(Batista-Duharte et al. 2013)	CpG ODN	CpG	B cell, dendritic cell, TLR9	(Lingnau, Riedl, and Gabain 2007; Zimmermann, Dalpke, and Heeg 2008; Cordeiro and Alonso 2016)
Saponin-Based Mixtures				**Microorganism-Derived Adjuvant**			
Adjuvant	**Component**	**Site of effect**	**References**	**Adjuvant**	**Component**	**Site of effect**	**References**
QS-21	Quil A derivative	TH1	(Batista-Duharte et al. 2013)	MPL	MPLA	TLR4 IL-2 IFN-γ	(Lingnau, Riedl, and Gabain 2007; Batista-Duharte et al. 2013; Baldrick et al. 2002; Casella and Mitchell 2008)
ISCOM	Liposome + saponin + cholesterol and phospholipid	TH1 TH2	(Pearse and Drane 2005; Morelli et al. 2012; Cordeiro and Alonso 2016)				
O/W Emulsion				**Particulate Antigen Delivery Systems**			
Adjuvant	**Component**	**Site of effect**	**References**	**Adjuvant**	**Component**	**Site of effect**	**References**
AS02	MPL, QS21, AS03	TLR4	(Batista-Duharte et al. 2013)	Virosome	"virus-like" phospholipid vesicles	Promote antigen presentation	(Maughan, Preston, and Williams 2015; Peterfalvi et al. 2019; Batista-Duharte et al. 2013; Wang and Xu 2013; Cordeiro and Alonso 2016; Glück, Burri, and Metcalfe 2005)

(Continued)

TABLE 14.2 (CONTINUED)
The Most Commonly Used Types of Adjuvants

O/W Emulsion				Particulate Antigen Delivery Systems			
Adjuvant	Component	Site of effect	References	Adjuvant	Component	Site of effect	References
AS03	Tween 80, DL α-tocopherol, squalene	IRE1α	(Maughan, Preston, and Williams 2015; Batista-Duharte et al. 2013; Wang and Xu 2020; Cordeiro and Alonso 2016)	Liposome	Cholesterol and phospholipid vesicles	B cell	(Cordeiro and Alonso 2016; Bernocchi, Carpentier, and Betbeder 2017; Gül and Dikmen-Yurdakök 2019)
AS04	Al(OH)$_3$ MPL	TLR4	(Maughan, Preston, and Williams 2015; Batista-Duharte et al. 2013; Wang and Xu 2020; Cordeiro and Alonso 2016)	Nanoparticle	Polymeric-based. Oil-based. Peptide-based.	Promote antigen presentation TH1 TH2 TLR, APC	(Skwarczynski and Toth 2011; Rudra et al. 2009; Skwarczynski and Toth 2016)
MF59	Tween 80, span85, squalene	ASC	(Maughan, Preston, and Williams 2015; Batista-Duharte et al. 2013; Wang and Xu 2020; Cordeiro and Alonso 2016)				

MPLA: Monophosphoryl lipid A; AS02: Adjuvant system 02, AS03: Adjuvant system 03; AS04: Adjuvant system 04; TLR: Toll-like receptor; APC: Antigen presenting cells; IRE1α: Endoplasmic reticulum stress sensor; ASC: Apoptosis-associated speck-like protein containing a CARD; TH: Helper T cell; IgG: Immunoglobulin G; ISCOM:Immune-Stimulating Complexes.

The types of adjuvants are:

1) Inorganic salts
2) Oligonucleotides (CpG DNA sequences)
3) Emulsions (e.g. MF59)
4) Saponin-based mixtures (e.g. QS-21 and ISCOMATRIX)
5) Microorganism-derived adjuvants
6) Cytokines
7) Particulate antigen delivery systems (liposomes, virosomes, nanoparticle-derived adjuvants)

14.3.1.1 Inorganic Salts

Aluminum compounds are known to be among the adjuvants used first in vaccines. It is a known fact that aluminum salt ($Al(OH)_3$ and $Al(PO)_4$) can trigger IgG with a relatively long-lasting immunity, ease of formulation and long-term reliability (Gül and Dikmen-Yurdakök 2019; Singh and Hagan 1999; Vecchi et al. 2012). There must be physical contact between an antigen and aluminum compounds, in other words, antigens must be absorbed to aluminum molecules in order for the aluminum compounds to have an adjuvant effect (Yurdakök and İnce 2008). Aluminum adjuvants slow down the diffusion of an antigen from an injection site, thereby saving time for the accumulation of inflammatory cells and enhancing the immune response. Therefore, it increases the uptake of the antigen by APCs (Gül and Dikmen-Yurdakök 2019).

The protein-containing antigens in a vaccine generally bind to the aluminum-derived adjuvants in two ways. The first mechanism is the binding of proteins to the aluminum-derived adjuvants via an electrostatic effect. The second mechanism is ligand exchange (Gül and Dikmen-Yurdakök 2019; Maughan, Preston, and Williams 2015). There is a substitution between hydroxyl or phosphate groups in the structure of the antigen and ligands in the aluminum-derived adjuvants (Maughan, Preston, and Williams 2015). It is reported that aluminum-derived adjuvants work on the four main stages of the immune response mechanism (Yurdakök and İnce 2008). The antigens adsorbed to aluminum are taken more effectively by dendritic cells (Gül and Dikmen-Yurdakök 2019).

An important problem for aluminum-derived adjuvants is that their use is limited to bacterial and viral vaccines requiring the production of neutralizing antibodies. In early studies with aluminum salts, aluminum compounds were found to be very successful in creating humoral immune responses including IgE production; however, they were not able to stimulate the cellular immune response (e.g. delayed type hypersensitivity) (Yurdakök and İnce 2008; Oleszycka and Lavelle 2014).

Aluminum, the source of aluminum-derived adjuvants, is a metal that is abundant in the environment. However, only small amounts of aluminum are absorbed through the intestinal barrier and the skin (Keith, Jones, and Chou 2002). Most of aluminum entering the body is excreted through the kidneys. Aluminum toxicity due to occupational exposure, neurological disease, and parental nutrition is related to renal and bone diseases (Tomljenovic and Shaw 2011). The most common side effect of aluminum-derived adjuvants is that it induces a local inflammation reaction and causes pain, swelling, redness, and local granuloma at the injection site. Clinical symptoms such as nausea, fever, suppression of the immune system, allergies may be observed following the local inflammation. This condition, which occurs in subcutaneous or intradermal injection of a vaccine instead of intramuscular injections, usually presents with mild clinical symptoms. In fact, aluminum-derived adjuvants reduce the prevalence and the severity of systemic adverse reactions by binding molecules and releasing them slowly. Thus, it reduces toxicity. Stimulating long-term brain inflammation and related neurological diseases, dementia, and various neurodegenerative diseases are among other toxic effects of aluminum-derived adjuvants (Gül and Dikmen-Yurdakök 2019).

The most important side effect of aluminum-derived adjuvants is the formation of local granulomas at the injection site. This situation develops especially when an intradermal or subcutaneous injection is administered instead of intramuscular. Local pain, inflammation, swelling, ulcer, and necrosis at the injection site may also be observed (Maughan, Preston, and Williams 2015). Other important side effects are tendency to allergy and potential neurotoxicities with increased IgE production. Aluminum is usually excreted from the body through the kidneys. In kidney dysfunction, aluminum accumulates in the body. High levels of aluminum affect the brain and bone tissues in particular. Aluminum intoxication has also been associated with Alzheimer's disease. Magnesium hydroxide ($Mg(OH)_2$), zinc sulphate ($ZnSO_4$), calcium phosphate ($Ca_3(PO_4)_2$), and iron and zirconium salts are used as alternative mineral salts (Yurdakök and İnce 2008; Maughan, Preston, and Williams 2015).

14.3.1.2 Oligonucleotides

Adjuvants containing the CpG dinucleotide directly stimulate B cells and dendritic cells. Thus, it promotes the production of cytokines and maturation/activation of antigen-presenting cells prior to an inflammation. CpG oligonucleotides are agonists of TLR found on the APCs (Klinman et al. 2009; Zimmermann, Dalpke, and Heeg 2008). These receptors are the source of the signals which trigger an immune response in the host when encountering a foreign antigen (Kindrachuk et al. 2009). They increase antigen-specific immune responses 5–500-fold (Klinman 2006). Nucleic acid vaccines (genetic vaccines) are based on the direct delivery of genetic material encoding the desired antigen to the host cell, and the creation of the immunity. They have some advantages such as better safety, the ability to specifically acquire immunity to an antigen, stimulating both B and T cell responses, relatively cheap cost and production (Zimmermann, Dalpke, and Heeg 2008; Gül and Dikmen-Yurdakök 2019). However, low stability and limited clinical application routes are some their disadvantages. On the other hand, areas of use of DNA and RNA vaccines are gradually increasing thanks to the production of various biomaterials with physical methods (such as electroporation, sonoporation and magnetofection) (Gül and Dikmen-Yurdakök 2019).

14.3.1.3   Emulsions

Emulsions are the second most frequently used and preferred adjuvants in humans and animals, subordinating to aluminum. This group includes water in oil (W/O; liquid antigen in paraffin) or oil in water (O/W) emulsions. They are considered to function by stimulating antibody production in plasma cells through forming a depot at the injection site and slow release of an antigen from there (Yurdakök and İnce 2008).

Emulsion-based adjuvants are commonly used in the development of vaccines (Spickler and Roth 2003). The most commonly used emulsions are oil-based emulsion adjuvants. An example of this is Freund's adjuvant. However, the use of Freund's adjuvant in humans was stopped due to its high toxic effects (granuloma, abscess, arthritis, amyloidosis, allergic reactions) (Stephenson et al. 2014).

Oil-based emulsion adjuvants contain a mixture of oil and aqueous phase which is stabilized with a surfactant. W/O(water/octanol) emulsions provide a strong and long-lasting immune effect. Mineral oil-based emulsions are known to be very efficient; however, they may sometimes cause local reactions with reactive antigens (Aucouturier, Dupuis, and Ganne 2001). Non-mineral oils are well-tolerated, but their immunogenicity is poor. W/O emulsions form a depot at the injection site, thereby causing the antigen to be released slowly. Thus, it may be used to reduce the dose, or the number of doses required for an effective antigen. MF59 is an O/W emulsion adjuvant. Squalene, which is included in its structure, is a triterpenoid hydrocarbon produced abundantly as a precursor of cholesterol and steroid hormones in humans. It is included in the structure of most of the adjuvants due to the fact that it is naturally available in the human body and it is biodegradable. In experimental studies on the less toxic MF59, it was suggested that it was a safe and effective vaccine adjuvant that was well tolerated in humans (Yurdakök and İnce 2008). MF59 has been used for an influenza vaccine for the elderly and young children (Morelli et al. 2012).

Unlike W/O emulsions, O/W emulsion-based adjuvants do not form an antigen depot at the injection site. Instead, oil droplets facilitate chemokine-induced immune cell uptake and the differentiation of macrophage and dendritic cells (Gül and Dikmen-Yurdakök 2019).

14.3.1.4   Saponins

Saponins are steroid or triterpenoid-form glycosides found in wild and cultivated plants, some primitive sea creatures, and bacteria. Saponins contain a steroidal or triterpenoid aglycon to which one or more sugar chains are bound (Sun, Xie, and Ye 2009). Saponins are extraordinary candidates for the development of a new adjuvant. It was found that many saponins had adjuvant effects on purified protein antigens (Song and Hu 2009). Their stimulation of the immune system in mammals has increased their use as a vaccine adjuvant. Saponin-based adjuvants modulate the cellular immune system, increase antibody production, and can be used at low doses for their efficacy. They have complex action pathways such as stimulating T-dependent antigens, inducing CD8$^+$ and enhancing the response to mucosal antigens (Gül and Dikmen-Yurdakök 2019). Saponins are generally safe; however, the safety changes depending on the route of administration, animal species, and some specific saponins. In addition, intravenous (IV) injections of saponins may cause toxicity due to hemolysis (Gül and Dikmen-Yurdakök 2019).

The saponin type obtained from Chilean soapbark tree (*Quillaja saponaria* Molina) shows strong adjuvant properties such as Quil A and its derivatives QS-21 (Sun, Xie, and Ye 2009; Ragupathi et al. 2010). They usually lead to immune cell proliferation and increased antibody production. There are studies showing that *Quillaja* saponins have mitogenic effects and cause T and B cell proliferation in this way. Although they have been used successfully in animal vaccines for a long time, they are highly toxic for humans. Besides forming severe local reactions and granulomas, they affect cholesterol in erythrocyte membrane and cause severe hemolysis. All *Quillaja* saponins are not still suitable for use in humans due to their high toxicity, hemolytic effects, and instable chemical structure (Yurdakök and İnce 2008; Ivanov et al. 2020).

Immune-stimulating complexes (ISCOM) are registered liposome adjuvants which contain saponin. The stability and small size of the ISCOM matrices help them remain in the circulation for a long time (Eratalay and Öner 2001). It was shown in several animal models that ISCOM induced both humoral (TH2 induction) and cellular immune response (TH1 induction), and potential cellular responses (such as CT) (Bengtsson, Morein, and Osterhaus 2011; Gül and Dikmen-Yurdakök 2019). A cell-mediated immune response is essential for effective vaccination against intracellular pathogens and chronic infections. ISCOMs can effectively stimulate both CD4+ and CD8+ T cell responses. The technology used leads to an increase in efficiency by decreasing the antigen dose and long-term immune responses (Gül and Dikmen-Yurdakök 2019). Within the scope of ISCOM, saponin-based *ISCOMATRIX*, which forms spontaneously then saponin, cholesterol, and phospholipid are mixed under controlled conditions, was developed as a versatile helper system which can increase both B cell and T cell responses. It is a system used to decrease the level of the dose administered and to stimulate the immune system (Gül and Dikmen-Yurdakök 2019; Pearse and Drane 2005). It has been reported that the most common adverse event was pain after administration (Pearse and Drane 2005).

14.3.1.5 Microorganism-Derived Adjuvants

Bacteria or fungi can be used as adjuvants due to their immunostimulant capacities. However, this is not possible because the use of microorganisms as adjuvants is very toxic to humans. Although the bacterial cell wall and lipopolysaccharides are not strong immunogens as antigens, they may enhance the immune response and show an adjuvant effect. They show their adjuvant effects by activating TLRs on the APCs (Yurdakök and İnce 2008).

Muramil dipeptide (MDP): N-acetyl muramil-L-alanyl-D-isoglutamine, also called muramil dipeptide (MDP), is the main part which is responsible for the adjuvant effect of microorganisms. MDP induces humoral immunity when used alone and it is a potent cellular immunity stimulant when used with liposomes or glycerol (Baldrick et al. 2002).

Lipopolysaccharides and monophosphoryl lipid A (MPLA): Another important adjuvant obtained from the gram negative bacterial cell wall is lipopolysaccharides. Lipid A is the part responsible for the adjuvant effect (Baldrick et al. 2002). In high pH environments, lipid A is hydrolyzed and monophosphoryl lipid A (MPLA) is formed. MPLA is as effective as lipid A; however, it is less toxic than lipid A. It was reported that it stimulated the cellular immune response by increasing the synthesis and the release of IL-2 and IFN-γ. MPLA is usually formulated with liposomes, emulsions, aluminum, or QS21 (Casella and Mitchell 2008).

14.3.1.6   Cytokines

Many cytokines are tried to be used as adjuvants because they stimulate the immune system. These small proteins play an important role in cell signaling, induction, and direction of cellular and humoral immune responses, and promote antibody production (Kishimoto, Taga, and Akira 1994). It was suggested that they increased the immune response following the use of various cytokines such as interferon (IFN-α, IFN-γ) and interleukin (IL-2, IL-12, IL-15, IL-18, IL-21) as adjuvants (Spickler and Roth 2003). However, they have stability problems due to their protein structure, and their half-life is short. Their areas of use are limited as they all show dose-dependent toxicity (especially hepatotoxicity and CYP-dependent drug interactions) (Yurdakök and İnce 2008).

14.3.1.7   Liposomes and Virosomes

Liposomes are cholesterol and phospholipid vesicles similar to cell membranes (Cordeiro and Alonso 2016; Bernocchi, Carpentier, and Betbeder 2017). Liposomes show adjuvant effects as depot systems that release an antigen slowly. These adjuvants affect antigens within the lumen or the membrane. They take part in the humoral immunity and in some cases are used with immunomodulators to increase their activity (Gül and Dikmen-Yurdakök 2019).

Liposomes are biodegradable, immunologically inert, non-toxic, and easy to produce. These properties give an advantage in their use as an adjuvant delivery system. A single liposome can hold more than one antigen, and long-term and high antibody response is obtained. However, the stability problems of liposomes pose an important obstacle (Eratalay and Öner 2001).

A virosome is a vaccine adjuvant and carrier system for the subunit vaccines. They have a multifunctional activity thanks to their structure and composition (Moser, Amacker, and Zurbriggen 2011). Virosomes, which are "virus-like" phospholipid vesicles containing viral proteins (Cordeiro and Alonso 2016). Virosomes are non-proliferating viral glycoprotein-coated liposomes which can be used to deliver vaccine antigens directly into a host cell (Moser, Amacker, and Zurbriggen 2011). They accelerate the antigen uptake of the APCs (Glück, Burri, and Metcalfe 2005; Wilschut 2009). Despite increasing use of peptide vaccines, designation stage may cause declines in the immunogenic efficacy of the peptide vaccines. Virosomes are developed as the adjuvant or carrier systems to increase the immunogenic efficacy of the synthetic peptide vaccines. The synthetic peptides on the virosomes are safe and immunogenic for humans even at low doses (Moser, Amacker, and Zurbriggen 2011).

14.3.1.8 Peptide-Based Vaccine and Particulate Antigen Delivery Systems

Vaccination plays a major role in the prevention of many fatal diseases. The conventional vaccination strategies have been effective in reducing morbidity and mortality due to infectious disease for many years. However, vaccination has a high side-effect potential because of excessive antigen load. For instance, the presence of proteins that do not induce protective immune response in the vaccine formulations also leads to allergic reactions. The allergic and other potential side effects of the vaccines directed the specialists to design peptide vaccines made up of small peptide fragments containing epitopes that may induce T and B cell mediated immune response were designed in the recent years (Li et al. 2014).

Peptide-based vaccines are designed to induce immune response (cellular and humoral) mediated by cytotoxic T cells and B cells (Kametani et al. 2015). T helper cells produce two types of immune response (Francis 2018; Skwarczynski and Toth 2011). TH1 produces proinflammatory response whereas TH2 produces anti-inflammatory response (this process occasionally triggers an allergic response). The appropriate peptide sequence length and binding affinity of the peptide vaccines to the APC and MHC cells play a key role in the production of T cell response to trigger immune response (Skwarczynski and Toth 2011). The peptides that are used in the production of these vaccines are the short fragment peptide molecules with a certain amino acid sequence that represents an antigen-specific epitope. If synthetic peptide sequences are longer than a length that T cells can recognize, then these synthetic peptides should be trimmed by proteases and peptidases to be converted to a shorter fragment peptide molecules to be recognized by the MHC cells on the APC surface or bound onto the MHC cells through an appropriate process (Li et al. 2014). Peptide-based vaccines contain multiple peptide epitopes corresponding to the subunits of a single pathogen or several epitopes of multiple distinct pathogens (Skwarczynski and Toth 2011). The key point of the process is to select the right epitope to induce the specific immune response. In the selection of these molecules, the initial stage is to determine the immunodominant sections of the epitopes that can induce protective immune response against a certain antigen in the course of humoral and/or cellular immunity. Immunodominant epitopes are selected from B cells, cytotoxic, and helper T-cell content. Also, the capacity of the epitopes to induce TH is crucial as well as induction of CT or B cells (Li et al. 2014).

The immune system presents antigens as a short peptide that can bind to MHC cells. A vaccine peptide should include at least one histocompatibility motif (agretope; T cell epitope) with MHC cells as an antigen (Yano et al. 2005). The interaction between MHC cell and peptide is of critical importance. The significant issue regarding this interaction is structural heterogeneity of MHC cells that are expressed on the antigen-presenting cell surface during the presentation of the peptide vaccine to T cells by the antigen-presenting cells (Celis 2002). Heterogeneity of MHC cells may affect the immune response process (Li et al. 2014). Vaccine peptides should have B cell epitope and capability to induce B cell receptors for induction of humoral immune response in the organism (Yano et al. 2005).

Most of the peptide-based vaccines are produced synthetically. The production of peptide vaccines has become simple, easy, reproducible, rapid, and affordable thanks to the technological improvements in the recent years. The complications associated with biological contamination are eliminated by obtaining peptide vaccines by chemical synthesis. These vaccines include no bacterial contaminants and demonstrate no oncogenic potential (Van der Burg et al. 2006; Skwarczynski and Toth 2016). These vaccines are usually water soluble and stable under storage conditions. Peptide vaccines are usually designed to demonstrate target-specific efficacy. Peptide vaccines cause no allergic or other side effects (Skwarczynski and Toth 2016). However, peptide vaccines have also some disadvantages to solve beside their advantages. At the isolation stage, immunogenic activity of the peptide vaccines weakens, and they require adjuvants or carriers (Li et al. 2014). The APCs play a key role in triggering immune response. Peptides stimulate the APCs, however short fragment peptides do not manifest high immunogenic impact (Skwarczynski and Toth 2011). The vaccines based on short fragment peptides may have weak immunological efficacy since short fragment peptides show low binding affinity to MHC cells on the APC surface (Van der Burg et al. 2006). The appropriate peptide length and binding affinity to MHC cells are effective in the induction of T cell response (Bijker et al. 2007). Therefore, modification procedures are performed to elevate binding affinity of the peptides (Van der Burg et al. 2006). Peptides require conformational changes to affect B cells. Additionally, peptides are easily broken down in vivo by many enzymes, thereby, only a limited portion of the peptides are taken by the APCs (Yang and Kim 2015). The epitopes at the C- and N- terminals of the amino acid chain are modified by facilitating the intake of peptides by the APCs to prevent inadequacy in the vaccine formulation and carrier systems conjugated with target-specific peptide epitope in the advanced studies of vaccine formulation (Skwarczynski and Toth 2011; Yang and Kim 2015).

Peptide-based vaccines have solely weak immunogenic impact and development of innovative adjuvant, and appropriate delivery systems are needed to increase long-duration immune response (Bijker et al. 2007; Rudra et al. 2009). The particulate vaccine delivery systems that were developed after nanotechnological approaches became applicable in the field of vaccine formulation during recent years; they also function as an immunological adjuvant at the same time. The particulate delivery systems aim at overcoming the deficiencies of the previous adjuvants. Peptide immunogens included in the nanoparticles are presented to the APCs more effectively, protect against in vivo enzymatic degradation, and cause increased immune response without need for adjuvants (Rudra et al. 2009).

14.3.1.8.1 Nanoparticle adjuvants

In vaccine development, a safe and highly efficient adjuvant deficiency (immunostimulant) is considered the biggest obstacle (Tornesello et al. 2020).

In peptide sequence antigen structures, modification does not occur very successfully without affecting the immunogenic properties. This problem has been overcome by developing a peptide antigen in a nanoparticle structure. Peptide antigens are hydrophilic; therefore, they are obtained by self-conjugating these molecules using hydrophobic partial polymer, lipid, and some special peptides (Tornesello et al. 2020)

Nanoparticle systems studied in the peptide vaccines:

1) Polymeric-based nanoparticles
2) Oil-based nanoparticles
3) Peptide-based nanoparticles

14.3.1.8.1.1   Polymer-Based Self-Assembled Nanostructures

Nanoparticles are manufactured using albumin, collagen, starch, chitosan, and dextran out of natural polymers and polymethylmethacrylate, polyesters, polyanhydrides, and polyamides among synthetic polymers (Li et al. 2014). There are biodegradable or non-biodegradable polymers. Non-biodegradable polymers may cause unexpected effects by accumulation in the body. In the vaccine studies, the characteristics such as toxic effects of the polymer on the organism, antigen release speed capacity, stability status under storage conditions, and stability in the in vivo conditions should be taken into account in making a decision for an ideal polymer carrier system (Skwarczynski and Toth 2011, 2016). The comprehensive toxicity tests for several synthetic polymers such as polyesters, polylactic acid (PLA), polyglycolic acid, and their copolymers poly(lactic-co-glycolic acid) (PLGA) have been carried out and they are FDA-approved for use in humans (Li et al. 2014; Cordeiro and Alonso 2016). The most commonly used biodegradable polymers are PLA, PLGA, polyglutamic acid (PGA), polycaprolactone (PCL), and polyhydroxybutyrate. PLGA is the most frequently used polymer in the nanoparticle studies (Li et al. 2014). Skwarczynski and Toth (2011) have reported in their study that MUC-1 peptide vaccine assembled into PLGA nanoparticle carrier system accompanied with adjuvant MPLA created immune response by inducing T cells. However, it has been noted in the same article that need for use of adjuvant in the PLGA-based systems still continues (Skwarczynski and Toth 2011).

Advantages:

1) Polymeric-based nanoparticles perform transport, presentation, and uptake of the antigen to the APCs
2) They induce both cellular and humoral response
3) Long-lasting continuous release of the encapsulated antigen occurs
4) The speed of antigen release is controlled by the breakdown speed of the polymer matrix. The composition of the polymer and nanoparticle size affect the breakdown speed of the polymer matrix. The formation of the polymer complex is affected by hydrophobic interactions, electrostatic forces, hydrogen bonds, and van der Waals interactions

This structure is used to protect the antigen from enzymatic degradation. These structures also have natural adjuvant effects. The structure of polymers can be easily changed. The most commonly used biodegradable synthetic polymers in vaccine applications include PLGA, PLA, and PCL (Zhao et al. 2017).

Beside these advantages, the stability of peptide antigens may be degraded at the stage of encapsulation and loss of immunogenicity may occur. Since hydrophobic characteristics of the polymers affect their distribution in the body, use of the easily excretable hydrophilic polymers is preferred (Li et al. 2014). The disadvantages include that the lengths of polymers may not be controlled during polymerization reactions. This may undesirably alter the size of the vaccine particles. In the binding of an antigen to the polymer structure in an epitope, byproducts at variable degrees may occur. Some polymer-based particles may biodegrade and accumulate in tissues. It is necessary to pay attention to in vivo toxicity potentials of these materials (Zhao et al. 2017).

Chitosan, out of the natural polymers, is an ideal polymer thanks to its non-toxic characteristics, biocompatibility, and biodegradability in the organism. It has been encountered that peptide vaccine formulations containing chitosan increased the immune response. However, water-soluble derivatives of chitosan were developed since this polymer indicates low water solubility (Li et al. 2014).

14.3.1.8.1.2 Oil-Based Self-Assembled Nanostructures

All adjuvants in commercial vaccines are in an oil-based formulation, except for aluminum- and virus-like particle-based adjuvants. Oils take the role of "dangerous signal" in the immune system by activating TLR receptors on the surface of the APCs. Oil-based vaccines are used to improve the amphiphilic properties of lipid-peptide conjugates. The length, number, and position of the conjugate oils play an important role in the administration of a vaccine. Although lipid-peptide conjugates can induce a strong immune response, potential toxicities such as membrane degradation caused by lipopeptides should be considered (Zhao et al. 2017).

14.3.1.8.1.3 Peptide-Based Self-Assembled Nanostructures

Although, many adjuvants (oil emulsions, TLR ligands, ISCOMs) have been researched to elevate immunogenic efficacy of the peptide epitopes, the improvement of nanoparticle and self-assembling systems directed the formulation studies of the novel peptide vaccines to focus on the use of these systems (Rudra et al. 2009). Peptides are heteropolymers consisting of amino acids. Peptides can form α-helical and β-sheet structures. β-sheet-structured conformations of peptides form nanofibers. Unlike vaccines with aluminum-derived adjuvant, nanofiber-shaped peptides do not cause cytotoxicity (Zhao et al. 2017).

Rudra et al. have reported that peptide epitopes assembled into a short synthetic peptide nanofiber (AC-QQKFQFQFEQQ-Am; Q11) induced high immune response without need for an adjuvant (Rudra et al. 2009).

Self-assembling concept-generated nanosize vaccine candidates utilize the well-documented aggregation properties of certain peptides. The advantages of the self-assembling peptide materials include molecular specificity, nanoscaled positioning of the ligands, multivalency, multifunctionality, and biocompatibility (Skwarczynski and Toth 2011). Yano et al. have reported that use of nanoparticles increased the binding of the peptide epitopes to the specific receptors in the immune system when peptide epitopes are assembled into arginine-glycine-aspartate nanoparticles (Yano et al. 2005). However, peptide-based conjugates have lower immunogenic properties compared to lipid and polymer-based nanostructures; therefore, peptide-based conjugates require a use at higher doses (Zhao et al. 2017).

Briefly, use of nanotechnology in the studies of vaccine formulation is one of the strategic steps for induction of the desired immune response and elimination of side effects.

14.3.2 Adverse Effects Related to Adjuvants

Adjuvants have been used for years to increase the immune response in practical vaccination. The toxic mechanisms of adjuvant vaccines are poorly understood, despite their immunological success (Batista-Duharte et al. 2013). Adjuvants can mediate potentially toxic effects. The type and mechanism of an adjuvant is decisive in the balance between immunogenicity and toxicity. For example, oil emulsions cause local toxicity, while cytokine adjuvants lead to vascular leak syndrome (Batista-Duharte et al. 2014). When the immune responses destroy invasive microorganisms, they may cause tissue damage and expose clinical signs of some diseases. These adverse effects can occur with specific adjuvant and antigen interactions. Side effects can be classified as local or systemic. Significant local reactions are pain, local inflammation, swelling, injection site necrosis, lymphadenopathy, granuloma, ulcer, and sterile abscess. Systemic reactions include nausea, fever, arthritis, uveitis, eosinophilia, allergies, anaphylaxis, and organ-specific toxicity and immunotoxicity (Gül and Dikmen-Yurdakök 2019).

Local ulcers and regional necrosis may be observed in oil emulsion applications. The reason is that short-chain hydrocarbons with detergent-like effects in the formulation disrupt the structure of the fat layer in the cell membrane, and thus cause cell lysis. Some W/O emulsions, on the other hand, can cause the enzymatic breakdown of fat chains and the formation of toxic fatty acids. Saponins, one of their surfactants, also cause lysis in the cell membrane.

Hypersensitivity appears common after the use of an adjuvant. The reason is the large number of immunocompetent compounds (CD4+, CD8+ T cells, macrophages, etc.) which migrate to the vaccine site.

Adjuvants cause toxic reactions when they overstimulate the immune system. The release of proinflammatory cytokines after the administration of an adjuvant vaccine may lead to the hyperactivation of immunological mechanisms. Overdoses of IL-2, which is a recommended adjuvant and a cytokine, were associated with autoimmune diseases.

The use of some aluminum-containing vaccines, Freund's adjuvant, and other adjuvants can cause large granulomas. Granulomas are particularly associated with depot adjuvants. Granulomas may appear metastatic when an excessive amount of oil emulsion is injected into a single site.

Adjuvants may overstimulate the inflammatory mechanism by causing the production of TNT and other mediators at high levels. Different symptoms (acute phase response (APR), modification of hepatic metabolism (MHM), vascular leak syndrome (VLS)) may be observed depending on the amount of these mediators in the plasma and the last time when they are found in the plasma (Batista-Duharte, Lindblad, and Oviedo-Orta 2011; Batista-Duharte et al. 2014).

Acute phase response (APR) is a temporary syndrome which occurs due to an injury or damage. Recent studies have shown that the serum level of TNF-α is the key mediator of the APR. Proinflammatory cytokines mediate the APR, and it is characterized by a local inflammatory response. Leukocytosis, acute protein production by hepatocytes, fever, and changes in fat, protein, and carbohydrate metabolism are among the symptoms which appear during the APR (Batista-Duharte et al. 2014).

In many studies conducted on the effects of immunostimulants on drug metabolism, it was reported that they were especially effective on the liver cytochrome P450 enzyme activity.

Vascular leak syndrome (VLS) is an important dose-limiting effect observed after several cytokine treatments. Effects varying from tissue edema to multiple organ failure are observed with increased vascular permeability (Batista-Duharte et al. 2014).

New generation adjuvants include different types of chemical compounds with different mechanisms of action and different potential adverse effects. The aim of the studies has been to find more effective auxiliary substances with fewer negative effects (Gül and Dikmen-Yurdakök 2019).

14.3.2.1 The Effect of Dose and the Preferred Route of Administration

The route of administration and dose of an adjuvant vaccine is effective in drug toxicity. Vaccines at low or high doses cause failure in the immune response. After vaccination, antigen antibody complexes are formed during the immune response. When this structure is formed in excessive amounts, it accumulates in the tissue and causes tissue damage. This complex, which is formed in the body through the administration of repeated dose or high dose antigen, can accumulate primarily in the kidneys, arteries, and lungs, and damage them (Batista-Duharte et al. 2014). Mucosal administrations (intranasal, oral, sublingual, rectal, vaginal) and transcutaneous administrations generally lead to fewer toxic effects than parenteral administrations, except for some oral administrations of an adjuvant vaccine (Neutra and Kozlowski 2006; Batista-Duharte, Lindblad, and Oviedo-Orta 2011) Furthermore, some cytokines cause fewer toxic effects when used as mucosal adjuvants (Batista-Duharte et al. 2014).

14.3.3 Toxicology Studies

Adverse effects which occur due to the interaction between a vaccine and the immune system are investigated by in silico, in vitro, and in vivo studies (Brennan and Dougan 2005). The studies are conducted according to the protocols stated in the guidelines of *"World Health Organization (WHO) guidelines on nonclinical evaluation of vaccines"* and *"European Medicines Agency (EMA) guidelines on adjuvants in vaccines for human use"* (Sun, Gruber, and Matsumoto 2012; WHO 2013; EMEA 2018, 2005). The evaluation of the immune response depends on the evaluation of the antibody response or the cellular immune response via immunoassay and multiplex assay analysis (Wolf, Kaplanski, and Lebron 2010). In toxicology studies of adjuvants, acute dose toxicity and repeated dose toxicities are primarily analyzed (Glueck 2002).

14.3.3.1 Acute Dose Toxicity

Single dose studies are designed to describe the relation between the dose level and the immune response. In addition, single dose studies evaluate only the gross effects of toxicity such as mortality, body weight, clinical findings, and macroscopic examination (Brennan and Dougan 2005). Although these studies are also recommended by the EMEA-CPMP guidelines, they are not always conducted in vaccines which are developed for repeated dosing if repeated dosing studies do not arouse any concern related to toxicity (Brennan and Dougan 2005; Forster 2012).

In single dose vaccine administrations, acute dose toxicity analysis is performed before repeated dose studies due to ethical reasons for the use of toxic adjuvants in some animal species (monkey) in uncommon treatment routes (Brennan and Dougan 2005).

14.3.3.2 Repeated Dose Toxicity

In animals, repeated dose toxicity studies are performed through repeated administration of a vaccine.

In a repeated dose toxicity study, an animal study is designed on the same route of clinical administration of a vaccine. The number of doses determined in the animal model is determined in a way to be one more dose than that of the clinical administration (N + 1 rule). In the animal model, the toxicity study is planned to take into account the full clinical human dose of the vaccine or the highest possible dose (Sun, Gruber, and Matsumoto 2012).

Two animal species (rodent/non-rodent) are used in new adjuvant vaccine studies, while one animal species is used in vaccine studies (Brennan and Dougan 2005). If there is an adjuvant in a vaccine study, study groups are formed to determine adverse effects as i) the group receiving antigen + adjuvant, ii) the group receiving antigen alone, iii) the group receiving adjuvant alone, and iv) saline-control group.

In repeated dose toxicity studies, antemortem parameters (mortality, body weight, food consumption, ophthalmic examinations, urinalysis, hematology, serum biochemistry, immunogenicity serum antibody test) and safety pharmacology studies parameters (body temperature, electrocardiograph (ECG), FAQ assessment test for cardiovascular, respiratory, and central nervous systems) are analyzed. Furthermore, a necropsy is performed in the animal model to identify early toxic effects within two to seven days following the last vaccination and deferred toxic effects in the second to fourth weeks following the last vaccination. Physical examination and histopathology tests are performed for all organs, and they are weighed. Histopathology tests are usually performed on the brain, kidney, liver, reproductive organs, and immune organs (Wolf, Kaplanski, and Lebron 2010).

Table 14.3 shows other toxicity parameters applied to vaccine adjuvants in nonclinical studies according to the EMEA-CPMP guidelines for adjuvants listed (EMEA 2005).

TABLE 14.3

Other Toxicity Parameters Applied to Vaccine Adjuvants in Non-Clinical Studies

	Toxicity of Adjuvant Alone	Method	References
Local tolerance	Depending on the route of administration, an analysis of local irritation induced by adjuvants should be performed. For oral and intranasal administration local and regional tolerance need to be assessed. Adverse effects such as granuloma may develop later with injected vaccines.	e.g. As noted in EMEA and WHO guidelines, the studies conducted included: single dose toxicity and local tolerance in rabbits.	(EMEA 2005; WHO 2013; Sun, Gruber, and Matsumoto 2012; Wolf, Kaplanski, and Lebron 2010)
Induction of hypersensitivity and anaphylaxis	Adjuvants may lead to hypersensitivity by inducing IgE increase. Hypersensitivity test is performed in appropriate models.	e.g. As noted in EMA and WHO guidelines, the studies conducted included: passive cutaneous anaphylaxis assay [PCA], and the active systemic anaphylaxis assay [ASA].	(EMEA 2005; WHO 2013; Sun, Gruber, and Matsumoto 2012; Wolf, Kaplanski, and Lebron 2010)
Pyrogenicity	Adjuvants should be tested with respect to their possible pyrogenic effects.		(Sun, Gruber, and Matsumoto 2012)
Systemic toxicity	Adjuvants may cause toxicity in various organs in the body system ((heart, lung, brain, liver, kidney, reproductive organs, spleen, thymus, bone marrow, lymph nodes). Protocols are designed to establish dose-effect relationships and include repeated administration at intervals reflecting the proposed clinical use. Studies are conducted regarding repeat dose toxicity.	e.g. As noted in EMA and WHO guidelines, the studies conducted included: repeated dose toxicity in rats and rabbits.	(EMEA 2005; WHO 2013; Batista-Duharte et al. 2014; Sun, Gruber, and Matsumoto 2012; Wolf, Kaplanski, and Lebron 2010)

(Continued)

TABLE 14.3 (CONTINUED)
Other Toxicity Parameters Applied to Vaccine Adjuvants in Non-Clinical Studies

	Toxicity of Adjuvant Alone	Method	References
Reproduction toxicity	If vaccination is to be performed in pregnant women, this study is conducted to identify the effect of the adjuvant on the reproductive system. An animal model is established before the study and administration of doses for the vaccine should be considered before and after pregnancy.	e.g. As noted in EMA and WHO guidelines, the studies conducted included: reproduction toxicity in rats.	(EMEA 2005; WHO 2013; Wolf, Kaplanski, and Lebron 2010)
Genotoxicity	Adjuvants might be derived from biological as well as from synthetic origin. In biologically derived adjuvants, genotoxicity study may not be conducted. Synthetically derived adjuvants require genotoxicity studies (e.g. gene mutation, chromosomal aberrations, DNA damage).	e.g. For in vitro genotoxicity tests human peripheral lymphocytes were used and for in vivo genotoxicity test rats were used. Comet assay was performed on rats and human lymphocyte culture.	(EMEA 2005; WHO 2013; Wolf, Kaplanski, and Lebron 2010; Yüzbaşıoğlu et al. 2013)
Toxicity of adjuvant in combination with the proposed antigen			
Local tolerance	Antigen injection in combination with adjuvants and injection containing adjuvant alone are compared to determine the severity of local reactions. Optimal dose ratio of adjuvant and antigen regarding the benefits and risks is investigated.	e.g. As noted in EMA and WHO guidelines, the studies conducted included: single dose toxicity and local tolerance in rabbits.	(EMEA 2005; WHO 2013; Sun, Gruber, and Matsumoto 2012; Wolf, Kaplanski, and Lebron 2010)
Repeated dose	According to the proposed clinical program, a dosing program is used in a laboratory study. For the reliability of a repeat dose toxicity study, the number of vaccine administrations during the study should be higher than the planned number of vaccine administrations in humans (N + 1).	e.g. As noted in EMA and WHO guidelines, the studies conducted included: repeated dose toxicity in rats and rabbits.	(EMEA 2005; WHO 2013; Sun, Gruber, and Matsumoto 2012; Wolf, Kaplanski, and Lebron 2010)

Moreover, to determine the effect of an adjuvant on the clinical response, non-clinical immunogenicity tests are used:

1. Dose-response study: Studies determining the effect of vaccine antigen at different doses and adjuvant at different doses on response.
2. Comparison study: The effect of a new adjuvant is compared with that of an antigen alone or a vaccine antigen adjuvanted with a known adjuvant.

14.4 CONCLUSION

This review mentions adjuvants, which create a stronger immune response and enhance the effect of a vaccine when included in the vaccine compared to the vaccine alone, and have an important role in vaccine development. In particular, it was emphasized that adjuvants are effective in enhancing the humoral and/or cellular immune response against antigens in the vaccine. In studies, it was stated that according to the desired effect during vaccination, adjuvants meeting the need may be selected among many adjuvants whose mechanism of action is known. Experimental studies on adjuvants determine not only the effects of adjuvants on the immune response, but also the level of adverse effects that may develop in the organism during the period of use. The variety in adjuvants is a result of the effort to develop adjuvants that will leave minimum toxic effect in the organism, while trying to achieve maximum immune response. Clinical trials are ongoing as new generation adjuvants developed also contain a wide variety of chemicals. The aim of the studies is to find the most effective excipients by determining the potential negative effects of new generation adjuvants with different mechanisms of action and developing the formulation to minimize these effects. Therefore, in this review, adverse effects that may occur in the organism with the use of vaccine adjuvants were evaluated. This review also referred to the role of immune cells in the immune system in the occurrence of adverse effects, and the local and systemic toxic effects which may occur with excessive stimulation of the immune cells. The potential toxic effects of each adjuvant on the organism were also evaluated based on the types of vaccine adjuvants. For adjuvant species, toxicity studies such as acute dose toxicity, repeat dose toxicity, and genotoxicity which are included in the in EMEA and WHO guidelines were included in this review. Thus, marketing activities and the final license of new vaccines will be supported by providing an immunogenic effect/side effect approach in vaccine adjuvants in light of the information in the literature and guidelines.

REFERENCES

Akdoğan, M. and M. Yöntem. 2018. "Cytokines." *Online Turkish Journal of Health Sciences* 3(1): 36–45. https://doi.org/10.26453/otjhs.350321.

Aucouturier, J., L. Dupuis, and V. Ganne. 2001. "Adjuvants designed for veterinary and human vaccines." *Vaccine* 19: 2666–2672.

Aytar, M. and G. Başbülbül. 2019. "Rekombinant Aşılar." *Elektronik Mikrobiyoloji Dergisi TR* 17(1): 1–10.

Baldrick, P., D. Richardson, G. Elliott, and A.W. Wheeler. 2002. "Safety evaluation of mono-phosphoryl lipid A (MPL): An immunostimulatory adjuvant." *Regulatory Toxicology and Pharmacology* 35: 398–413. https://doi.org/10.1006/rtph.2002.1541.

Batista-Duharte, A., E.B. Lindblad, and E. Oviedo-Orta. 2011. "Progress in understanding adjuvant immunotoxicity mechanisms." *Toxicology Letters* 203: 97–105. https://doi.org/10.1016/j.toxlet.2011.03.001.

Batista-Duharte, A., D. Portuondo, Z.I. Carlos, and O. Pérez. 2013. "An approach to local immunotoxicity induced by adjuvanted vaccines." *International Immunopharmacology* 17(3): 526–536. https://doi.org/10.1016/j.intimp.2013.07.025.

Batista-Duharte, A., D. Portuondo, O. Pérez, and Z.I. Carlos. 2014. "Systemic immunotoxic-ity reactions induced by adjuvanted vaccines." *International Immunopharmacology* 20: 170–180. https://doi.org/10.1016/j.intimp.2014.02.033.

Baxter, D. 2007. "Active and passive immunity, vaccine types, excipients and licensing." *Occupational Medicine* 57: 552–556. https://doi.org/10.1093/occmed/kqm110.

Bengtsson, K.L., B. Morein, and A.D.M.E. Osterhaus. 2011. "ISCOM technology-based matrix M ™ adjuvant: Success in future vaccines relies on formulation." *Expert Review of Vaccines* 10(4): 401–403. https://doi.org/10.1586/erv.11.25.

Bernocchi, B., R. Carpentier, and D. Betbeder. 2017. "Nasal nanovaccines." *International Journal of Pharmaceutics* 530: 128–138. https://doi.org/10.1016/j.ijpharm.2017.07.012.

Beyaz, F. 2004. "Development, Functions and Histochemical Properties of T Lymphocytes." *Journal of The Faculty of Veterinary Medicine Erciyes University* 11(1): 61–66.

Bijker, M.S., C.J.M. Melief, R. Offringa, and S.V.D. Burg. 2007. "Design and development of synthetic peptide vaccines: Past, present and future." *Expert Review of Vaccines* 6(4): 591–603.

Bird, J.J., D.R. Brown, A.C. Mullen, N.H. Moskowitz, M.A. Mahowald, J.R. Sider, T.F. Gajewski, C. Wang, and S.L. Reiner. 1998. "Helper T cell differentiation is controlled by the cell cycle." *Immunity* 9: 229–237.

Brennan, F.R. and G.D. 2005. "Non-Clinical safety evaluation of novel vaccines and adju-vants: New products, new strategies." *Vaccine* 23: 3210–3222. https://doi.org/10.1016/j.vaccine.2004.11.072.

Van Der Burg, S.H., M.S. Bijker, M.J.P., Welters, R. Offringa, and C.J.M. Melief. 2006. "Improved peptide vaccine strategies, creating synthetic artificial infections to maxi-mize immune efficacy." *Advanced Drug Delivery Reviews* 58: 916–930. https://doi.org/10.1016/j.addr.2005.11.003.

Casella, C.R. and T.C. Mitchell. 2008. "Putting endotoxin to work for us: Monophosphoryl lipid A as a safe and effective vaccine adjuvant." *Cellular and Molecular Life Sciences* 65: 3231–3240. https://doi.org/10.1007/s00018-008-8228-6.

Celis, E. 2002. "Getting peptide vaccines to work: Just a matter of quality control ?" *The Journal of Clinical Investigation* 110(12): 1765–1768. https://doi.org/10.1172/JCI200 217405.Optimizing.

Cordeiro, A.S., and M.J. Alonso. 2016. "Recent advances in vaccine delivery." *Pharmaceutical Patent Analyst* 5(1): 49–73.

Defranco, A.L., R.M. Locksley and M. Robertson. 2007. "Overview of the immune response." In *Immunity: The immune response in infectious and inflammatory disease*. Oxford: New Science Press Ltd, 1–18.

Elgert, K.D. 2009a. "Cell and orrgans of the immune system." In *Immunology: Understanding the immune system*. New Jersey: John Wiley&Sons, 27–59.

Elgert, K.D. 2009b. "Innate immunity." In *Immunology: Understanding the immune system*. New Jersey: John Wiley&Sons, 61–104.

Elgert, K.D. 2009c. "Introduction to the immune system." In *Immunology: Understanding the immune system*, 1–25. New Jersey: John Wiley&Sons.

EMEA. 2005. "Guideline on adjuvants in vaccines for human use." *Committee for Medicinal Products for Human Use (CHMP)*, 1–18.

EMEA. 2018. "Guideline on clinical evaluation of vaccines." *Committee for Medicinal Products for Human Use (CHMP), no. EMEA/CHMP/VWP/164653/05 Review* 1: 1–21.

Engels, N. and J. Wienands. 2018. "Memory control by the B cell antigen receptor Niklas Engels." *Immunological Reviews* 283: 150–160. https://doi.org/10.1111/imr.12651.

Eratalay, A. and F. Öner. 2001. "Aşılar ve Aşı Adjuvanları." *FABAD J. Pharm. Sci* 25: 21–33.

Forster, R. 2012. "Study designs for the nonclinical safety testing of new vaccine products." *Journal of Pharmacological and Toxicological Methods* 66: 1–7. https://doi.org/10.1016/j.vascn.2012.04.003.

Francis, M.J. 2018. "Recent advances in vaccine technologies." *The Veterinary Clinics of North America: Small Animal Practice* 48: 231–241. https://doi.org/10.1016/j.cvsm.2017.10.002.

Garçon, N., L. Segal, F. Tavares, and M.V. Mechelen. 2011. "The safety evaluation of adjuvants during vaccine development: The AS04 experience." *Vaccine* 29: 4453–4459. https://doi.org/10.1016/j.vaccine.2011.04.046.

Germolec, D., R. Luebke, A. Rooney, K. Shipkowski, R. Vandebriel, and H.V. Loveren. 2017. "Immunotoxicology: A brief history, current status and strategies for future immunotoxicity assessment." *Current Opinion in Toxicology* 5: 55–59.

Glueck, R. 2002. "Pre-clinical and clinical investigation of the safety of a novel adjuvant for intranasal immunization." *Vaccine* 20(Suppl. 1): 42–44.

Glück, R., K.G. Burri, and I. Metcalfe. 2005. "Adjuvant and antigen delivery properties of virosomes." *Current Drug Delivery* 2: 395–400.

Goran, A., F. Schmiedeke, C. Bachert, B.M. Bröker, and S. Holtfreter. 2020. "Allergy — A new role for T cell superantigens of staphylococcus aureus?" *Toxins* 12(176): 1–21. https://doi.org/10.3390/toxins12030176.

Gül, B. and B. Dikmen-Yurdakök. 2019. "Vaccine Adjuvants and Their Undesirable Effects." *Bulletin of Veterinary Pharmacology and Toxicology Association* 10(2): 91–105.

Hanna, Jacob, J.H. Buckner, O. Mandelboim, J. Hanna, T. Gonen-gross, J. Fitchett, T. Rowe, et al. 2004. "Novel APC-like properties of human NK cells directly regulate T cell activation find the latest version: Novel APC-like properties of human NK cells directly regulate T cell activation." *The Journal of Clinical Investigation* 114(11): 1612–1623. https://doi.org/10.1172/JCI200422787.1612.

Hogen Esch, H., D.T. O'Hagan, and C.B. Fox. 2018. "Optimizing the utilization of aluminum adjuvants in vaccines: You might just get what you want." *NPJ Vaccines* 51(September): 1–11. https://doi.org/10.1038/s41541-018-0089-x.

Ivanov, K., E. Garanina, A. Rizvanov, and S. Khaiboullina. 2020. "Inflammasomes as targets for adjuvants." *Pathogens* 9(252): 1–12.

Justiz Vaillant, A.A., and K. Ramphul. 2020. Immunoglobulin. Treasure Island: Stat Pearls. https://www.ncbi.nlm.nih.gov/books/NBK513460/.

Kaech, S.M., E.J. Wherry, and R. Ahmed. 2002. "Effector and memory T-cell differentiation: Implications for vaccine development." *Nature* 2 (April): 251–262. https://doi.org/10.1038/nri778.

Kametani, Y., A. Miyamoto, B. Tsuda, and Y. Tokuda. 2015. "B cell epitope-based vaccination therapy." *Antibodies* 4: 225–239. https://doi.org/10.3390/antib4030225.

Keith, L.S., D.E. Jones, and C.S.J. Chou. 2002. "Aluminum toxicokinetics regarding infant diet and vaccinations." *Vaccine* 20(Suppl. 3): 13–17.

Kindrachuk, J., H. Jenssen, M. Elliott, R. Townsend, A. Nijnik, S.F. Lee, V. Gerdts, L.A. Babiuk, S.A. Halperin, and R.E.W. Hancock. 2009. "A novel vaccine adjuvant comprised of a synthetic innate defence regulator peptide and CpG oligonucleotide links innate and adaptive immunity." *Vaccine* 27: 4662–4671. https://doi.org/10.1016/j.vaccine.2009.05.094.

Kishimoto, T., T. Taga, and S. Akira. 1994. "Cytokine signal transduction." *Cell* 76: 253–262.

Klinman, D.M. 2006. "Adjuvant activity of CpG oligodeoxynucleotides." *International Reviews of Immunology* 25: 135–154. https://doi.org/10.1080/08830180600743057.

Klinman, D.M., S. Klaschik, T. Sato, and D. Tross. 2009. "CpG Oligonucleotides as adjuvants for vaccines targeting infectious diseases." *Advanced Drug Delivery Reviews* 61(3): 248–255. https://doi.org/10.1016/j.addr.2008.12.012.

Klostermen, L. 2009. "What is the immune system?" In Ang, K. (Ed.), *Immune system*. New York: Marshall Covendish Benchmark, 1–18.

Kohler, S., and A. Thiel. 2009. "Life after the thymus: CD31+ and CD31- human naive CD4+ T-cell subsets." *Blood* 113(4): 769–774. https://doi.org/10.1182/blood-2008-02-139154.

Krakauer, T. and C. Russo. 2001. "Serum cytokine levels and antibody response to influenza vaccine in the elderly." *Immunopharmacology and Immunotoxicology* 23(1): 35–41. https://www.ncbi.nlm.nih.gov/books/NBK513460/.

Lebien, T.W. and T.F. Tedder. 2008. "B lymphocytes : How they develop and function." *Blood* 112(5): 1570–1580. https://doi.org/10.1182/blood-2008-02-078071.

Li, W., M.D. Joshi, S. Singhania, K.H. Ramsey, and A.K. Murthy. 2014. "Peptide vaccine: Progress and challenges." *Vaccines* 2: 515–536. https://doi.org/10.3390/vaccines2030515.

Lingnau, K., K. Riedl, and A.V. Gabain. 2007. "IC31 ® and IC30, novel types of vaccine adjuvant based on peptide delivery systems." *Expert Rev. Vaccines* 6(5): 741–746.

Lund, F.E. 2008. "Cytokine-Producing B lymphocytes — key regulators of immunity." *Current Opinion in Pharmacology* 20: 332–338. https://doi.org/10.1016/j.coi.2008.03.003.

Luster, M.I. 2013. "A historical perspective of immunotoxicology." *Journal of Immunotoxicology* 6901: 1–6. https://doi.org/10.3109/1547691X.2013.837121.

Maughan, C.N., S.G. Preston, and G.R. Williams. 2015. "Particulate inorganic adjuvants: Recent developments and future outlook." *Journal of Pharmacy and Pharmacology* 67(3): 426–449. https://doi.org/10.1111/jphp.12352.

Maurer, M., S. Altrichter, O. Schmetzer, J. Scheffel, M.K. Church, M. Metz, R.J. Ludwig, and M.K. Church. 2018. "Immunoglobulin E-mediated autoimmunity." *Frontiers in Immunology* 9: 1–17. https://doi.org/10.3389/fimmu.2018.00689.

Morelli, A.B., D. Becher, S. Koernig, A. Silva, D. Drane, and E. Maraskovsky. 2012. "ISCOMATRIX: A novel adjuvant for use in prophylactic and therapeutic vaccines against infectious diseases." *Journal of Medical Microbiology* 61: 935–943. https://doi.org/10.1099/jmm.0.040857-0.

Moser, C., M. Amacker, and R. Zurbriggen. 2011. "Influenza virosomes as a vaccine adjuvant and carrier system." *Expert Reviews* 10(4): 437–446.

Mosmann, T.R. and R.L Coffman. 1989. "TH1 and TH2 cells: Different patterns of lymphokine secretion lead to different functional properties." *Ann. Rev. Immunol.* 7: 145–173.

Neutra, M.R. and P.A. Kozlowski. 2006. "Mucosal vaccines: The promise and the challenge." *Nature Immunology* 6: 148–158. https://doi.org/10.1038/nri1777.

Oleszycka, E. and E.C. Lavelle. 2014. "Immunomodulatory properties of the vaccine adjuvant alum." *Current Opinion in Immunology* 28(1): 1–5. https://doi.org/10.1016/j.coi.2013.12.007.

Ozato, K., H. Tsujimura, and T. Tamura. 2002. "Toll-like receptor signaling and regulation of cytokine gene expression in the immune system." *Bio Techniques* 33: 66–75.

Parker, D.C. 1993. "T cell- dependent B cell activation." *Annual Reviews Further* 11: 332–360.

Pearse, M.J. and D. Drane. 2005. "ISCOMATRIX adjuvant for antigen delivery." *Advanced Drug Delivery Reviews* 57: 465–474. https://doi.org/10.1016/j.addr.2004.09.006.

Peterfalvi, A., E. Miko, T. Nagy, B. Reger, D. Simon, A. Miseta, and L. Szereday. 2019. "Much more than a pleasant scent: A review on essential oils supporting the immune system." *Molecules* 24: 1–16.

Pulendran, B., and R. Ahmed. 2011. "Immunological mechanisms of vaccination." *Nature Immunology* 12(6): 509–517. https://doi.org/10.1038/ni.2039.

Ragupathi, G., P. Damani, K. Deng, M.M. Adams, J. Hang, C. George, P.O. Livingston, and D.Y. Gin. 2010. "Preclinical evaluation of the synthetic adjuvant SQS-21 and its constituent isomeric saponins." *Vaccine* 28: 4260–4267. https://doi.org/10.1016/j.vaccine.2010.04.034.

Rudra, J.S., Y.F. Tian, J.P. Jung, and J.H. Collier. 2009. "A self-assembling peptide acting as an immune adjuvant." *PNAS* 107(2): 622–627. https://doi.org/10.1073/pnas.0912124107.

Seder, R.A., P.A. Darrah, and M. Roederer. 2008. "T-cell quality in memory and protection: Implications for vaccine design." *Nature* 8(April): 247–259. https://doi.org/10.1038/nri2274.

Singh, M. and D.O. Hagan. 1999. "Advances in vaccine adjuvants." *Nature Biotechnology* 17: 1075–1081.

Skwarczynski, M. and I. Toth. 2011. "Peptide-Based subunit nanovaccines." *Current Drug Delivery* 8: 282–289.

Skwarczynski, M. and I. Toth. 2016. "Peptide-based synthetic vaccines." *Chemical Science* 7: 842–854. https://doi.org/10.1039/c5sc03892h.

Sompayrac, L.M. 2019. "An overview 1." In *How the immune system works*. Oxford: John Wiley&Sons, 1–12.

Song, X. and S. Hu. 2009. "Adjuvant activities of saponins from traditional chinese medicinal herbs." *Vaccine* 27: 4883–4890. https://doi.org/10.1016/j.vaccine.2009.06.033.

Songu, M. and H. Katılmış. 2012. "Immune system and protection from infections." *Journal of Medical Updates* 2(1): 31–42. https://doi.org/10.2399/jmu.2012001006.

Spickler, A.R. and J.A. Roth. 2003. "Adjuvants in veterinary vaccines: Modes of action and adverse effects." Journal of Veterinary Internal *Medicine* 17: 273–281.

Stephenson, R., H. You, D. Mcmanus, and I. Toth. 2014. "Schistosome vaccine adjuvants in preclinical and clinical research." *Vaccines* 2: 654–685. https://doi.org/10.3390/vaccines2030654.

Sun, H., Y. Xie, and Y. Ye. 2009. "Advances in saponin-based adjuvants." *Vaccine* 27: 1787–1796. https://doi.org/10.1016/j.vaccine.2009.01.091.

Sun, Y., M. Gruber, and M. Matsumoto. 2012. "Overview of global regulatory toxicology requirements for vaccines and adjuvants." *Journal of Pharmacological and Toxicological Methods* 65(2): 49–57. https://doi.org/10.1016/j.vascn.2012.01.002.

Tang, P.C., Y. Zhang, M.K. Chan, W.W. Lam, J.Y. Chung, W. Kang, K. To, and H. Lan. 2020. "The emerging role of innate immunity in chronic kidney diseases." *International Journal of Molecular Sciences* 21: 1–19.

Tomljenovic, L. and C.A. Shaw. 2011. "Aluminum vaccine adjuvants: Are they safe?" *Current Medicinal Chemistry* 18: 2630–2637.

Tornesello, A.L., M. Tagliamonte, M.L. Tornesello, F.M. Buonaguro, and L. Buonaguro. 2020. "Nanoparticles to improve the efficacy of peptide-based cancer vaccines." *Cancers* 12: 1–20.

Vallhov, H., N. Kupferschmidt, S. Gabrielsson, S. Paulie, M. Strømme, A.E. Garcia-Bennett, and A. Scheynius. 2012. "Adjuvant properties of mesoporous silica particles tune the development of effector T cells." *Small* 8(13): 2116–2124. https://doi.org/10.1002/smll.201102620.

Vecchi, S., S. Bufali, D.A.G. Skibinski, D.T. O'Hagan, and M. Singh. 2012. "Aluminum adjuvant dose guidelines in vaccine formulation for preclinical evaluations." *Journal of Pharmaceutical Sciences* 101(1): 17–20. https://doi.org/10.1002/jps.22759.

Verdier, F. 2002. "Non-clinical vaccine safety assessment." *Toxicology* 174: 37–43.

Wang, Z. and J. Xu. 2020. "Better adjuvants for better vaccines: Progress in adjuvant delivery systems, modifications, and adjuvant – antigen codelivery." *Vaccines* 8(1): 128.

Watts, T.H. 2005. "TNF / TNFR family members in costimulation of T cell responses." *Annual Review of Immunology* 23: 23–68. https://doi.org/10.1146/annurev.immunol. 23.021704.115839.

WHO. 2013. "Guidelines on the nonclinical evaluation of vaccine adjuvants and adjuvanted vaccines." 64th Meeting of the WHO Expert Committee on Biological Standardization, 1–55.

Wilschut, J. 2009. "Influenza vaccines: The virosome concept." *Immunology Letters* 122: 118–121. https://doi.org/10.1016/j.imlet.2008.11.006.

Wolf, J.J., C.V. Kaplanski, and J.A. Lebron. 2010. "Nonclinical safety assessment of vaccines and adjuvants." In Vaccine *adjuvants, methods in molecular biology*. Springer Science + Businesss Media, 29–40. https://doi.org/10.1007/978-1-60761-585-9.

Yang, H. and D.S. Kim. 2015. "Peptide immunotherapy in vaccine development: From epitope to adjuvant." In *Advances in protein chemistry and structural biology*, 1st ed., Vol. 99. Elsevier Inc, 1–14. https://doi.org/10.1016/bs.apcsb.2015.03.001.

Yano, A., A. Onozuka, Y. Asahi-ozaki, S. Imai, N. Hanada, Y. Miwa, and T. Nisizawa. 2005. "An ingenious design for peptide vaccines." *Vaccine* 23: 2322–2326. https://doi.org/10.1 016/j.vaccine.2005.01.031.

Yurdakök, K. and T. İnce. 2008. "Aşı Adjuvanları." *Cocuk Sağlığı ve Hastalıkları Dergisi* 51: 225–239.

Yüzbaşıoğlu, D., F. Ünal, F. Koç, S. Öztemel, H. Aksoy, S. Mamur, and F. Demirtaş Korkmaz. 2013. "Genotoxicity assessment of vaccine adjuvant squalene." *Food and Chemical Toxicology Journal* 56: 240–246. https://doi.org/10.1016/j.fct.2013.02.034.

Zeitlin, L., R.A. Cone, T.R. Moench, and K.J. Whaley. 2000. "Preventing infectious disease with passive immunization." *Microbes and Infection* 2: 701–708. https://doi.org/10.1 016/S1286-4579(00)00355-5.

Zhao, G., S. Chandrudu, M. Skwarczynski, and I. Toth. 2017. "The application of self-assembled nanostructures in peptide-based subunit vaccine development." *European Polymer Journal* 1–34. https://doi.org/10.1016/j.eurpolymj.2017.02.014.

Zimmermann, S., A. Dalpke, and K. Heeg. 2008. "CpG oligonucleotides as adjuvant in therapeutic vaccines against parasitic infections." *International Journal of Medical Microbiology* 298: 39–44. https://doi.org/10.1016/j.ijmm.2007.07.011.

15 Role of Epigenetics in Immunity and Immune Response to Vaccination

Necip Ozan Tiryakioğlu

Necip Ozan Tiryakioğlu will always be remembered for his hardworking, scientific seriousness, and productivity. We would like to extend our condolences to his family, lovers, and the scientific community.

CONTENTS

15.1 THE IMMUNE SYSTEM

An effective immune system is the result of interaction and interplay between the innate immune system and the adaptive immune system.

Innate immunity functions as general resistance, which provides continuous but non-specific protection against pathogens. This general resistance involves natural protections such as skin, low gastric pH, lysozyme, mucus, cilia, and the natural microflora. The innate immune system does not provide a specific and reinforced response to repeated pathogen exposure and is therefore described as not having "long term memory" (Turvey and Broide 2010).

Additional components of innate immunity include interferons and collagen-containing C-type lectins (collectins). Interferons are categorized as Type I and Type II. Type I interferons are synthesized by host cells in response to viral infection and give rise to increased antiviral activity in neighboring cells (De Andrea et al. 2002). Collagen-containing C-type lectins on the other hand are present in serum and on mucosal surfaces. Collectins act by binding to membrane oligosaccharides or lipids of microorganisms. This binding may cause a direct elimination of the micro-organisms by membrane destabilization or result in an indirect response such as facilitating the phagocytosis of infectious microorganisms by cell aggregation (Atochina-Vasserman 2012). The innate immune system also has three complement pathways classified as the classical pathway, the properdin pathway, and the lectin pathway. The classical pathway is induced by the binding of immunoglobulin M (IgM) or some immunoglobulin G (IgG) antibodies to the surface antigens of microorganisms. The other complement pathways, the properdin, and the lectin pathways do not require antibody binding for activation. Instead they are induced by the accumulation of certain membrane-binding proteins on microbial membranes. The combinatory action of these three pathways is called the complement cascade which triggers three important functions of the immune system: 1) phagocytosis, 2) inflammation, and 3) rupturing of bacterial cell walls (Rus, Cudrici, and Niculescu 2005).

Another crucial part of innate immunity is the pattern recognition receptors (PPRs), which are produced by innate immune system cells and function as molecular detectors for certain pathogen-associated molecules. The first group of molecules recognized by PPRs are pathogen-associated molecular patterns (PAMPs) (Amarante-Mendes et al. 2018; Akira, Uematsu, and Takeuchi 2006). PAMPs are not found naturally in mammalian cells and comprise of molecules which are necessary for microbial survival. The second group of molecules recognized by PPRs are the damage-associated molecular patterns (DAMPs). DAMPs are molecules released by host cells upon cell damage (Rajaee, Barnett, and Cheadle 2018). PPRs cooperate with each other to initiate expression of certain genes, triggering cellular immune responses to eliminate pathogens. They also induce the release of inflammatory cytokines (Takeda, Kaisho, and Akira 2003). The final components of innate immunity are mononuclear phagocytes and granulocytic cells. These cells also establish the link between innate and adaptive immunity. Mononuclear phagocytes are produced by bone marrow and can later differentiate into monocytes. Differentiated monocytes later migrate in tissues and further differentiate into macrophages or dendritic cells (Hume et al. 2002). Dendritic cells are crucial for the link between the innate and adaptive immune systems (Hoebe, Janssen, and Beutler 2004). Granulocytes are categorized as neutrophils, eosinophils, basophils, and mast cells (Breedveld et al. 2017). Neutrophils are the most numerous and active of phagocytic cells. Eosinophils show less phagocytic activity and are more specialized against parasites. Basophil constitutes approximately 1% of circulating leukocytes (Stone, Prussin, and Metcalfe 2010). They are particularly prominent in allergic reactions and release histamine among other molecules when they receive damage (Siracusa et al. 2013). Mast cells reside in tissues and contain relatively high amounts of histamine and heparin which are released upon their activation (Rivera et al. 2008). A fully functional immune system and therefore an effective immune response relies on the interaction between the innate and adaptive immune systems. Innate immunity functions as a quick response and may in some cases be able to eliminate the pathogen by itself. In other cases, the activation of the adaptive immune system by the innate immune system is required to eliminate the pathogen. The main functional difference between these systems is the response time. The first encounter with a certain pathogen evokes a quick response from the innate immune system and a slower but more systemic response from the adaptive system. The adaptive immune system adapts to antigens from pathogens or vaccination over repeated exposure, hence its name, and provides a more rapid response with each exposure due to its "memory" (Nicholson 2016). The two types of adaptive immunity are antibody-mediated immunity and cell-mediated immunity. The B cells and antibodies constitute the antibody-mediated immunity while the T cells compose cell-mediated immunity (Janeway et al. 2001). The antibody-mediated immunity, also called as humoral immunity, functions via activated B cells and antibodies. B cells originate from bone marrow and express B cell receptors which bind specific antigens and induce a response. Since these antigens do not require T cell activation to activate B cells, they are called T-independent antigens. Bacterial polysaccharides and lipopolysaccharides are such T-independent antigens and can induce antibody production by B cell without the help of T cells (Janeway et al. 2001). The activation of B cells without T-helper cells induce a weaker response in comparison to the activation with T-helper cells (Kurosaki, Kometani, and Ise 2015; McHeyzer-Williams et al. 2012). The response induced via T-helper cells is more effective, especially in terms of long-term immune memory, which is the main goal of vaccine-induced immunizations (Goldsby et al. 2003). The binding of antigens to B cell receptors and the release of cytokines from T cells stimulate the maturation of B cells, leading to more specific antibody production. Mature B cells later produce clones which produce IgM. IgM is the foremost antibody produced after a first-time encounter with a specific antigen (Capolunghi et al. 2013). Further downstream in the immune response the production of antibodies shifts to IgGs from IgMs. This event is called immunoglobulin class switching or isotype switching (Market and Papavasiliou 2003). IgG is the main antibody type in circulation and more effective at antibody neutralization. These features make IgGs crucial for vaccine immunization. The production of memory cells and IgGs facilitates a quicker and widespread response for the future encounters with the pathogen.

The second type of adaptive immunity is cell-mediated immunity. Cell-mediated immunity functions via T cells which are released into circulation following their maturation in the thymus. T cells are categorized as $CD4^+$ cells and $CD8^+$ cells according to the type of T cell receptor (TCR) they express. The helper T cells express CD4 receptors while cytotoxic T cells express CD8 TCR (Margolick, Markham, and Scott 2006). There are two types of helper T cells, Th1 and Th2. Th1 cells are involved in cell-mediated immunity and Th2 cells are involved in antibody-mediated immunity (O'Garra and Arai 2000). In contrast to B cells, T cells require antigen processing by antigen-presenting cells for antigen recognition. Following activation and clonal expansion, memory T cells are produced to induce a rapid immune response for subsequent infections (Pennock et al. 2013). Following their formation, memory T cells can provide immunity for approximately ten years (Hammarlund et al. 2003).

15.2 IMMUNIZATION

The two types of immunity, active or passive immunity, can be achieved via vaccination. Active immunity via vaccination occurs as a result of exposure to antigens while passive immunity is achieved through administration of antibodies.

The success of immunization depends on various factors including the antigen, the route of administration, and the dose.

Active immunization is achieved by the introduction of an antigen's most immunogenic form possible to elicit a long-term immune response without the pathogenic effects. While most viral vaccines are based on live attenuated viruses, many bacterial vaccines are based on acellular components, including harmless toxin components.

The detection of immunization agents by the innate immune system requires the recognition of immune response-inducing regions of the antigens called epitopes. Following the activation of innate immunity these antigens will be represented on the surface of antigen presenting cells (APCs). In case of a viral antigen, APCs will present the antigen to cytotoxic T cells and induce the cell-mediated immunity. In case of a bacterial antigen, APCs present the antigen to helper T cells and induce humoral immunity (Figure 15.1).

Passive immunization relies on pre-formed antibodies. These antibodies can be obtained commercially or from a donor.

Both active and passive immunization show their effect through inhibition of a pathogen's attachment molecule or toxin, activation of complement system, and induction of cell cytotoxicity.

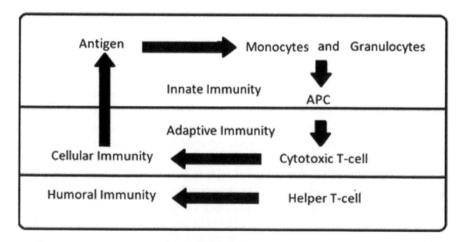

FIGURE 15.1   Immune response to antigen.

15.3   EPIGENETICS

Epigenetics is a rapidly evolving field and studies the functional changes which do not involve sequence alterations. The Greek prefix epi- indicates alterations "in addition to" the traditional genetics based on sequence variations. Most epigenetic changes function as a form of gene regulation, effecting gene activity and expression. Epigenetic modifications are essential for many cellular processes including development, differentiation, and cell metabolism.

There are several mechanisms contributing to gene regulation via epigenetic modifications, including covalent modifications of histones, DNA methylation, and non-coding RNAs.

15.3.1   COVALENT MODIFICATIONS OF HISTONES AND DNA

Covalent modifications are crucial for development where they regulate the differentiation of stem cells to different cell types with various functions (Teif et al. 2014). One of the best-studied examples of this effect is the differentiation of neural stem cells into oligodendrocytes through chromatin remodeling with histone methylation (Hallgrimsson and Hall 2011). There are two mechanisms for chromatin remodeling: 1) post translational modifications and 2) DNA methylation.

Post translational modifications are achieved through the modifications of histone amino acids. These modifications alter the chromatin structure which in turn effect transcription. The most common and widely studied histone modifications are acetylation, phosphorylation, and methylation.

15.3.1.1   Histone Acetylation

Histone acetylation is the most common and most studied form of histone modifications. The regulation of acetylation is achieved by opposing actions of two enzyme families: histone acetyltransferases (HATs) and histone deacetylases (HDACs). HATs transfer an acetyl group to lysine side chains which results in weaker interaction between DNA and histones due to neutralization of lysine's charge. The decrease in interaction between DNA and histones leads to increased gene transcription. The action of HDACs shows the opposing effect by catalyzing the removal of acetyl groups from lysine residues. Deacetylation restores the positive charge of lysine residues which stabilizes the interaction between positively charged histones and negatively charged DNA. The increased interaction contributes to the transcriptional repression of certain genes.

15.3.1.2   Histone Phosphorylation

In contrast to acetylation, phosphorylation involves more than one type of amino acid, namely serines, threonines, and tyrosines (Rossetto, Avvakumov, and Côté 2012). Phosphorylation is catalyzed by histone kinases where a phosphate group from ATP is transferred to the amino acid side chain. Phosphorylation increases the negative charge of the histone resulting in structural changes which are reversed by phosphatases by removal of phosphate groups from amino acid side chains on histones (Rossetto, Avvakumov, and Côté 2012). Although phosphorylation has been associated in general with DNA repair events, recent studies have shown phosphorylation also plays a role in gene regulation similar to acetylation (Kotova et al. 2011).

15.3.1.3   Histone Methylation

Histone methylation can occur on either lysines or arginines. In contrast to acetylation and phosphorylation, methylations do not change the charge of the histones and are not limited with the addition of single chemical group. Lysines can accept up to three methyl groups while arginines can accept up to two methyl groups.

As with other types of covalent histone modifications, histone methylations influence transcriptional activity by altering the chemical interaction between histones and DNA (Ng et al. 2009). The decrease in the interaction results in an increase in transcriptional activity due to more easily accessible DNA. The change in the transcriptional activity is determined by the type modified amino acids and the number of methyl groups added (Ng et al. 2009).

The modification reaction is catalyzed by histone methyltransferases where methyl groups from S-adenosyl methionines (SAM) are transferred to amino acid residues on H3 and H4 histones (Bannister and Kouzarides 2011).

15.3.1.4   DNA Methylation

The second mechanism of covalent modifications is DNA methylation. DNA methylation occurs when methyl groups from SAMs are transferred to cytosines on DNA molecules via DNMTs to form a 5-methylcytosine (Figure 15.2). Methylation is crucial for many cellular processes including development, imprinting, and X-chromosome inactivation. This process usually results in functional changes for the methylated DNA segment. In humans 70% of all CG dinucloetides are methylated (Strichman-Almashanu et al. 2002).

CpG islands are segments of approximately 1000 base pairs long, GC-rich DNA with a minimum length of 200 bp and a minimum GC percentage of 50%. They are located in nearly 40% of mammalian promoters which are often promoters of housekeeping genes (Fatemi et al. 2005). CpG sites near housekeeping gene promoters are not methylated to prevent the silencing of these genes. On the other hand, CpG islands near inactive genes are methylated resulting in silencing of gene expression (Mohn et al. 2008) (Figure 15.3). Under normal physiologic conditions, gene silencing due to methylation contributes to development and differentiation (Weber et al. 2007). In terms of pathogenic conditions, changes in DNA methylation are usually associated with tumorigenesis. Deregulation of DNA methylation patterns is a

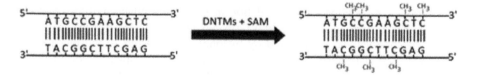

FIGURE 15.2   CpG Island methylation near promotor leads to transcriptional silencing.

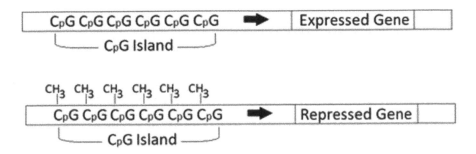

FIGURE 15.3 Unmethylated and methylated states of a dsDNA segment.

frequent event in many cancer types. The deregulation is usually observed as hypermethylation of tumor suppressor genes and hypomethylation of oncogenes leading to downregulation of tumor suppressors and upregulation of oncogenes (Skvortsova, Stirzaker, and Taberlay 2019).

15.3.2 Non-Coding RNAs

Non-coding RNAs (ncRNA) are RNA molecules that do not code for a protein and therefore are not translated into proteins. While showing a wide range of functionality, most of the ncRNAs have regulatory or housekeeping roles. Epigenetically functional ncRNAs include microRNAs (miRNA), long non-coding RNAs (lncRNA) and circular RNAs (circRNA). In addition to their functional diversity, ncRNAs can also be classified according to their size as short ncRNAs and long ncRNAs (Zaratiegui, Irvine, and Martienssen 2007). ncRNAs with maximum length of 200 nucleotides (nt) are considered short ncRNAs while ncRNAs longer than 200 nt are classified as long ncRNAs.

15.3.2.1 MicroRNAs

The most extensively studied group of short ncRNAs are miRNAs. miRNAs are single stranded ~20 nt long regulatory RNAs which mainly function as post-transcriptional regulators of gene expression.

miRNAs are transcribed by RNA polymerase II as a hairpin loop structure, called the pri-miRNA. Following transcription, pri-miRNAs are capped and polyadenilated (Cai, Hagedorn, and Cullen 2004). Pri-miRNAs undergo a nuclear processing where the pri-miRNA is cleaved by microprocessor complex, comprised of DiGeorge syndrome critical region 8 (DGCR8) and Drosha proteins (Gregoryi, Chendrimada, and Shiekhattar 2006; Conrad et al. 2014). The cleavage product is called a pre-miRNA. Nuclear processing is followed by the export of pre-miRNA to cytoplasm by Exportin. Pre-miRNA is cleaved by an enzyme called dicer to yield a double stranded miRNA complex. Following the cleavage by dicer, one strand of this miRNA duplex is loaded into the RNA-induced silencing complex (RISC) to interact with its target (Kim and Kim 2012; Park et al. 2011). Gene silencing by miRNAs has two modes depending on the miRNA target complementarity. Perfect or near perfect complementarity between the miRNA and its target mRNA induces the cleavage and degradation of target mRNA. In case of a non-perfect complementarity, the target mRNA is silenced through inhibition of translation (Lim et al. 2005).

15.3.2.2 Circular RNAs

Circular RNAs are formed through a process called backsplicing during which splice acceptor and splice donor sites of a pre-mRNA are covalently joined to produce a circular transcript (Figure 15.4) (Barrett, Wang, and Salzman 2015).

The lack of a 5′ or 3′ end makes these RNAs more stable due to their resistance to exonuclease-mediated degradation. One study has shown that their half-life is at least 2.5 times longer than the half-life of their linear RNAs (Enuka et al. 2016).

Several circRNAs show tissue-specific expression, indicating a tissue-dependent function role for ncRNAs (Salzman et al. 2013). This is also supported by the discrepancy between circRNA and corresponding mRNA levels (Salzman et al. 2012).

One of first circRNAs to be characterized is the mouse circRNA: Sry. Findings indicate a very interesting function for this circRNA. Evidently, Sry represses miR-138 activity by binding 16 miR-138 molecules and acting as a miRNA sponge (Hansen et al. 2013). ciRS-7 is another circRNA which functions as a miRNA sponge. A recent study has shown that ciRS-7 can bind more than 70 molecules of miR-7 (Memczak et al. 2013).

Transcriptional or post-transcriptional regulation has also been suggested as a possible function for circRNAs. It has been shown that some circRNAs contain intron and are localized to the nucleus. These circRNAs can also interact with U1 small nuclear ribonucleoprotein to promote transcription (You et al. 2015).

15.3.2.3 Long Non-Coding RNAs

ncRNAs longer than 200 nt are classified as long ncRNAs. In a fashion similar to pre-miRNAs lncRNAs are also transcribed by RNA polymerase II and undergo 5′ capping and polyadenylation. The human genome encodes for approximately 16,000 lcnRNAs

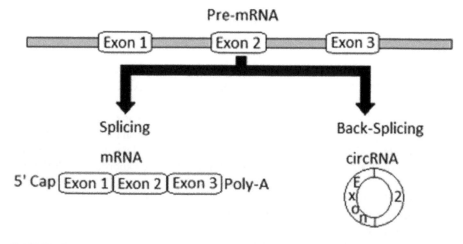

FIGURE 15.4   circRNA formation through backsplicing.

which give rise to almost 30,000 different transcripts. Despite being considered new additions to the field of RNA biology, the function of some lncRNAs has been known for almost 30 years. One of the lncRNAs whose function has been identified in the early 1990s is lncRNA Xist. Xist plays a central role in X-chromosome inactivation (Brown et al. 1992). Recent studies have demonstrated several new functions for lncRNAs including antiviral response in addition to regulation of development and differentiation (Fatica and Bozzoni 2014; Fortes and Morris 2016). A common mechanism for exerting such functions for lncRNAs is to act as post transcriptional regulators by altering mRNA/protein stability and translation (Yoon, Abdelmohsen, and Gorospe 2013).

Certain lncRNAs interact with chromatin modification proteins to regulate gene expression depending on the type modifying protein complex (Morlando et al. 2014; Marchese and Huarte 2014). In terms of regulation of epigenetic mechanisms lncRNAs can also interact with methyltransferases to induce transcriptional repression (Schmitz et al. 2010).

Another way of gene regulation by lncRNAs is achieved via interaction with transcription factors. One of the best studied examples of this, is the binding of GAS5 lncRNA to glucocorticoid receptor (GR). This binding impairs GR's interaction with glucocorticoid response elements (GRE) and suppresses the expression of GRE-containing genes (Kino et al. 2010).

15.4 EPIGENETIC REGULATION OF THE RESPONSE TO IMMUNIZATION

In an immunological context, "heterogeneity" is used to describe two different phenomena and both of them are central to the immune response to vaccination. In the more frequently used sense of the term, immune heterogeneity describes intracellular diversity of the immune cells, resulting in diverse antigen recognition and an overall more effective immune response. In the less frequently used sense of the term, immune heterogeneity describes the interpersonal diversity of the immune response to vaccination. Differential epigenetic regulation is central to immune heterogeneity and therefore plays a crucial role in immune response to vaccination.

Despite the fact that an effective adaptive immune response depends on the diversity of immune cells, a strictly controlled cell differentiation is also required to preserve homeostatic balance. This means that the regulation of the immune system should adhere to strict conditions for differentiation while allowing a flexible regulation of mature cells to provide an effective immune response against various pathogenic factors. Epigenetic regulation of gene expression is an efficient way of controlling differentiation while also allowing flexible adaptation of these cells. This regulation occurs reciprocally as exemplified by the cytokines, which can alter the epigenetic status of immune cells.

15.4.1 COVALENT MODIFICATIONS IN REGULATION OF IMMUNE RESPONSE

Recent studies have shown that epigenetic mechanisms are crucial for development of immune cells (Busslinger and Tarakhovsky 2014). One such study is conducted by Komori et al. where they identified a methylation signature associated with 39 upregulated and hypomethylated genes. This methylation signature is unique to memory CD4 cells and correlates with activation-induced gene expression (Komori et al. 2015). Similar findings have also been obtained by additional studies, where it was shown that related orphan receptor C, P-selectin, and E-selectin genes were hypomethylated in CD4$^+$ memory T cells, in contrast to their hypermethylated states in naive CD4$^+$ T cells (Syrbe et al. 2004; Schmidl et al. 2011). Changes in DNA methylation patterns are also observed in CD8$^+$ T cells during differentiation to memory cells in certain genes after exposure to viral antigens (Youngblood et al. 2011). Evidently alterations to chromatin accessibility play a key role in gene regulation in immune system. Macrophage toll-like receptor (TLR) signaling results in acetylated lysines on histones interacting with cytokine genes. This modification in turn leads to upregulation of these genes by facilitating its interaction with transcription-enhancing elements (Aung et al. 2006; Nicodeme et al. 2010). In contrast to this, a thienotriazolodiazepine derivative, JQ1, suppresses the expression of cytokines by impairing the aforementioned interaction. This suppression has been shown to confer protection against exposure to heat-killed *Salmonella typhimurium* in mice (Belkina et al. 2013). Direct covalent histone modifications can also have an inhibitory effect on gene expression in macrophages. Histone deacetylase 11 modifies the histone structure, giving rise to a more condensed chromatin. This condensation inhibits the interaction of transcription factors with Interleukin 10 (IL-10) gene and suppresses its expression (Villagra et al. 2009).

The histone methyltransferase HRX, methylates GATA3, and IL-4 gene loci, upregulating GATA3 and IL-4 which results in the production of specific Th2 cell cytokines (Yamashita et al. 2006). Interestingly, Zimmermann et al. identified a group of CpG sites whose hypomethylation results in decreased humoral response to influenza vaccination (Zimmermann et al. 2016). In accordance with this finding, a comparison of influenza vaccine responses in different age groups revealed increased epigenetic remodeling in subjects who are over 50 years of age and responded to the influenza vaccination (Gensous et al. 2018).

Some findings even suggest the use of demethylation agents as vaccine potentiator. The cancer-testis/cancer germline antigen, NY-ESO-1, is targeted by vaccines against ovarian cancer. But the efficacy of the vaccine suffers from low NY-ESO-1 expression. When decitabine was used as a demethylation agent to induce global hypomethylation, NY-ESO-1 expression and therefore the efficacy of vaccine was increased (Odunsi et al. 2014).

15.4.2 Non-Coding RNAs in Regulation of Immune Response

Since ncRNAs exert their function as non-translated transcripts, researchers have focused on profiling of ncRNAs in different immune cell types and/or in different immunological conditions, revealing differential expression of various ncRNAs. Neilson et al. have investigated miRNA expression profiles of developing T cells and have shown that in miRNAs levels vary significantly between different developmental stages (Neilson et al. 2007).

As the best studied group of ncRNAs, numerous functions have been assigned to various miRNAs. One such miRNA is miR-155 which affects differentiation of CD4$^+$ T lymphocytes into Th1 cells, the development of regulatory T cells (Rodriguez et al. 2007). miR-155 also plays a role in the development of B cell memory. miR-155 targets the transcription factor PU.1 and PU.1 upregulation facilitates IgG1 switch. In cells with low miR-155 expression, IgG1 generation is significantly impaired (Vigorito et al. 2007).

There are several other miRNAs which can influence the development of B cell development. Studies have shown that increased miR-212/132 expression inhibits B cell progenitor survival, while miR-181, miR-17-92 cluster promotes the differentiation of B cells (Zheng et al. 2018).

Another member of the miR-181 family, miR-181a modulates the threshold for TCR signaling and in turn regulating the sensitivity of T cells to peptide antigens (Li et al. 2007). miR-181a also alters the maturation of T cells as shown in a study conducted by Chen et. al. The findings of this study indicate that miR-181a overexpression results in a decrease in the number of CD8$^+$ T cells (C. Chen et al. 2004).

miR-451 is frequently found in extracellular vesicles and it alters the immune response to whole-virus vaccines against influenza. Its internalization into macrophages downregulates type I interferon (IFN) and interleukin 6 expression in response to whole-virus vaccines (Okamoto et al. 2018). Gannavaram et al. have illustrated additional roles for miRNA in vaccine immunity. miR-21 expression determines the early vaccine immunity induced by LdCen−/− immunization. First, they generated centrin gene-deleted *Leishmania donovani* parasites (LdCen−/−) as live attenuated vaccines against leishmaniasis. Their findings suggested that exposure to this antigen suppresses miR-2, which in turn facilitates IL-12-mediated activation of adaptive immunity and development of a Th1 immunity (Gannavaram et al. 2019). In another study conducted to investigate the effect of H1N1 vaccination on serum miRNA expression in children, miR-142-3p was found to be downregulated following vaccination. Nevertheless, the results could not be replicated in a validation cohort (Drury, Pollard, and O'Connor 2019).

Due to their stability and presence in various body fluids, recent research has focused on the use of ncRNAs as biomarkers as a minimally invasive way of assessing the safety and efficacy of vaccine candidates. One such study was conducted by Atherton et al. to evaluate the usefulness of miRNAs as vaccine response biomarkers. They compared sera miRNA profiles before and after immunization with different respiratory syncytial virus vaccine types in mice. This comparison revealed possible roles for certain miRNAs, including miR-467f and miR-106a. The upregulation miR-467f after vaccination seems to be associated with an increase in the frequency of RSV-specific CD8 T cells and RSV-specific memory T cell precursors, while miR-106a upregulation after vaccination seems to suppress IL-10 expression and regulate Th2-type responses (Atherton et al. 2019).

Functional sequence alterations in miRNAs, especially changes to the seed region have been implied in numerous diseases. In order to investigate the effects of such sequence variations in response to hepatitis B vaccine, Xion et al. have compared genotype frequencies in vaccine responders and non-responders. Their results indicate that rs2910164 variation in miR-146a and rs7372209 variation in miR-26a-1 increased the risk of not responding to hepatitis B vaccine. Furthermore, these variations were also associated with different antibody levels following immunization (Xion et al. 2013). Despite being more numerous and regulating a plethora of pathways, the role of miRNAs in the regulation of immune response is more substantiated in comparison to circRNAs and lncRNAs due to their relatively simple mechanism of action. In each case miRNAs downregulate their targets by post-transcriptional regulation. In comparison, circRNAs and lncRNAs have much more diverse mechanisms of action. Many recent studies, investigating the role of circRNAs in immune system, focus on macrophages. Several circRNAs have been shown to influence polarization and antigen presentation in macrophages (Ng et al. 2016; Zhangi, Zhang, and Lv 2017). One of those circRNAs is circ-RasGEF1B, which stabilizes intercellular adhesion molecule 1 transcripts and effects antigen presentation by promoting cell-cell interactions. Additionally, circ-RasGEF1B has also been shown to be induced by bacterial lipopolysaccharide (LPS) and play an important role in LPS response since its depletion leads to downregulation of LPS-induced molecules (Ng et al. 2017).

Virus infections alter circRNA synthesis. RNA binding protein nuclear factor 110 (NF110) is required for the formation of functional circRNA-protein complexes (circ-RNP) in the nucleus and viral infection induced export of NF110 inhibits circ-RNP formation (Li et al. 2017). It has also been shown that circRNAs also influence the cellular response to simian virus 40 (SV40) where circRNAs act as miRNA decoys and participate in the regulation of SV40-associated pathways (Shi et al. 2017).

lncRNAs are functionally the most diverse group of ncRNAs and therefore have been implicated in various immune response pathways, including differentiation of DCs and T cells, cytokine production, and activation of CD8[+] T cells (Zhang et al. 2013; Wang et al. 2014; Imamura et al. 2014; Wang et al. 2015).

For instance, in T cells the activity of the transcription factor, nuclear factor of activated T cells (NFAT), is inhibited by the lncRNA, noncoding repressor of NFAT (NRON). This results in the suppression of IL-2 production (Willingham et al. 2005). In another study, Wang et al. have shown that lnc-DC activates STAT3 while also blocking its dephosphorylation and contributes to differentiation of DCs (Wang et al. 2014).

Ilott et al. identified 40 differentially expressed lncRNAs in monocytes. Two of these lncRNAs, IL-1β-RBT46 and IL-1β-eRNA, were involved in regulation of LPS-induced gene expression in a similar way to circ-RasGEF1B (Ilott et al. 2014). lncRNAs were also shown to not only regulate LPS-induced expression but also be regulated by LPS themselves. One such lncRNA is Ptprj-as1 which is induced in response to LPS bone marrow-derived mouse macrophages (Dave et al. 2013).

Many lncRNAs are directly associated with immune response to infections and vaccinations. In a study conducted using whole-transcriptome analysis to identify differentially expressed lncRNAs in response to severe acute respiratory syndrome coronavirus (SARS-CoV) infection in mice, 509 lncRNAs have shown differential expression, where approximately 95% of them had a change of 2.5 fold or more (Peng et al. 2010). In 2019 Diogenes et al. conducted a study where they used a meta-analytical approach to analyze data obtained from over 2000 blood transcriptome results from vaccine studies to reveal the significance of lncRNAs in response to immunization. The measurements in different time points revealed a time-dependent induction of certain lncRNA signatures. On the first day following vaccination, signatures associated with monocytes, TLR signaling, and antigen presentation were induced while a B cell and CD4+ T cell associated signature was induced on day seven following vaccination. The lncRNA differentiation antagonizing non-protein coding RNA (DANCR) had shown B and T cell specific expression while the lncRNA apoptosis associated transcript in bladder cancer (AATBC) was specifically upregulated in monocytes. DANCR is involved in the regulation of histone modifications, sponging of miR216a to inhibit differentiation, and sponging of miR345-p to promote cell growth (Mao et al. 2017; L. Chen et al. 2020; Zhu et al. 2020). AATBC is proliferation-associated lncRNA. It has been shown to play a role in cell proliferation in urinary bladder cancer (UBC) and is a candidate UBC biomarker, in addition to being reported as a regulator of pinin in promotion of metastasis (Zhao et al. 2015; Afshar et al. 2019; Tang et al. 2020). In a more recent study the comparison between transcriptome data of immune cells from influenza A virus (IAV) patients during infection and after recovery has revealed the lncRNA IVRPIE (Inhibiting IAV Replication by Promoting IFN and ISGs Expression) as a key factor in antiviral innate immunity. In vitro ectopic expression of IVRPIE inhibited IAV replication, whereas its silencing led to promotion of IAV replication. The inhibition of viral replication by IVRPIE can be attributed to upregulation interferon β1 and interferon-stimulated genes by IVRPIE (Zhao et al. 2020).

Effective response to pathogens and vaccinations relies on the diversity and adaptability of immune cells. Evidently epigenetics plays a very significant role in providing diversity and adaptability to immune cells. As exemplified in the previous sections the role of epigenetics in the immune system has very substantial implications including, but not limited to, improvement of immune response to pathogens, surveillance of response to vaccines, and the use epigenetic modifying agents as vaccine adjuvants.

REFERENCES

Afshar, S., S. Seyedabadi, M. Saidijam, P. Samadi, H. Mazaherilaghab, and A. Mahdavinezhad. 2019. "Long non-coding ribonucleic acid as a novel diagnosis and prognosis biomarker of bladder cancer." *Avicenna Journal of Medical Biochemistry* 7(1): 28–34.

Akira, S., S. Uematsu, and O. Takeuchi. 2006. "Pathogen recognition and innate immunity." *Cell* 124(4): 783–801.

Amarante-Mendes, G.P., S. Adjemian, L.M. Branco, L.C. Zanetti, R. Weinlich, and K.R. Bortoluci. 2018. "Pattern recognition receptors and the host cell death molecular machinery." *Frontiers in Immunology* 9: 2379.

Atherton, L.J., P.A. Jorquera, A.A. Bakre, and R.A. Tripp. 2019. "Determining immune and miRNA biomarkers related to respiratory syncytial virus (RSV) vaccine types." *Frontiers in Immunology* 10: 2323.

Atochina-Vasserman, E.N. 2012. "S-nitrosylation of surfactant protein D as a modulator of pulmonary inflammation." *Biochimica et Biophysica Acta (BBA)-General Subjects* 1820(6): 763–769.

Aung, H.T., K. Schroder, S.R. Himes, K. Brion, W. van Zuylen, and A. Trieu. 2006. "LPS regulates proinflammatory gene expression in macrophages by altering histone deacetylase expression." *FASEB J* 20: 1315–1327.

Bannister, A.J. and T. Kouzarides. 2011. "Regulation of chromatin by histone modifications." *Cell Research* 21(3): 381–395.

Barrett, S.P., P.L. Wang, and J. Salzman. 2015. "Circular RNA biogenesis can proceed through an exon-containing lariat precursor." *elife* 4: e07540.

Belkina, A.C., B.S. Nikolajczyk, and G.V. Denis. 2013. "BET protein function is required for inflammation: Brd2 genetic disruption and BET inhibitor JQ1 impair mouse macrophage inflammatory responses." *Journal Immunology* 190: 3670–3678.

Breedveld, A., T. Groot Kormelink, M. van Egmond, and E.C. de Jong. 2017. "Granulocytes as modulators of dendritic cell function." *Journal of Leukocyte Biology* 102(4): 1003–1016.

Brown, C.J., B.D. Hendrich, J.L. Rupert, R.G. Lafreniere, Y. Xing, J. Lawrence, and H.F. Willard. 1992. "The human XIST gene: Analysis of a 17 kb inactive X-specific RNA that contains conserved repeats and is highly localized within the nucleus." *Cell* 71(3): 527–542.

Busslinger, M., and A. Tarakhovsky. 2014. "Epigenetic control of immunity." *Cold Spring Harbor Perspectives in Biology* 6(6): a019307.

Cai, X., C.H. Hagedorn, and B.R. Cullen. 2004. "Human microRNAs are processed from capped, polyadenylated transcripts that can also function as mRNAs." *RNA* 10(12): 1957–1966.

Capolunghi, F., M.M. Rosado, M. Sinibaldi, A. Aranburu, and R. Carsetti. 2013. "Why do we need IgM memory B cells?" *Immunology Letters* 152(2): 114–120.

Chen, C.Z., L. Li, H.F. Lodish, and D.P. Bartel. 2004. "MicroRNAs modulate hematopoietic lineage differentiation." *Science* 303(5654): 83–86.

Chen, L., Z. Song, J. Wu, Q. Huang, Z. Shen, X. Wei, and Z. Lin. 2020. "LncRNA DANCR sponges miR-216a to inhibit odontoblast differentiation through upregulating c-Cbl." *Experimental Cell Research* 387(1): 111751.

Conrad, T., A. Marsico, M. Gehre, and U.A. Ørom. 2014. "Microprocessor activity controls differential miRNA biogenesis in vivo." *Cell Reports* 9(2): 542–554.

Dave, R.K., M.E. Dinger, M. Andrew, M. Askarian-Amiri, D.A. Hume, and S. Kellie. 2013. "Regulated expression of PTPRJ/CD148 and an antisense long noncoding RNA in macrophages by proinflammatory stimuli." *PLoS One* 8(6): e68306.

De Andrea, M., R. Ravera, D. Gioia, M. Gariglio, and S. Landolfo. 2002. "The interferon system: An overview". *European Journal of Paediatric Neurology* 6(Suppl. A): A41–A46, discussion A55–8. doi:10.1053/ejpn.2002.057.

Drury, R.E., A.J. Pollard, and D. O'Connor. 2019. "The effect of H1N1 vaccination on serum miRNA expression in children: A tale of caution for microRNA microarray studies." *PloS One* 14(8): e0221143.

Enuka, Y., M. Lauriola, M.E. Feldman, A. Sas-Chen, I. Ulitsky, and Y. Yarden. 2016. "Circular RNAs are long-lived and display only minimal early alterations in response to a growth factor." *Nucleic Acids Research* 44(3): 1370–1383.

Fatemi, M., M.M. Pao, S. Jeong, E.N. Gal-Yam, G. Egger, D.J. Weisenberger, and P.A. Jones. 2005. "Footprinting of mammalian promoters: Use of a CpG DNA methyltransferase revealing nucleosome positions at a single molecule level." *Nucleic Acids Research* 33(20): e176–e176.

Fatica, A. and I. Bozzoni. 2014. "Long non-coding RNAs: New players in cell differentiation and development." *Nature Reviews Genetics* 15(1): 7–21.

Fortes, P. and K.V. Morris. 2016. "Long noncoding RNAs in viral infections." *Virus Research* 212: 1–11.

Gannavaram, S., P. Bhattacharya, A. Siddiqui, N. Ismail, S. Madhavan, and H. Nakhasi. 2019. "miR-21 expression determines the early vaccine immunity induced by LdCen−/− immunization." *Frontiers in Immunology* 10: 2273.

Gensous, N., C. Franceschi, BB. Blomberg, C. Pirazzini, F. Ravaioli, D. Gentilini, A.M. Di Blasio, et al. 2018. "Responders and non-responders to influenza vaccination: A DNA methylation approach on blood cells." *Experimental Gerontology* 105: 94–100. https://doi.org/10.1016/j.exger.2018.01.019.

Goldsby, R., T.J. Kindt, B.A. Osborne, and J. Kuby. 2003. "Chapter 2: Cells and organs of the immune system." In *Immunology*, 5th ed. New York: W. H. Freeman and Company, 24–56.

Gregory, R.I., T.P. Chendrimada, and R. Shiekhattar. 2006. "MicroRNA biogenesis: Isolation and characterization of the microprocessor complex." In *MicroRNA protocols*. Tottowa, NJ: Humana Press. 33–47.

Hallgrimsson, B. and B. Hall. 2011. "Nervous system development." In *Epigenetics*. UC Press, ISBN: 9780520267091.

Hammarlund, E., M.W. Lewis, S.G. Hansen, L.I. Strelow, J.A. Nelson, G.J. Sexton, J.M. Hanifin, and M.K. Slifka. 2003. "Duration of antiviral immunity after smallpox vaccination." *Nature Medicine* 9: 1131–1137.

Hansen, T.B., T.I. Jensen, B.H. Clausen, J.B. Bramsen, B. Finsen, C.K. Damgaard, and J. Kjems. 2013. "Natural RNA circles function as efficient microRNA sponges." *Nature* 495(7441): 384–388.

Hoebe, K., E. Janssen, and B. Beutler. 2004. "The interface between innate and adaptive immunity." *Nature Immunology* 5(10): 971–974.

Hume, D.A., I.L. Ross, S.R. Himes, R.T. Sasmono, C.A. Wells, and T. Ravasi. 2002. "The mononuclear phagocyte system revisited." *Journal of Leukocyte Biology* 72(4): 621–627.

Ilott, N.E., J.A. Heward, B. Roux, E. Tsitsiou, P.S. Fenwick, L. Lenzi, I. Goodhead, et al. 2014. "Long non-coding RNAs and enhancer RNAs regulate the lipopolysaccharide-induced inflammatory response in human monocytes." *Nature Communications* 5: 3979. https://doi.org/10.1038/ncomms4979.

Imamura, K., N. Imamachi, G. Akizuki, M. Kumakura, A. Kawaguchi, K. Nagata, A. Kato, et al. 2014. "Long noncoding RNA NEAT1-dependent SFPQ relocation from promoter region to paraspeckle mediates IL8 expression upon immune stimuli." *Molecular Cell* 53(3): 393–406.

Janeway, C., P. Travers, M. Walport, and M. Shlomchik. 2001. *Immunobiology*, 5th ed. New York: Garland Science.

Kim, Y. and V.N. Kim. 2012. "MicroRNA factory: RISC assembly from precursor microRNAs." *Molecular Cell* 46(4): 384–386.

Kino, T., D.E. Hurt, T. Ichijo, N. Nader, and G.P. Chrousos. 2010. "Noncoding RNA gas5 is a growth arrest–and starvation-associated repressor of the glucocorticoid receptor." *Science Signaling* 3(107): ra8–ra8.

Komori, H.K., T. Hart, S.A. LaMere, P.V. Chew, and D.R. Salomon. 2015. "Defining CD4 T cell memory by the epigenetic landscape of CpG DNA methylation." *The Journal of Immunology* 194(4): 1565–1579.

Kotova, E., N. Lodhi, M. Jarnik, A.D. Pinnola, Y. Ji, and A.V. Tulin. 2011. "Drosophila histone H2A variant (H2Av) controls poly (ADP-ribose) polymerase 1 (PARP1) activation in chromatin." *Proceedings of the National Academy of Sciences* 108(15): 6205–6210.

Kurosaki, T., K., Kometani, and W. Ise. 2015. "Memory B cells." *Nature Reviews Immunology* 15(3): 149–159.

Li, Q.J., J. Chau, P.J. Ebert, G. Sylvester, H. Min, G. Liu, R. BraichManoharan, et al. 2007. "miR-181a is an intrinsic modulator of T cell sensitivity and selection." *Cell* 129(1): 147–161. https://doi.org/10.1016/j.cell.2007.03.008.

Li, X., C.X. Liu, W. Xue, Y. Zhang, S. Jiang, Q.F. Yin, J. Wei, R.W. Yao, L. Yang, and L.L. Chen. 2017. "Coordinated circRNA biogenesis and function with NF90/NF110 in viral infection." *Molecular Cell* 67: 214–227. https://doi.org/10.1016/j.molcel.2017.05.023.

Lim, L.P., N.C. Lau, P. Garrett-Engele, A. Grimson, J.M. Schelter, J. Castle, D.P. Bartel, P.S. Linsley, and J.M. Johnson. 2005. "Microarray analysis shows that some microRNAs downregulate large numbers of target mRNAs." *Nature* 433(7027): 769–773.

Mao, Z., H. Li, B. Du, K. Cui, Y. Xing, X. Zhao, and S. Zai. 2017." LncRNA DANCR promotes migration and invasion through suppression of lncRNA-LET in gastric cancer cells." *Bioscience Reports* 37(6): BSR20171070. https://doi.org/10.1042/BSR20171070

Marchese, F.P. and M. Huarte. 2014. "Long non-coding RNAs and chromatin modifiers: Their place in the epigenetic code." *Epigenetics* 9(1): 21–26.

Margolick, J.B., R.B. Markham, and A.L. Scott. 2006. "Infectious disease epidemiology: Theory and practice." Chapter 10. In Nelson, K.E., and Masters, C.F. (Eds), *The immune system and host defense against infections.* Boston: Jones and Bartlett, 317–343.

Market, E. and F.N. Papavasiliou. 2003. "V(D)J recombination and the evolution of the adaptive immune system." *PLoS Biology* 1(1): e16.

McHeyzer-Williams, M., S. Okitsu, N. Wang, and L. McHeyzer-Williams. 2012. "Molecular programming of B cell memory." *Nature Reviews Immunology* 12(1): 24–34.

Memczak, S., M. Jens, A. Elefsinioti, F. Torti, J. Krueger, A. Rybak, L. Maier, et al. 2013. "Circular RNAs are a large class of animal RNAs with regulatory potency." *Nature* 495(7441): 333–338.

Mohn, F., M. Weber, M. Rebhan, T.C. Roloff, J. Richter, M.B. Stadler, M. Biebel, and D. Schübeler. 2008. "Lineage-specific polycomb targets and de novo DNA methylation define restriction and potential of neuronal progenitors." *Molecular Cell* 30(6): 755–766.

Morlando, M., M. Ballarino, A. Fatica, and I. Bozzoni. 2014. "The role of long noncoding RNAs in the epigenetic control of gene expression." *ChemMedChem* 9(3): 505–510.

Neilson, J.R., G.X. Zheng, C.B. Burge, and P.A. Sharp. 2007. "Dynamic regulation of miRNA expression in ordered stages of cellular development." *Genes and Development* 21: 578–589.

Ng, S.S., W.W. Yue, U. Oppermann, and R.J. Klose. 2009. "Dynamic protein methylation in chromatin biology." *Cellular and Molecular Life Sciences* 66(3): 407.

Ng, W.L., G.K. Marinov, E.S. Liau, Y.L. Lam, Y.Y. Lim, and C.K. Ea. 2016. "Inducible RasGEF1B circular RNA is a positive regulator of ICAM-1 in the TLR4/LPS pathway." *Rna Biology* 13: 861–71.

Ng, W.L., G.K. Marinov Y.M. Chin Y.Y. Lim, and C.K. Ea. 2017. "Transcriptomic analysis of the role of RasGEF1B circular RNA in the TLR4/LPS pathway." *Scientific Reports* 7: 12227.

Nicholson, L.B. 2016. "The immune system." *Essays in Biochemistry* 60(3): 275–301.

Nicodeme, E., K.L. Jeffrey, U. Schaefer, S. Beinke, S. Dewell, and C.W. Chung. 2010. "Suppression of inflammation by a synthetic histone mimic." *Nature* 2010; 468: 1119–1123.

O'Garra, A. and N. Arai. 2000. "The molecular basis of T helper 1 and T helper 2 cell differentiation." *Trends in Cell Biology* 10(12): 542–550.

Odunsi, K., J. Matsuzaki, S.R. James, P. Mhawech-Fauceglia, T. Tsuji, A. Miller, W. Zhang, et al. 2014. "Epigenetic potentiation of NY-ESO-1 vaccine therapy in human ovarian cancer." *Cancer Immunology Research* 2(1): 37–49. https://doi.org/10.1158/2326-6066.CIR-13-0126.

Okamoto, M., Y. Fukushima, T. Kouwaki, T. Daito, M. Kohara, H. Kida, and H. Oshiumi. 2018. "MicroRNA-451a in extracellular, blood-resident vesicles attenuates macrophage and dendritic cell responses to influenza whole-virus vaccine." *Journal of Biological Chemistry* 293(48): 18585–18600.

Park, J.E., I. Heo, Y. Tian, D.K. Simanshu, H. Chang, D. Jee, D.J. Patel, and V.N. Kim. 2011. "Dicer recognizes the 5′ end of RNA for efficient and accurate processing." *Nature* 475(7355): 201–205.

Peng, X., L. Gralinski, C.D. Armour, M.T. Ferris, M.J. Thomas, S. Proll, B.G. Trerheway, et al. 2010. "Unique signatures of long noncoding RNA expression in response to virus infection and altered innate immune signaling." *MBio* 1(5): e00206-10. doi:10.1128/mBio.00206-10.

Pennock, N.D., J.T. White, E.W. Cross, E.E. Cheney, B.A. Tamburini, and R.M. Kedl. 2013. "T cell responses: Naive to memory and everything in between." *Advances in Physiology Education* 37(4): 273–283.

Rajaee, A., R. Barnett, and W.G. Cheadle. 2018. "Pathogen-and danger-associated molecular patterns and the cytokine response in sepsis." Surgical Infections 19(2): 107–116.

Rivera, J., N.A. Fierro, A. Olivera, and R. Suzuki. 2008. "New insights on mast cell activation via the high affinity receptor for IgE." *Advances in Immunology* 98: 85–120.

Rodriguez, A., E. Vigorito, S. Clare, M.V. Warren, P. Couttet, D.R. Soond, S. van Dongen, et al. 2007. "Requirement of bic/ microRNA-155 for normal immune function." *Science* 316: 608–611.

Rossetto, D., N. Avvakumov, and J. Côté. 2012. "Histone phosphorylation: A chromatin modification involved in diverse nuclear events." *Epigenetics* 7(10): 1098–1108.

Rus, H., C., Cudrici, and F. Niculescu. 2005. "The role of the complement system in innate immunity." *Immunologic Research* 33(2): 103–112. doi:10.1385/IR:33:2:103.

Salzman, J., C. Gawad, P.L. Wang, N. Lacayo, and P.O. Brown. 2012. "Circular RNAs are the predominant transcript isoform from hundreds of human genes in diverse cell types." *PloS one* 7(2).

Salzman, J., R.E. Chen, M.N. Olsen, P.L. Wang, and P.O. Brown. 2013. "Cell-type specific features of circular RNA expression." *PLoS Genetics* 9(12): e1003777.

Schmidl, C., L. Hansmann, R. Andreesen, M. Edinger, P. Hoffmann, and M. Rehli. 2011. "Epigenetic reprogramming of the RORC locus during in vitro expansion is a distinctive feature of human memory but not naïve Treg." *European Journal of Immunology* 41(5): 1491–1498. https://doi.org/10.1002/eji.201041067.

Schmitz, K.M., C. Mayer, A. Postepska, and I. Grummt. 2010. "Interaction of noncoding RNA with the rDNA promoter mediates recruitment of DNMT3b and silencing of rRNA genes." *Genes & Development* 24(20): 2264–2269.

Shi, J., N. Hu, J. Li, Z. Zeng, L. Mo, J. Sun, M. Wu, and Y. Hu. 2017. "Unique expression signatures of circular RNAs in response to DNA tumor virus SV40 infection." *Oncotarget* 8: 98609–98622. https://doi.org/10.18632/oncotarget.21694.

Siracusa, M.C., B.S. Kim, J.M. Spergel, and D. Artis. 2013. "Basophils and allergic inflammation." *Journal of Allergy and Clinical Immunology* 132(4): 789–801.

Skvortsova, K., C. Stirzaker, and P. Taberlay. 2019. "The DNA methylation landscape in cancer." *Essays in Biochemistry* 63(6): 797–811. https://doi.org/10.1042/EBC20190037.

Stone, K.D., C. Prussin, and D.D. Metcalfe. 2010. "IgE, mast cells, basophils, and eosinophils." *Journal of Allergy and Clinical Immunology* 125(2): S73–S80.

Strichman-Almashanu, L.Z., R.S. Lee, P.O. Onyango, E. Perlman, F. Flam, M.B. Frieman, and A.P. Feinberg. 2002. "A genome-wide screen for normally methylated human CpG islands that can identify novel imprinted genes." *Genome Research* 12(4): 543–554.

Syrbe, U., S. Jennrich, A. Schottelius, A. Richter, A. Radbruch, and A. Hamann. 2004. "Differential regulation of Pselectin ligand expression in naive versus memory CD4+ T cells: Evidence for epigenetic regulation of involved glycosyltransferase genes." *Blood* 104: 3243–3248.

Takeda, K., T. Kaisho, and S. Akira. 2003. "Toll-like receptors." *Annual Review of Immunology* 21: 335–376.

Tang, T., L. Yang, Y. Cao, M. Wang, S. Zhang, Z. Gong, F. Xiong, et al. 2020. "LncRNA AATBC regulates Pinin to promote metastasis in nasopharyngeal carcinoma." *Molecular Oncology.* https://doi.org/10.1002/1878-0261.12703.

Teif, V.B., D.A. Beshnova, Y. Vainshtein, C. Marth, J.P. Mallm, T. Höfer, and K. Rippe. 2014. "Nucleosome repositioning links DNA (de)methylation and differential CTCF binding during stem cell development." *Genome Research* 24(8): 1285–1295.

Turvey, S.E. and D.H. Broide. 2010. "Innate immunity." *The Journal of Allergy and Clinical Immunology* 125(Suppl 2): S24–S32.

Vigorito, E., K.L. Perks, C. Abreu-Goodger, S. Bunting, Z. Xiang, S. Kohlhaas, and P.P. Das. 2007. "microRNA-155 regulates the generation of immunoglobulin class-switched plasma cells." *Immunity* 27(6): 847–859. https://doi.org/10.1016/j.immuni.2007.10.009.

Villagra, A., F. Cheng, H.W. Wang, I. Suarez, M. Glozak, and M. Maurin. 2009. "The histone deacetylase HDAC11 regulates the expression of interleukin 10 and immune tolerance" *Nature Immunology* 10: 92–100.

Wang, P., Y. Xue, Y. Han, L. Lin, C. Wu, S. Xu, Z. Jiang, J. Xu, Q. Liu, and X. Cao. 2014. "The STAT3-binding long noncoding RNA lnc-DC controls human dendritic cell differentiation." *Science* 344(6181): 310–313. https://doi.org/10.1126/science.1251456.

Wang, Y., H. Zhong, X. Xie, C.Y. Chen, D Huang, L. Shen, H. Zhang, Z.W. Chen, and G. Zeng. 2015. "Long noncoding RNA derived from CD244 signaling epigenetically controls CD8+ T-cell immune responses in tuberculosis infection." *Proceedings of the National Academy of Sciences* 112(29): E3883–E3892. https://doi.org/10.1073/pnas.1501662112

Weber, M., I. Hellmann, M.B. Stadler, L. Ramos, S. Pääbo, M. Rebhan, and D. Schübeler. 2007. "Distribution, silencing potential and evolutionary impact of promoter DNA methylation in the human genome." *Nature Genetics* 39(4): 457.

Willingham, A.T., A.P. Orth, S. Batalov, E.C. Peters, B.G. Wen, P. Aza-Blanc, and P.G. Schultz. 2005. "A strategy for probing the function of noncoding RNAs finds a repressor of NFAT." *Science* 309(5740): 1570–1573.

Xion, Y., S. Chen, R. Chen, W. Lin, and J. Ni. 2013. "Association between microRNA polymorphisms and humoral immunity to hepatitis B vaccine." *Human Vaccines & Immunotherapeutics* 9(8): 1673–1678.

Yamashita, M., K. Hirahara, R. Shinnakasu, H. Hosokawa, S. Norikane, M.Y. Kimura, A. Hasegawa, and T. Nakayama. 2006. "Crucial role of MLL for the maintenance of memory T helper type 2 cell responses." *Immunity* 24: 611–622.

Yoon, J.H., K. Abdelmohsen, and M. Gorospe. 2013. "Posttranscriptional gene regulation by long noncoding RNA." *Journal of Molecular Biology* 425(19): 3723–3730.

You, X., I. Vlatkovic, A. Babic, T. Will, I. Epstein, G. Tushev, G. Akbalik, et al. 2015. "Neural circular RNAs are derived from synaptic genes and regulated by development and plasticity." *Nature Neuroscience* 18(4): 603.

Youngblood, B., K.J. Oestreich, S.J. Ha, J. Duraiswamy, R.S. Akondy, E.E. West, and Z. Wei. 2011. "Chronic virus infection enforces demethylation of the locus that encodes PD-1 in antigen-specific CD8+ T cells." *Immunity* 35(3): 400–412.

Zaratiegui, M., D.V. Irvine, and R.A. Martienssen. 2007. "Noncoding RNAs and gene silencing." *Cell* 128(4): 763–776.

Zhang, H., C.E. Nestor, S. Zhao, A. Lentini, B. Bohle, M. Benson, and H. Wang. 2013. "Profiling of human CD4+ T-cell subsets identifies the TH2-specific noncoding RNA GATA3-AS1." *Journal of Allergy and Clinical Immunology* 132(4): 1005–1008.

Zhangi, Y., X. Li, M. Zhang, and K. Lv. 2017. "Microarray analysis of circular RNA expression patterns in polarized macrophages." *International Journal of Molecular Medicine* 39: 373–379.

Zhao, F., T. Lin, W. He, J. Han, D. Zhu, K. Hu, Li, W., Z. Zheng, J. Huang, and W. Xie. 2015. "Knockdown of a novel lincRNA AATBC suppresses proliferation and induces apoptosis in bladder cancer." *Oncotarget* 6(2): 1064–1078. https://doi.org/10.18632/oncotarget.2833.

Zhao, L., M. Xia, K. Wang, C. Lai, H. Fan, H. Gu, P. Yang, and X. Wang 2020. "A long non-coding RNA IVRPIE promotes host antiviral immune responses through regulating interferon β1 and ISG expression." *Frontiers in Microbiology* 11: 260. https://doi.org/10.3389/fmicb.2020.00260.

Zheng, B., Z. Xi, R. Liu, W. Yin, Z. Sui, B. Ren, H. Miller, Q. Gong, and C. Liu. 2018. "The function of MicroRNAs in B-cell development, lymphoma, and their potential in clinical practice." *Frontiers in Immunology* 9: 936. https://doi.org/10.3389/fimmu.2018.00936.

Zhu, C.Y., C.R. Fan, Y.L. Zhang, Q.X. Sun, M.J. Yan, W. Wei, G.F. Liu, and J.J. Liu. 2020. "LncRNA DANCR affected cell growth, EMT and angiogenesis by sponging miR-345-5p through modulating Twist1 in cholangiocarcinoma." *European Review for Medical and Pharmacological Sciences* 24(5): 2321–2334. https://doi.org/10.26355/eurrev_202003_20498.

Zimmermann, M.T., A.L. Oberg, D.E. Grill, I.G. Ovsyannikova, I.H. Haralambieva, R.B. Kennedy, and G.A. Poland. 2016. "System-wide associations between DNA-methylation, gene expression, and humoral immune response to influenza vaccination." *PLoS One* 11(3): e0152034. https://doi.org/10.1371/journal.pone.0152034.

16 "Omics" Technologies in Vaccine Research

Sezer Okay

CONTENTS

16.1 INTRODUCTION

Since the utilization of vaccinia virus against smallpox by Edward Jenner in 1796, vaccines have been saving lives as successful and cost-effective vehicles to improve public health. Later, in 1881, Louis Pasteur developed the rabies vaccine, and his well-known principles, "isolate, inactivate, and inject" provided a basis for the development of traditional vaccines such as the ones against rubella, measles, and mumps as well as the Bacillus Calmette-Guerin (BCG) vaccine against tuberculosis, obtained via 230 times serial passage of *Mycobacterium bovis* in bile medium (Serruto and Rappuoli 2006; Seib, Zhao, and Rappuoli 2012; Bidmos et al. 2018; Ismail, Ahmad, and Azam 2020).

In the empirical vaccine development, first the pathogen is isolated, and then the candidate agents are modified to improve their efficacy, safety, and tolerability, without a requirement to comprehend the details of the immune responses to the vaccine. Although many important diseases were eradicated using these types of vaccines, the empirical approach is insufficient today for the development of successful vaccines against some of the challenging pathogens which are unculturable in vitro, and sometimes antigenic determinants vary extensively or broader applications of these types of vaccines are limited due to molecular mimicry (Kennedy and Poland 2011; Poland et al. 2011; Ismail, Ahmad, and Azam 2020).

The barriers to empirical vaccine development also include: (i) long-lasting and effective immunity cannot be obtained in some individuals, (ii) the pathogen may develop mutations, (iii) protection may not be obtained via neutralizing antibodies alone, (iv) resulting antibodies may build harmful immune complexes, (v) some individuals may fail to respond to the vaccine due to an immature immune system or acquired maternal immunity may interfere, (vi) in nature, benign agents causing mild infection but providing sufficient immunity are not present, (vii) immunogenic quality of the pathogen may be impaired during attenuation, (viii) the attenuated form may revert to its wild type which is more dangerous, and (ix) the attenuated form is still dangerous for immune-compromised individuals (Poland et al. 2011).

Vaccines had been produced quite a long time using conventional techniques, but crucial progress was made in vaccine development by the utilization of molecular biology and genetics methods to obtain recombinant vaccines. Two examples of successful recombinant vaccines are the hepatitis B and acellular pertussis (aP) vaccines. The hepatitis B vaccine includes a non-infectious viral subunit, hepatitis B surface antigen (HBsAg), produced recombinantly and highly purified, and the recombinant aP vaccine includes one or more components of *Bordetella pertussis*, namely, filamentous haemagglutinin (FHA), pertactin, fimbrial proteins types 2 and 3, as well as detoxified pertussis toxin (Serruto et al. 2009; Yılmaz et al. 2016).

In 1995, the genome of *Haemophilus influenzae* was sequenced for the first time for a bacterial species. Thus, emergence of microbial genomics added a new perspective to vaccinology. The genome of a microorganism covers a complete set of putative antigens, and bioinformatic analyses make it possible to identify which protein is more potent for use in vaccine development. The term "reverse vaccinology" was proposed for discovery of novel antigens as vaccine candidates via analysis of the genome sequences (Serruto et al. 2009; He et al. 2010; Bidmos et al. 2018). The whole genome of a pathogen is screened via in silico analyses in reverse vaccinology to find out candidate proteins such as outer membrane proteins with potential antigenic property (Figure 16.1). These identified proteins are generally produced using recombinant DNA technology, formulated as vaccines, and mice are vaccinated with these formulations to determine the provided immune responses and protection capacity (Seib, Zhao, and Rappuoli 2012). In addition to genomics, the approaches such as transcriptomics and proteomics make it possible to investigate the set of antigens produced by a pathogen under specified conditions via utilization of the mRNA and protein samples of the microorganism, respectively (Rinaudo et al. 2009).

Today, an increased number of published bacterial genomes are available, and many bioinformatics tools are present for detailed analyses. Genome sequences of many isolates of the same species, including pathogenic and non-pathogenic ones, can be compared with each other via pan-genome analysis and potential antigens can be determined (Kanampalliwar 2020) (Figure 16.1). Moreover, reverse vaccinology can also be used for identification of immunogenic epitopes of specific proteins. This approach has many advantages including utilization for non-cultivable microorganisms, non-abundant antigens, and for antigens which are non-immunogenic during infection (Soltan et al. 2020).

In the second phase of reverse vaccinology, called reverse vaccinology 2.0 (RV 2.0), the approach has been directed to isolation and recombinant production of heavy and light (κ or λ) variable regions of immunoglobulin genes. Antigen-specific monoclonal antibodies (mAbs) are obtained from antibody-secreting cells which are immortalized using techniques such as myeloma fusions. These mAbs are effective for a variety of viruses by targeting and neutralizing them (Bidmos et al. 2018). In RV 2.0, the antigens are also selected according to their high-resolution structures to find out the ones recognized by neutralizing antibodies with immunologic

Reverse Vaccinology

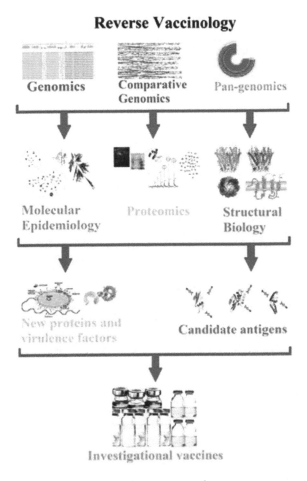

FIGURE 16.1 An overview of reverse vaccinology. One or more genomes of pathogens are analyzed to find out candidate genes for vaccine development. Genomic approaches can be integrated with the disciplines of molecular epidemiology, proteomics or structural biology. Eventually, new proteins and virulence factors as well as candidate antigens can be identified. Candidate antigens are used for the preparation of investigational vaccines (modified from Seib, Zhao and Rappuoli 2012).

performance. Thus, more homogeneous and stable antigens are produced efficiently making the immunization more affordable and practical (Dormitzer, Ulmer, and Rappuoli 2008). The chronological milestones in the methodologies for vaccine development are summarized in Table 16.1.

In 2007, the term "vaccinomics" was introduced covering the disciplines such as "omics" technologies, bioinformatics, systems biology, genetics, and immunology to study possible adverse reactions and biomarkers for use in the clinical trials, host-pathogen interactions, and host immune responses against vaccines for the rational design of safe and effective new vaccines both for entire populations and for specific

254   Synthetic Peptide Vaccine Models

TABLE 16.1
The Chronological Milestones in the Strategies for Vaccine Development (modified from Rappuoli et al. 2016).

Year	Milestone	Comment
1796	The beginning of vaccinology	The vaccinia virus was used against smallpox by Edward Jenner.
1995	Whole-genome of a bacterium sequenced first time	The genome of *Haemophilus influenzae* was published by **Fleischmann et al. (1995)**.
2000	The beginning of reverse vaccinology	The vaccine candidates for serogroup B strain of *Neisseria meningitides* (MenB) were reported by **Pizza et al. (2000)** discovered via genomics.
2002	Use of monoclonal antibodies for vaccine development	**Burton (2002)** proposed that monoclonal antibodies can be used for the development of more efficient vaccines via molecular characterization of antibody-pathogen-antigen interaction.
2007	Introduction of the term "vaccinomics"	**Poland (2007)** proposed the term "vaccinomics" as the interaction of many disciplines for vaccine development.
2008	The beginning of reverse vaccinology 2.0	**Dormitzer, Ulmer and Rappuoli (2008)** introduced the structural vaccinology for the structure-based antigen design.
2012	The first regulatory approval for a genome-based vaccine	European Medicines Agency (EMA) recommended approval of first vaccine for meningitis B.
2013	The first report of structure-based vaccine protective in animal model	**McLellan et al. (2013)** reported that the structure-based vaccine against Respiratory Syncytial Virus was protective in mice and macaques.

individuals via individualized medicine methodologies. Vaccinomics studies aim to define correlates of protection as well as deepen the knowledge on the protective immune pathways and effects of pathogens on immunity. The phenotypes of naïve individuals after receiving live viral vaccines can be observed as non-responders, normal responders, and hyper-responders according to humoral immune responses. In combination with cell-mediated immunity (CMI), skewed responses with high antibody-low CMI or vice versa, and early or late responders with different kinetics of immune activation can be observed in addition to non-responder and hyper-responder phenotypes (Poland 2007; Poland, Ovsyannikova, and Jacobson 2009; Haralambieva and Poland 2010; Poland and Oberg 2010; Kennedy and Poland 2011).

Vaccines provide production by means of adaptive immunity, and the innate immunity acts as an intermediate between antigens found in the vaccine and the adaptive immunity of the vaccinee. Also, identification of molecular signatures induced by vaccination is important to define the elements underlying the adaptive immune responses. Thus, the immune responses providing protection will be predicted to evaluate the potency of vaccines or to identify the unresponsive individuals.

Interactions among the vaccine, innate immunity, and adaptive immunity can be investigated in detail via systems biology approaches (Buonaguro and Pulendran 2011; Petrizzo et al. 2012). It is also important to investigate the repertoire of B cell and T cell receptors needed for adaptive immune responses. The dynamic immune repertoire can be identified using high-throughput sequencing technologies via computational and systems immunology strategies (Miho et al. 2018).

Vaccines have saved millions of lives; however, they may also bring about some adverse reactions. Usually vaccines produce mild local inflammation, swelling, fever, or redness. Since vaccines should pass safety tests in clinical trials before commercialization, severe adverse reactions should not be common. Nevertheless, life-threatening rare allergic reactions may appear upon vaccination due to the genetic differences among the individuals. The term "adversomics" was proposed for the integration of systems biology and immunogenomics to find out potential factors in the adverse reactions to vaccines at the molecular level (Whitaker, Ovsyannikova, and Poland 2015).

Many omics approaches are used in vaccine research having interactions with each other. Figure 16.2 summarizes the main omics techniques utilized for vaccine development. In omics technologies, an excess amount of data is acquired, which should be processed and analyzed to reach a conclusion. Bioinformatic tools and

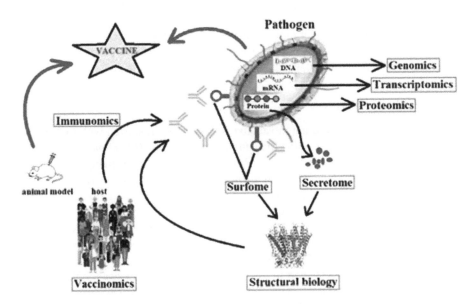

FIGURE 16.2 Schematic overview of omics approaches utilized in vaccine research. Antigen candidates are predicted using genomics, transcriptomics and proteomics techniques. Sub-proteome analysis can also be performed for a specific subcellular compartment such as surface proteins (surfome) or secreted proteins (secretome). The protective epitopes are predicted using structural biology. The complete set of antigens recognized by the host is investigated via immunomics, and the responses of humans to a vaccine are evaluated via vaccinomics (modified from Bagnoli et al. 2011).

databases storing and organizing the data for both immune reactivity and pathogens are the means to evaluate the huge amount of data obtained from omics approaches. The amount of acquired data from genomics, transcriptomics, and proteomics studies increases day by day. Therefore, utilization of the bioinformatic tools covering the most comprehensive information is important to obtain the most exact result (Sette and Rappuoli 2010).

Rashid et al. (2017) followed an interdisciplinary computational framework of genomics, subtractive proteomic, and structural biology approaches using various bioinformatic tools to predict candidate antigens for *Pseudomonas aeruginosa*. First, the proteome of *P. aeruginosa* with subcellular localization was screened, and the proteins were analyzed to identify essentiality and virulence. Second, selected proteins were checked for their immuno-protective potential. The crystal structures of these proteins were analyzed to monitor the host-pathogen interactions. Next, broad spectrum immunogenic peptides were identified via epitope prediction, which can bind to various MHC alleles. Lastly, sequences of identified proteins were obtained from different virulent strains to find out the immune-protective consensus sequences among different strains of *P. aeruginosa*. As a result, two antibiotic efflux pumps, two components of chaperone-usher pathway, a penicillin-binding protein of bacterial cell wall, an extracellular component of type III secretion system, and three uncharacterized secretory proteins were determined as the potential candidate antigens.

The number and variety of omics technologies utilized in vaccine research have been increasing as time passes. However, the core approaches are genomics, transcriptomics, and proteomics, which were mentioned in this chapter.

16.2 GENOMICS

The genomic era is a revolution in vaccine development. It started with the shotgun sequencing technology producing the genome sequence of *Haemophilus influenzae* in 1995 by The Institute for Genomic Research (TIGR). Later, many technologies were developed for next-generation sequencing, such as massively parallel signature sequencing, polony sequencing, 454 pyrosequencing, reversible terminator sequencing, sequencing by oligonucleotide ligation detection (SOLiD), single-molecule real-time sequencing, ion torrent sequencing, or DNA nanoball sequencing (reviewed by Rajesh and Jaya 2017). Due to these advanced technologies, the number of published bacterial genomes increased considerably with 14,754 completed and 128,146 permanent draft genomes deposited in the Genomes Online Database (GOLD, https://gold.jgi.doe.gov) by June 2020. The organisms with completed genomes cover various bacterial pathogens, and they serve vaccine development to find out potential antigens (Serruto et al. 2009).

As an early example, Pizza et al. (2000) identified 570 open reading frames (ORFs) belonging to serogroup B strain of *Neisseria meningitides* (MenB) genome, and they were able to express 350 of them. Mice were immunized with the purified proteins, and 85 of them showed positive results. Next, seven proteins successful in immunologic evaluations were selected for further analysis, and two of them had bactericidal activity similar to that of outer membrane vesicles providing protection against homologous strains in humans.

Genomes of the pathogens can be searched for the motifs found in the known immunogenic proteins. Rosini et al. (2006) identified pilus-like structures based on LPXTG amino acid motif in eight genomes of *Streptococcus agalactiae* which is a group B streptococcus (GBS). One of the proteins (GBS59) was produced recombinantly and used for the immunization of mice. Male and female mice were vaccinated three times and mated. Their offspring were challenged, and 56% protection was obtained.

Although genome analysis of one strain provides valuable information about the putative antigenic proteins, it does not reveal the intra-species variations among different strains of pathogens. Mutations on the genes encoding these putative antigens may change the antigenic property or may not show the same efficacy against different strains. Therefore, a wider point of view is needed for the antigen discovery. A pan-genome is the global repertoire of genes belonging to a species obtained from its different strains, which includes three parts: (i) the core-genome with invariable and conserved genes in all strains, (ii) the dispensable genome including genes found in some strains, and (iii) the strain-specific genes found only in one strain (Kaushik and Sehgal 2008; Serruto et al. 2009).

Maione et al. (2005) performed comparative genome analysis of eight GBS isolates and identified 312 surface proteins. These proteins were produced recombinantly and used to immunize the mice. Four of the proteins conferred protective potential, and the combination of these proteins provided high protection against various strains. Similarly, Moriel et al. (2013) compared the ten complete and 33 draft genomes of *Acinetobacter baumannii*, and identified 62 antigens as vaccine candidates. Of these, 20 proteins were predicted to have a beta-barrel structure causing a problem for protein solubility. The remaining 42 proteins were identified as 18 outer membrane lipoproteins, ten haemagglutinins and adhesins, nine toxins and enzymes, two solenoid repeat proteins, and three hypothetical proteins.

Reverse vaccinology approaches were used to investigate the potential vaccine candidates for *Histophilus somni*. Genome sequences of different clinical *H. somni* isolates were obtained via next-generation sequencing. The genome regions coding for proteins were predicted, and surface exposure as well as antigenicity scores of 20 proteins were determined. The candidate proteins in the investigated isolates were compared and one of them, having conserved proteins, was selected for vaccine development. Cloning of the genes for 18 proteins was successful, and 13 proteins gave positive signals in western blot analysis with convalescent bovine serum (Madampage et al. 2015).

Another strategy to discover potential antigens and virulence factors is comparison of the genomes of pathogenic and non-pathogenic strains of the same species via comparative genome analysis. Thus, the genes responsible for pathogenesis can be identified (Serruto et al. 2009). The biological trends can be identified and characterized via comparison of large-scale genomes to understand a particular phenomenon or to highlight an interesting exception. The comparative genomics can be used to discover the patterns common among bacteria and the increasing number of published genomes strengthens it. These patterns cover the distribution of structural properties, relative amounts of specific genes, and the differential expression of the encoded proteins under varying conditions (Barocchi, Censini, and Rappuoli 2007).

Moriel et al. (2010) compared the genomes of pathogenic and non-pathogenic strains of extraintestinal pathogenic *Escherichia coli* (ExPEC) reporting 19 genomic islands and 230 antigens only found in the pathogenic strain. These proteins were produced recombinantly and used for the vaccination of mice. They found that nine proteins conferred protection against bacterial challenge in mice.

In vaccine development, selected antigen candidates should not have homologues in the hosts. Therefore, the genome of the host is also analyzed in addition to the genome of the pathogen. A comparative and subtractive genomic analysis was conducted for *Mycoplasma genitalium* to identify potential antigen candidates. A total of 79 proteins of *M. genitalium* were identified with no similarity to human proteins, and 67 of them were predicted as potential drug and vaccine targets. Of these, 16 membrane proteins were reported as candidates for vaccine development (Butt et al. 2012).

16.3   TRANSCRIPTOMICS

Differential regulation of specific genes or pathways upon interaction with antigens as well as regulation of immune responses upon encountering a pathogen can be identified using transcriptomics. Results obtained by several independent studies can be integrated via meta-analysis using public databases. Thus, a common profile of the host responses, such as the expression of genes related to transcription factors, cytokine production, or signal transduction, against a pathogen can be revealed. Eventually, observed transcriptional processes provide a comprehensive understanding of host defense mechanisms to an infection at the level of diverse cells and compartments (Buonaguro and Pulendran 2011; Holtfreter et al. 2016).

Global transcriptome analysis can be used in two ways to identify the candidate antigens. First, the quantitative expression level of each gene under a specific condition is provided by transcriptome analysis. Since these transcript levels are mostly correlated with the amount of the encoded proteins, in silico analyses can be used to predict the highly expressed and surface-exposed proteins. Thus, the genes with low expression can be excluded for the recombinant production of the proteins. Second, comparative transcriptome analysis can be performed under different disease conditions to predict the cellular pathways and proteins related to the infection, which have high potential in vaccine discovery (Grandi 2006).

In order to identify candidate antigens from *Ctenocephalides felis* for vaccine development against cat flea, transcriptomics data were evaluated from unfed adult fleas. RNA-seq analysis produced 59,558 transcripts and 11,627 unigenes, of which 1620 unigenes were predicted in the exoproteome, 177 of them encoding proteins with transmembrane regions or signal peptides. The gene ontology (GO) analysis of these unigenes showed that 96% of the proteins belonged to cell membrane or extracellular space. Among them, six proteins were selected as candidate antigens, combining with proteomics data (Contreras et al. 2018).

Efficacy of the candidate antigens is desired to be valid for the majority of the target population. Therefore, polymorphisms should be considered in the selection process of the antigens. For this purpose, single nucleotide polymorphisms (SNPs) are analyzed for a set of transcripts to find out the genetic diversity. However, it is important to be cautious that the bioinformatic tool used for the analysis may identify the variants having polymorphism at one or more positions as different transcripts (Contreras et al. 2018). Additionally, polymorphisms in the genes related with immune responses, such as the genes encoding toll-like receptors (TLRs), might also have influence on the success of the vaccine (Buonaguro and Pulendran 2011).

In addition to RNA-seq, microarray is another methodology for the analysis of global gene expression. In microarray analysis, different conditions are used for the growth of microorganisms, mRNAs are isolated, cDNAs are synthesized and hybridized. Thus, altered gene expression in response to change in the growth conditions is determined and affected metabolic pathways and regulatory elements are revealed. Also, transcriptome of a pathogen grown in a cell, tissue, or animal model deciphers the host-pathogen interactions (Serruto and Rappuoli 2006).

Transcriptome analysis can also be performed to find out the signature of the immune responses raised against a vaccine. Gaucher et al. (2008) investigated the immune responses induced by yellow fever vaccine via microarray analysis. Expression levels of 594 genes were reported to be changed significantly after vaccination. The highest regulation of gene expression was observed at days three and seven post-vaccination. The genes with differential expression were reported as transcription factors as well as the ones associated with immune cells, TLRs, interferon pathway, complement system, and components of the inflammasome.

MicroRNAs (miRNAs) are among the important regulators of gene expression at posttranscriptional level. Therefore, expression profiles of miRNAs at specific conditions such as host-pathogen interaction confer valuable information about the regulation of genes involved in the immune responses (Buonaguro and Pulendran 2011).

16.4 PROTEOMICS

Proteomics is the high-throughput analysis of the complete set of proteins of an organism, a cell, or the part of a cell. In this approach, the protein sample is isolated from the cells or the specific subcellular compartment according to the intended type of proteome such as secretome or surfome. Following the protein isolation, gel-based or gel-free techniques are used for the identification of each protein in the sample. In gel-based approaches, the proteins are separated by either conventional one-dimensional gel electrophoresis (1-DE) or two-dimensional gel electrophoresis (2-DE) for a higher resolution. In 2-DE, the proteins are separated according to their isoelectric point (pI) values in the first dimension, called isoelectric focusing (IEF). In the second dimension, proteins on the IEF strips are separated according to their molecular weight (MW) using sodium dodecyl sulphate polyacrylamide gel electrophoresis (SDS-PAGE), and following the staining of the gel using a dye such as Coomassie brilliant blue, the proteins are visualized as fine spots (Figure 16.3). The protein samples can be labeled prior to 2-DE using spectrally resolvable fluorophores for

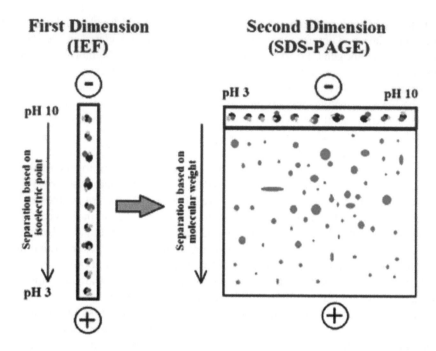

FIGURE 16.3 The basic principles of two-dimensional gel electrophoresis (2-DE). In the first dimension, protein samples are separated according to their isoelectric point (pI) via isoelectric focusing (IEF). The proteins on IEF strips are separated according to their molecular weight (MW) using sodium dodecyl sulphate polyacrylamide gel electrophoresis (SDS-PAGE) (modified from Raak et al. 2018).

given masses and charges comigrating with proteins on 2D-gel, a technique called 2D difference in-gel electrophoresis (2D-DIGE). Desired bands in 1-DE or spots in 2-DE are isolated from the gel, and mass spectrometric techniques such as matrix-assisted laser desorption ionization-time of flight (MALDI-TOF) and tandem mass spectrometry (MS/MS) are used for the peptide identification. Gel-based techniques are well-established and used widely. However, they have some limitations such as application problems for certain classes of proteins, and quantitative reproducibility. Gel-free approaches are used to overcome these problems. In shotgun proteomics, proteins are digested using proteases to obtain complex peptide fractions, and they are resolved via different high-throughput fractionation approaches such as ion-exchange chromatography, reversed-phase chromatography, 2D liquid chromatography, or OFFGEL electrophoresis (Kaushik and Sehgal 2008; Walters and Mobley 2010; Scherp et al. 2010; Abdallah et al. 2012; Baiwir et al. 2018).

A detailed picture of the protein-to-protein interactions related to the immune responses against infections, virulence of the pathogen, or host-pathogen interactions can be obtained via proteomics, and "interactome" of the cell is revealed. Proteins undergo modifications to gain their function or they are found in different locations of the cell. These certain groups of proteins can be isolated and identified using proteomics techniques. For instance, phosphorylated proteins playing roles in the signal transduction are identified via the phosphoproteomics technique, and the "kinome" of the cell is identified (Buonaguro and Pulendran 2011) or secreted proteins can be identified via "secretome" analysis (Bidmos et al. 2018).

Although many proteins were identified and characterized, there are still many uncharacterized and hypothetical proteins which might be encountered during the bioinformatic analysis of the omics data. Hypothetical proteins can be prioritized as a vaccine candidate if they have differential gene and protein expressions in different hosts infected by different bacterial and viral pathogens. Some other problems in data search are lack of the genome of investigated organism and alternate splice forms for eukaryotic proteins as well as the presence of retired sequences and alternate species or strain representations in many databases (McGarvey et al. 2009).

In an attempt to find out candidate antigens from *Ctenocephalides felis*, soluble plasma membrane proteins of adult fleas were investigated using proteomics approaches. The reverse phase-liquid chromatography-mass spectrometry analysis identified 218 flea and 82 cat proteins from the Neoptera and Carnivora Uniprot databases, respectively. The proteomics data were combined with those obtained from transcriptomics analysis, and seven different proteins were identified to be in the plasma membrane. Six proteins were determined as candidate antigens, and five of them were successfully produced in *Escherichia coli*. The recombinant proteins were formulated with oil-based adjuvant, and cats were vaccinated with these formulations. Different developmental stages of flea were evaluated for vaccine efficacy, and flea infestations were controlled up to 46% in vaccinated cats (Contreras et al. 2018).

In order to identify the immunogenic proteins, outer membrane proteins (OMPs) from *Stenotrophomonas maltophilia* culture were isolated by Xu et al. (2018) and separated using 2-DE. Proteins were identified via MALDI-TOF MS and nano liquid chromatography (nanoLC) coupled Fourier transform ion cyclotron resonance (FTICR) MS/MS. Two OMPs were selected according to the results of immunoblotting analysis, one reacts strongly (OmpA) and the other reacts weakly (Smlt4123) with the anti-*S. maltophilia* serum from rabbits. Recombinant OmpA and Smlt4123 proteins were formulated with Freund's incomplete adjuvant and mice were vaccinated with these formulations. Sera from the mice vaccinated with Smlt4123 formulation reduced the bacterial count significantly but OmpA formulation did not. The bacterial load in the blood of mice vaccinated with Smlt4123 formulation was also lower than the control group.

Immunoproteome analysis including immunoblotting of 2-DE gels provides important data about the immunogenic potential of the candidate proteins. Altındiş et al. (2009) isolated the total soluble proteins from two *Bordetella pertussis* strains, and the protein samples were separated via 2-DE followed by an immunoblotting using the sera from the mice obtained against inactivated or live cells of *B. pertussis* strains. A total of 25 proteins were identified as giving reactions to the sera from mice, and 21 of them were reported to be the novel *B. pertussis* antigens. Later, the surface proteome of these strains was also studied via immunoproteome approaches, and 27 immunogenic spots were identified (Tefon et al. 2011).

Immunoproteomic analysis of *Mycobacterium immunogenum* for its cellular and secreted proteins was performed via a combination of 2-DE, MALDI-TOF, and immunoblotting analyses. A total of 33 immunoreactive proteins were identified, and eight of them matched with the homologues of the known antigens from mycobacteria while 11 proteins were found as hypothetical and 14 proteins were matched with the proteins from other bacteria. The major protein spot on the 2D-gel of secretome analysis was isolated and used on the murine alveolar macrophages to evaluate its potential for induction of innate immune response. The proinflammatory cytokines IL-6, IL-18, IL-1β, and TNF-α were found to be up-regulated, and anti-inflammatory cytokine IL-10 was down-regulated (Gupta et al. 2009).

REFERENCES

Abdallah, C., E. Dumas-Gaudot, J. Renaut, and K. Sergeant. 2012. "Gel-based and gel-free quantitative proteomics approaches at a glance." *International Journal of Plant Genomics* 2012: 494572. doi:10.1155/2012/494572.

Altındiş, E., B.E. Tefon, V. Yıldırım, E. Ozcengiz, D. Becher, M. Hecker, and G. Ozcengiz. 2009. "Immunoproteomic analysis of *Bordetella pertussis* and identification of new immunogenic proteins." *Vaccine* 27(4): 542–548. doi:10.1016/j.vaccine.2008.11.020.

Bagnoli, F., B. Baudner, R.P.N. Mishra, E. Bartolini, L. Fiaschi, P. Mariotti, V. Nardi-Dei, P. Boucher, and R. Rappuoli. 2011. "Designing the next generation of vaccines for global public health." *OMICS* 15(9): 545–566. doi:10.1089/omi.2010.0127.

Baiwir, D., P. Nanni, S. Müller, N. Smargiasso, D. Morsa, E. De Pauw, and G. Mazzucchelli. 2018. "Gel-free proteomics." Chapter 5. In de Almeida,. A.M., Eckersall, D., and Miller, I. (Eds), *Proteomics in domestic animals: From farm to systems biology.* Springer International Publishing AG, 55–101. doi:10.1007/978-3-319-69682-9_5.

Barocchi, M.A., S. Censini, and R. Rappuoli. 2007. "Vaccines in the era of genomics: The pneumococcal challenge." *Vaccine* 25: 2963–2973. doi:10.1016/j.vaccine.2007.01.065.

Bidmos, F.A., S. Siris, C.A. Gladstone, and P.R. Langford. 2018. "Bacterial vaccine antigen discovery in the reverse vaccinology 2.0 era: Progress and challenges." *Frontiers in Immunology* 9: 2315. doi:10.3389/fimmu.2018.02315.

Buonaguro, L. and B. Pulendran. 2011. "Immunogenomics and systems biology of vaccines." *Immunological Reviews* 239(1): 197–208. doi:10.1111/j.1600-065X.2010.00971.x.

Burton, D.R. 2002. "Antibodies, viruses and vaccines." *Nature Reviews Immunology* 2(9): 706–713. doi:10.1038/nri891.

Butt, A.M., S. Tahir, I. Nasrullah, M. Idrees, J. Lu, and Y. Tong. 2012. "*Mycoplasma genitalium*: A comparative genomics study of metabolic pathways for the identification of drug and vaccine targets." *Infection, Genetics and Evolution* 12(1): 53–62. doi:10.1016/j.meegid.2011.10.017.

Contreras, M., M. Villar, S. Artigas-Jerónimo, L. Kornieieva, S. Mytrofanov, and J. de la Fuente. 2018. "A reverse vaccinology approach to the identification and characterization of *Ctenocephalides felis* candidate protective antigens for the control of cat flea infestations." *Parasites and Vectors* 11(1): 43. doi:10.1186/s13071-018-2618-x.

Dormitzer, P.R., J.B. Ulmer, and R. Rappuoli. 2008. "Structure-based antigen design: A strategy for next generation vaccines." *Trends in Biotechnology* 26(12): 659–667. doi:10.1016/j.tibtech.2008.08.002.

Fleischmann, R.D., M.D. Adams, O. White, R.A. Clayton, E.F. Kirkness, A.R. Kerlavage, C.J. Bult, et al. 1995. "Whole-genome random sequencing and assembly of *Haemophilus influenzae* Rd." *Science* 269(5223): 496–512. doi:10.1126/science.7542800.

Gaucher, D., R. Therrien, N. Kettaf, B.R. Angermann, G. Boucher, A. Filali-Mouhim, J.M. Moser, et al. 2008. "Yellow fever vaccine induces integrated multilineage and polyfunctional immune responses." *The Journal of Experimental Medicine* 205(13): 3119–3131. doi:10.1084/jem.20082292.

Grandi, G. 2006. "Genomics and proteomics in reverse vaccines." Chapter 20. In Humphery-Smith, I., and Hecker, M. (Eds), *Microbial proteomics: Functional biology of whole organisms.* John Wiley & Sons, Inc, 379–393. doi:10.1002/0471973165.ch20.

Gupta, M.K., V. Subramanian, and J.S. Yadav. 2009. "Immunoproteomic identification of secretory and subcellular protein antigens and functional evaluation of the secretome fraction of *Mycobacterium immunogenum*, a newly recognized species of the *Mycobacterium chelonae-Mycobacterium abscessus* group." *Journal of Proteome Research* 8: 2319–2330. doi:10.1021/pr8009462.

Haralambieva, I.H. and G.A. Poland. 2010. "Vaccinomics, predictive vaccinology and the future of vaccine development." *Future Microbiology* 5: 1757–1760. doi:10.2217/fmb.10.146.

He, Y., R. Rappuoli, A.S. De Groot, and R.T. Chen. 2010. "Emerging vaccine informatics." *Journal of Biomedicine and Biotechnology* 2010: 218590. doi:10.1155/2010/218590.

Holtfreter, S., J. Kolata, S. Stentzel, S. Bauerfeind, F. Schmidt, N. Sundaramoorthy, and B.M. Bröker. 2016. "Omics approaches for the study of adaptive immunity to *Staphylococcus aureus* and the selection of vaccine candidates." *Proteomes* 4(1): 11. doi:10.3390/proteomes4010011.

Ismail, S., S. Ahmad, and S.S. Azam. 2020. "Vaccinomics to design a novel single chimeric subunit vaccine for broad-spectrum immunological applications targeting nosocomial Enterobacteriaceae pathogens." *European Journal of Pharmaceutical Sciences* 146: 105258. doi:10.1016/j.ejps.2020.105258.

Kanampalliwar, A.M. 2020. "Reverse vaccinology and its applications." Chapter 1. In Tomar, N. (Ed.), Immunoinformatics. *Methods in Molecular Biology.* Springer Science+Business Media, LLC, 1–16. doi:10.1007/978-1-0716-0389-5_1.

Kaushik, D.K. and D. Sehgal. 2008. "Developing antibacterial vaccines in genomics and proteomics era." *Scandinavian Journal of Immunology* 67: 544–552. doi:10.1111/j.1365-3083.2008.02107.x.

Kennedy, R.B. and G.A. Poland. 2011. "The top five 'game changers' in vaccinology: Toward rational and directed vaccine development." *OMICS* 15(9): 533–537. doi:10.1089/omi.2011.0012.

Madampage, C.A., N. Rawlyk, G. Crockford, Y. Wang, A.P. White, R. Brownlie, J. Van Donkersgoed, C. Dorin, and A. Potter. 2015. "Reverse vaccinology as an approach for developing *Histophilus somni* vaccine candidates." *Biologicals* 43(6): 444–451. doi:10.1016/j.biologicals.2015.09.001.

Maione, D., I. Margarit, C.D. Rinaudo, V. Masignani, M. Mora, M. Scarselli, H. Tettelin, et al. 2005. "Identification of a universal Group B streptococcus vaccine by multiple genome screen." *Science* 309(5731): 148–150. doi:10.1126/science.1109869.

McGarvey P.B., H. Huang, R. Mazumder, J. Zhang, Y. Chen, C. Zhang, S. Cammer, et al. 2009. "Systems integration of biodefense omics data for analysis of pathogen-host interactions and identification of potential targets." *PLoS ONE* 4(9): e7162. doi:10.1371/journal.pone.0007162.

McLellan, J.S., M. Chen, M.G. Joyce, M. Sastry, G.B. Stewart-Jones, Y. Yang, B. Zhang, et al. 2013. "Structure-based design of a fusion glycoprotein vaccine for respiratory syncytial virus." *Science* 342(6158): 592–598. doi:10.1126/science.1243283.

Miho, E., A. Yermanos, C.R. Weber, C.T. Berger, S.T. Reddy, and V. Greiff. 2018. "Computational strategies for dissecting the high-dimensional complexity of adaptive immune repertoires." *Frontiers in Immunology* 9: 224. doi:10.3389/fimmu.2018.00224.

Moriel, D.G., I. Bertoldi, A. Spagnuolo, S. Marchi, R. Rosini, B. Nesta, I. Pastorello, et al. 2010. "Identification of protective and broadly conserved vaccine antigens from the genome of extraintestinal pathogenic *Escherichia coli*." *Proceedings of the National Academy of Sciences of the United States of America* 107(20): 9072–9077. doi:10.1073/pnas.0915077107.

Moriel, D.G., S.A. Beatson, D.J. Wurpel, J. Lipman, G.R. Nimmo, D.L. Paterson, and M.A. Schembri. 2013. "Identification of novel vaccine candidates against multidrug-resistant *Acinetobacter baumannii*." *PLoS ONE* 8(10): e77631. doi:10.1371/journal.pone.0077631.

Petrizzo, A., M. Tornesello, F.M. Buonaguro, and L. Buonaguro. 2012. "Immunogenomics approaches for vaccine evaluation." *Journal of Immunotoxicology* 9(3): 236–240. doi:1 0.3109/1547691X.2012.707698.

Pizza, M., V. Scarlato, V. Masignani, M.M. Ciuliani, B. Arico, M. Comanducci, G.T. Jenning, et al. 2000. "Identification of vaccine candidates against serogroup B meningococcus by whole-genome sequencing." *Science* 287(5459): 1816–1820. doi:10.1126/science.287.5459.1816.

Poland, G.A. 2007. "Pharmacology, vaccinomics, and the second golden age of vaccinology." *Clinical Pharmacology and Therapeutics* 82(6): 623–626. doi:10.1038/sj.clpt.2007.6100379.

Poland G.A. and A.L. Oberg. 2010. "Vaccinomics and bioinformatics: Accelerants for the next golden age of vaccinology." *Vaccine* 28(20): 3509–3510. doi:10.1016/j.vaccine.2010.03.031.

Poland, G.A., I.G. Ovsyannikova, and R.M. Jacobson. 2009. "Application of pharmacogenomics to vaccines." *Pharmacogenomics* 10(5): 837–852. doi:10.2217/pgs.09.25.

Poland, G.A., I.G. Ovsyannikova, R.B. Kennedy, I.H. Haralambieva, and R.M. Jacobson. 2011. "Vaccinomics and a new paradigm for the development of preventive vaccines against viral infections." *OMICS* 15(9): 625–636. doi:10.1089/omi.2011.0032.

Raak, N., R.A. Abbate, A. Lederer, H. Rohm, and D. Jaros. 2018. "Size separation techniques for the characterisation of cross-linked casein: A review of methods and their applications." *Separations* 5: 14. doi:10.3390/separations5010014.

Rajesh, T. and M. Jaya. 2017. "Next-generation sequencing methods." Chapter 7. In Gunasekaran, P., Noronha, S., and Pandey, A. (Eds), *Current developments in biotechnology and bioengineering: Functional genomics and metabolic engineering*. Elsevier B.V., 143–158. doi:10.1016/B978-0-444-63667-6.00007-9.

Rappuoli, R., M.J. Bottomley, U. D'Oro, O. Finco, and E. De Gregorio. 2016. "Reverse vaccinology 2.0: Human immunology instructs vaccine antigen design." *The Journal of Experimental Medicine* 213(4): 469–481. doi:10.1084/jem.20151960.

Rashid, M.I., A. Naz, A. Ali, and S. Andleeb. 2017. "Prediction of vaccine candidates against *Pseudomonas aeruginosa*: An integrated genomics and proteomics approach." *Genomics* 109: 274–283. doi:10.1016/j.ygeno.2017.05.001.

Rinaudo, C.D., J.L. Telford, R. Rappuoli, and K.L. Seib. 2009. "Vaccinology in the genome era." *The Journal of Clinical Investigation* 119: 2515–2525. doi:10.1172/JCI38330.

Rosini, R., C.D. Rinaudo, M. Soriani, P. Lauer, M. Mora, D. Maione, A. Taddei, et al. 2006. "Identification of novel genomic islands coding for antigenic pilus-like structures in *Streptococcus agalactiae*." *Molecular Microbiology* 61(1): 126–141. doi:10.1111/j.1365-2958.2006.05225.x.

Scherp, P., G. Ku, L. Coleman, and I. Kheterpal. 2010. "Gel-based and gel-free proteomic technologies." Chapter 13. In Gimble, J.M., and Bunnell, B.A. (Eds), *Adipose-derived stem cells: Methods and protocols, methods in molecular biology*, Vol 702. 163–190. doi:10.1007/978-1-61737-960-4_13.

Seib, K.L., X. Zhao, and R. Rappuoli. 2012. "Developing vaccines in the era of genomics: A decade of reverse vaccinology." *Clinical Microbiology and Infection* 18 (Suppl. 5): 109–116. doi:10.1111/j.1469-0691.2012.03939.x.

Serruto, D.and R. Rappuoli. 2006. "Post-genomic vaccine development." *FEBS Letters* 580: 2985–2992. doi:10.1016/j.febslet.2006.04.084.

Serruto, D., L. Serino, V. Masignani, and M. Pizza. 2009. "Genome-based approaches to develop vaccines against bacterial pathogens." *Vaccine* 27: 3245–3250. doi:10.1016/j. vaccine.2009.01.072.

Sette, A. and R. Rappuoli. 2010. "Reverse vaccinology: Developing vaccines in the era of genomics." *Immunity* 33(4): 530–541. doi:10.1016/j.immuni.2010.09.017.

Soltan, M.A., D. Magdy, S.M. Solyman, and A. Hanora. 2020. "Design of *Staphylococcus aureus* new vaccine candidates with B and T cell epitope mapping, reverse vaccinology, and immunoinformatics." *OMICS* 24(4): 195–204. doi:10.1089/omi.2019.0183.

Tefon, B.E., S. Maass, E. Ozcengiz, D. Becher, M. Hecker, and G. Ozcengiz 2011. "A comprehensive analysis of *Bordetella pertussis* surface proteome and identification of new immunogenic proteins." *Vaccine* 29(19): 3583–3595. doi:10.1016/j.vaccine.2011.02.086.

Walters M.S. and H.L.T. Mobley. 2010. "Bacterial proteomics and identification of potential vaccine targets." *Expert Reviews in Proteomics* 7(2): 181–184. doi:10.1586/EPR.10.12.

Whitaker, J.A., Ovsyannikova, I.G., and G.A. Poland. 2015. "Adversomics: A new paradigm for vaccine safety and design." *Expert Review of Vaccines* 14(7): 935–947. doi:10.1586 /14760584.2015.1038249.

Xu, G., X. Tang, X. Shang, Y. Li, J. Wang, J. Yue, and Y. Li. 2018. "Identification of immunogenic outer membrane proteins and evaluation of their protective efficacy against *Stenotrophomonas maltophilia*." *BMC Infectious Diseases*, 18(1): 347. doi:10.1186/ s12879-018-3258-7.

Yılmaz, Ç., A. Apak, E. Özcengiz, and G. Özcengiz. 2016. "Immunogenicity and protective efficacy of recombinant iron superoxide dismutase protein from *Bordetella pertussis* in mice models." *Microbiology and Immunology* 60: 717–724. doi:10.1111/1348-0421.12445.

17 Computer-Aided Epitope Identification and Design of Epitope Mimetics

Greg Czyryca

CONTENTS

17.1 INTRODUCTION

In silico tools for epitope prediction are well established, and excellent relevant reviews are available [e.g. Soria-Guerra et al. (2015)]. As of mid-2020, multiple peptide-based cancer vaccines are advancing through the final stages of clinical trials. Writing yet another review seemed redundant and unnecessary. However, there is a gap in this landscape of successful science. The existing (and yet again – successful!) predictive methods are primarily concerned with epitopes capable of eliciting T cell responses. Examples of research involving disorganized peptides eliciting antibodies are less numerous. Then, peptidic mimetics of structured epitopes belong to a completely exotic area of science, overlapping with yet another emerging area: protein/peptide design. The author therefore decided primarily to devote this chapter to the subject of *in silico* design of structured epitope mimetics. Hopefully, the reader will find this useful.

17.2   RATIONALE FOR EPITOPE-BASED VACCINES

The evolution of vaccines from crude biological mixtures, through attenuated or inactivated whole pathogen forms, to whole-protein subunit vaccines, has not stopped there. Epitope design is the logical next step, expected to deliver vaccines with further improved precision, safety, and functionality.

To illustrate the need for such improved precision, consider, for example, the sharing of peptide sequences between the human papillomavirus (HPV) and the human proteome (Kanduc and Shoenfeld 2019), presenting a tangible risk of cross-reactivity and autoimmune reactions. Another example of an undesirable antibody elicitation is the antibody-dependent enhancement (ADE), caused by certain prospective full-length S protein SARS vaccines, but not present when antibodies are elicited more selectively, toward the receptor-binding domain only (Du et al. 2009). Preventing the elicitation of certain antibodies *via* a precise selection of epitopes is therefore a worthy goal. Elicitation of antibodies against transient epitopes is yet another example of the potential benefits of rational epitope design. A case in point is HIV gp120. HIV has evolved a conformational masking mechanism for the purpose of hiding the epitopes of its envelope protein gp120 from the host's immune system (Kwong et al. 2002). The conformational equilibria reduce the epitopes' effective concentration and thus their immunogenicity. Conversely, a conformationally stable epitope mimetic, representing an immunogenic "snapshot" of the protein's dynamic state, would offer a better chance of eliciting the desired immune response.

The choice of methodology for epitope design is defined by the component of the immune system that is targeted. A very concise review of relevant immunology topics will therefore be useful.

17.3   INNATE AND ADAPTIVE IMMUNE SYSTEM: HUMORAL AND CELLULAR RESPONSE

Innate immunity is the host organism's first line of defense against invading pathogens. The components of the innate system recognize the most general molecular characteristics of pathogens. Innate immunity is often considered only tangential to vaccine development, although the reader may want to refer to a review "Innate immunity and vaccines" by Platt and Wetzler (2013), or to an analysis concerned with the BCG vaccine, by Covián et al. (2019).

The evolutionarily more advanced *adaptive immune system* employs two components: the *humoral*, and the *cell-based* response. The humoral system ("humoral" indicates something related to body fluids) relies on soluble proteins called *antibodies*, or synonymously, *immunoglobulins*. Immunoglobulins with structurally diverse *variable loops* are produced by genetically diverse *B cells*, also known as *B lymphocytes*. The specificity of antibodies toward their targets – pathogens' structural elements called *antigens* – is acquired as a result of a quasi-evolutionary process triggered by antigen binding, *via* a B cell receptor (BCR) that combines a membrane-anchored antibody expressed by a given B cell with a transmembrane protein tasked with passing a signal to proliferate. That evolutionary process employs two mechanisms: *clonal selection* and *somatic hypermutation* for further fine-tuning of the affinity. From the point of view of epitope design, it is important to note:

Adaptation *via* change, selection, and proliferation of B lymphocyte germ lines produces structurally highly diverse antibodies, with efficiently optimized binding affinities to their respective antigens.

A similar evolutionary mechanism sculpts the recognition elements in T lymphocytes that are responsible for cellular immune response, with one fundamental difference: instead of free antigen recognition, pathogens' proteins are degraded into short peptidic fragments by the attacked cell's proteolytic machinery, these fragments are complexed with *major histocompatibility complex* (MHC) proteins, and such complexes are presented at the surface of the cell, where they are recognized by T cell receptors (TCRs).

17.4 DIFFERENT COMPONENTS OF THE IMMUNE SYSTEM TARGET EPITOPES OF A DIFFERENT NATURE

17.4.1 LINEAR EPITOPES

Short peptidic fragments, resulting from protein degradation and recognized by T lymphocytes, unavoidably lack an intrinsic structural organization. Compatible MHC proteins are required to form stable complexes with these peptides. "Stable" is the key word here, because the peptide-MHC complex subsequently needs to form a ternary complex with a T cell receptor. A weak binding between the peptide and MHC would impose too large an entropic penalty on the formation of such a ternary complex. MHC proteins therefore possess well-defined cavities, capable of accepting peptidic fragments, forming high affinity complexes with them, and assisting them in assuming stable spatial organization (Figure 17.1).

Epitopes consisting of continuous peptidic sequences are called *linear epitopes*. Short, linear epitopes, often lacking structural organization, are particularly (but not exclusively) relevant to cellular immune response, mediated by T cells.

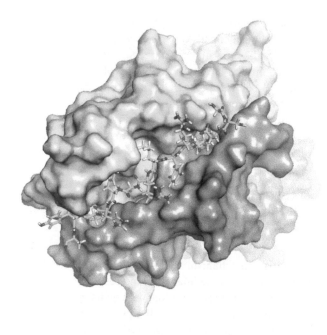

FIGURE 17.1 A heterodimeric MHC protein complexing a linear epitope. PDB structure 1AQD.

17.4.2 CONFORMATIONAL EPITOPES

The adaptability of B cell genomes allows for an extreme structural diversity of antibodies' antigen-binding hypervariable loops (Figure 17.2). Antibodies are typically elicited against complex *conformational epitopes* **consisting of structurally organized, discontinuous peptidic sequences**. An extreme example of a rare, broadly neutralizing anti-HIV antibody is shown in Figure 17.3. The antibody forms a distinct "probe" compatible with the viral gp120 glycoprotein conformational epitope.

Antibodies' ability to bind to intricate conformational epitopes does not preclude an opposite scenario: recognition of disorganized linear epitopes. Examples of paratopes (antibodies' antigen-binding regions) forming MHC-like cavities to bind disorganized linear epitopes can be found (Figure 17.4). To further complicate attempts to generalize, examples of structurally organized linear epitopes can also be invoked, e.g. heptad repeats present in several viral fusion proteins. Nevertheless, three main categories of use cases for computer-aided epitope design can be defined: (1) inherently disorganized peptidic epitopes for eliciting T cell response, (2) disorganized peptides for eliciting B cell response, and (3) structurally organized mimetics of either conformational or linear epitopes in their native geometries, also for eliciting B cell response. While antibodies specific to MHC-peptide complexes are known (see, e.g. Wittman et al. 2006), they arguably have little relevance to computer-aided vaccine engineering and will not be discussed here.

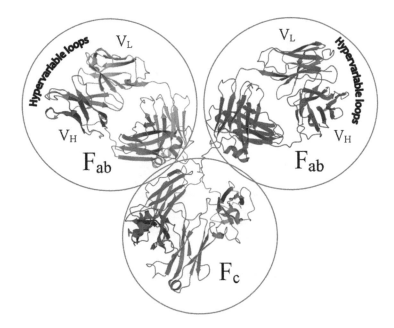

FIGURE 17.2 Anatomy of Immunoglobulin G (IgG), based on PDB 1IGY. Hypervariable loops are stretched on more conserved sheet structures within variable light (V_L) and heavy (V_H) subunits of the antigen-binding regions (F_{ab}).

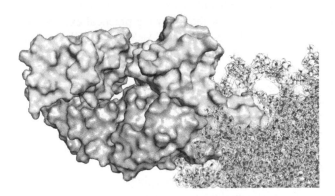

FIGURE 17.3 Extensive conformational epitope from prefusion HIV gp120 complexed with a broadly neutralizing antibody PGT122. PDB 5FYJ.

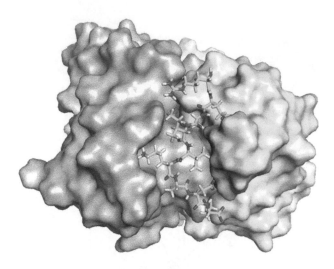

FIGURE 17.4 Example of an antibody binding to a linear, disorganized epitope. PDB 5EOQ.

17.5 GOALS AND MEANS OF EPITOPE PREDICTION AND DESIGN

17.5.1 EXERCISE – DESIGN A PEPTIDE-BASED SARS-CoV-2 VACCINE

Vaccines mimic natural biological processes associated with acquiring immunity. The awareness of the underlying biology of a given medical condition is a prerequisite for computer-aided epitope design. The need to start from seemingly self-evident things can best be illustrated by a real-life example. At the time this book is written, COVID-19, caused by the SARS-CoV-2 virus, spurs numerous efforts aimed at developing a vaccine. The objective of the following practical exercise will be to design peptides for eliciting antibodies against the SARS-CoV-2 spike protein (S) – arguably its most immunogenic feature. There is also a good chance that antibodies against S will be neutralizing, *i.e.* not only tagging the viral protein for the immune system, but also directly interfering with its biological function. S attaches the virus to its cellular receptor angiotensin-converting enzyme 2 (ACE2). Some antibodies elicited against the receptor-binding domain (RBD) of S should be capable of competing with ACE2, thus preventing the attachment and blocking the entire process of viral entry

The comprehensive Immune Epitope Database and Analysis Resource (IEDB, www.iedb.org), maintained under a contract funded by the National Institutes of Health (NIH), is the obvious initial step for epitope prediction. In addition to providing a collection of experimentally verified epitopes, IEDB is a web interface to several epitope prediction algorithms. Being an important resource, supported by NIH, IEDB is likely going to remain available to researchers in the foreseeable future. IEDB provides an interface to DiscoTope, an industry standard in B cell epitope prediction (Haste-Andersen, Nielsen, and Lund 2006; Kringelum, Lundegaard, and Nielsen 2012) (Figure 17.5).

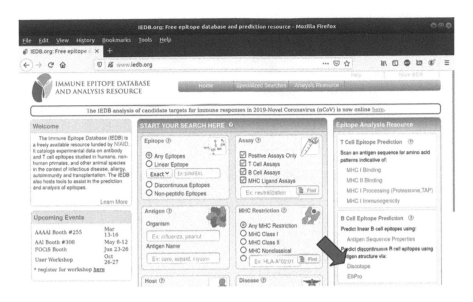

FIGURE 17.5 IEDB starting page.

No preparation of input files is needed. The IEDB implementation of DiscoTope retrieves a Protein Data Bank structure, and the user only needs to select a specific protein chain to analyze. For our purpose, the structural information will be imported from PDB entry 6VW1 – a structure of the SARS-CoV-2 receptor-binding domain complexed with ACE2 (Shang et al. 2020) (Figure 17.6).

SARS-CoV-2 receptor binding domain residues form chain E in this structure. DiscoTope then analyzes the chosen chain, taking into account amino acid statistics, spatial information, and surface exposure, and delivers its prediction (Figure 17.7).

At first glance, IEDB has performed impressively. DiscoTope identified a continuous 13-amino acid sequence that forms a distinct structural feature in the spike protein in the proximity of the receptor-binding domain. Indeed, it is not hard to imagine an antibody binding to this region of S. While DiscoTope failed to predict the epitope for an actual, S-directed, neutralizing antibody CR3022 (PDB 6W41), the other B cell epitope prediction tool available from IEDB – ElliPro (Ponomarenko et al. 2008), made a correct prediction here. So, what should happen next? Is the 14-mer from Figure 17.8 directly suitable to the role of a vaccine component? Perhaps a peptide of such length is capable (depending on its exact amino acid composition) of forming a complex with MHC proteins and eliciting a T cell response, but there is a catch: T cell response is not what we actually want in this case. T cell-inducing vaccines have their legitimate place in vaccinology, in cases when cell-directed response is appropriate, e.g. in cancer, malaria, or HIV infections; refer to a review by Gilbert (2012). For the purpose of protecting against acute viral infection, however, antibody protection is preferable, even though the cell-based arm of the immune system also

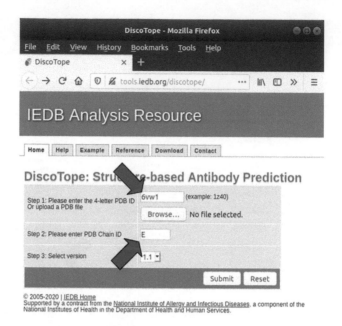

FIGURE 17.6   DiscoTope input window.

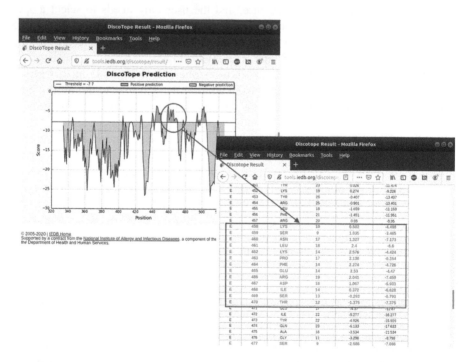

FIGURE 17.7   DiscoTope output. This continuous 13-amino acid epitope deserves a closer look.

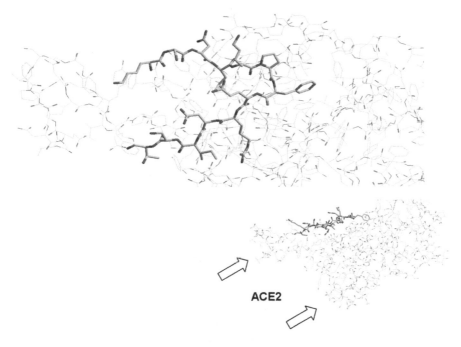

FIGURE 17.8 Epitope KSNLKPFERDIST predicted by DiscoTope 1.1 for the receptor-binding domain of SARS-CoV-2 spike protein (S).

plays a role in B cell activation (mediated by helper T cells) and in the final clearance of the pathogen.

For the purpose of stimulating B cells, peptides such as the one in Figure 17.8 are not suitable effectors, as they lack the means to retain their native structural organization. In its native state, our 13-mer owes its organization to the bulk of the spike protein's structure, of which it is an integral part. Extracted from that protein, the peptide will unavoidably become disorganized. Even if antibodies can be elicited against such a peptide, it is highly unlikely that such antibodies will be compatible with the native structure.

17.6   PRACTICAL GUIDE TO STRUCTURE-BASED DESIGN OF EPITOPE MIMETICS

The experience with IEDB provokes several observations regarding the practicality of the available state-of-the-art B cell epitope prediction tools. First, the algorithms deliver reasonably accurate predictions, although a competent structural biologist could readily reach similar conclusions without the use of bioinformatics tools. Second, the algorithm's advantage over human intelligence is even more disputable considering that the predictions are not fully reliable, and the results require a critical evaluation by a human researcher anyway. Third and foremost, these predictions are the end of the road within established science. The identified epitope sequences alone are generally not suitable to the role of vaccine components.

There is an exception to that last statement. MacRaild et al. (2018) have published a review centered on the abundance of intrinsically disordered antigens in biological organisms, which makes such antigens important targets in vaccine development. While the statement regarding the abundance is correct, a belief that it directly translates into a similar abundance of suitable epitopes is probably too optimistic, for two reasons. First, "disorganized protein" is an oxymoron. To perform its biological function, a protein needs to adopt a structural organization. Therefore, if a disorganized form exists, it has to be transient, or only some parts of the protein may lack organization. Second, while antibodies can be elicited against disorganized epitopes (refer again to Figure 17.4), it must be remembered that they emerge from the evolutionary process of clonal selection, guided by the epitope's affinity to the immunoglobulin component of the B cell receptor. To maximize this affinity, antibody structures capable of engulfing the disorganized epitope will be preferentially selected. As a result, when the disorganized epitope is a part of some larger native tertiary structure, an antibody elicited against the extracted epitope will likely be geometrically incompatible with the epitope's native environment.

Disorganized antigens should definitely not be dismissed. Opportunities for utilizing isolated loops or disorganized termini can occasionally present themselves. Such opportunities, however, should be considered an exception to the general, common sense principle:

> To elicit protective antibodies, B cell receptors should be stimulated by geometrically stable epitope mimetics, recreating the epitopes' native geometries.

This observation explains the scarcity of research involving epitope mimetics for targeting B cells, while eliciting T cell response is (as of 2020) an established science, with numerous vaccine candidates completing clinical trials. Design of three-dimensional peptidic assemblies is an emerging science.

17.6.1 APTAMERS

Relatively small, yet structured amino acid assemblies capable of protein-like binding to target molecules are collectively called aptamers (*aptus* – fit, and *meros* – part). Although the term originally referred to RNA molecules (Ellington and Szostak 1990), its usage has since been expanded to include peptides. In fact, it is appropriate and convenient to also apply the term to heteropolymers that are structurally stabilized by non-biological components. Aptamers constructed around structurally defined epitopes should, as a matter of principle, be capable of eliciting B cell response. Several practical methods for constructing epitope-containing aptamers will be discussed next, still in the context of our practical exercise: design of a SARS-CoV-2 epitope mimetic for eliciting B cell response.

17.6.2 Foregone Conclusion in Epitope Selection

Aptamers are not small molecules. As heteropolymers, they are not prohibitively difficult or expensive to synthesize, but they are not trivial either. Now, let us consider the possibility of unwanted attrition and waste of resources. Bioinformatics epitope prediction algorithms are inherently incapable of accounting for certain risk factors, and do not necessarily agree with each other. Then, design of epitope-carrying aptamers is not an exact science either; not every structure ultimately proves sufficiently geometrically stable, unforeseen synthetic problems can also arise. Ultimately, an aptamer may prove not immunogenic enough. While some trial and error is unavoidable, a reduction of the branching factor at the earliest epitope selection stage is highly desirable, as the most productive means of pruning the combinatorial space.

Let us again consider the epitope shown in Figure 17.8 and the associated risk factors. It is uncertain if, in the actual biological system, the identified epitope would not be masked by some glycan. Its position in relation to other spike protein chains within the homotrimeric supramolecular structure is, without additional structural information, unknown; it can be sterically obstructed; it can be transient. We may expend substantial effort and resources to engineer its mimetics, only to later learn that the antibodies they elicit are not cross-reactive with the native spike protein. These uncertainties can be eliminated by selecting epitopes belonging to the areas of the viral protein experimentally proven as capable of complexing some ligand, be it an antibody, or the cellular receptor.

> If a certain region of a protein is capable of binding a ligand, it is a very safe bet that the same region is accessible to other ligands as well.

This is a *foregone conclusion* that may only occasionally fail in cases involving allosteric regulation. With X-ray crystallography, NMR and cryo-EM studies becoming increasingly routine, the availability of structures of *liganded* proteins is sufficient to prioritize such empirical data for practical purposes over bioinformatics predictions.

Figure 17.9 presents such process of manual selection. The work started with the structure of SARS-CoV-2 receptor-binding domain in a complex with its cellular receptor ACE2 (PDB 6VW1). There is an immediately identifiable structural feature within RBD that interacts with a partially disorganized fragment of ACE2. Because of this lack of organization in the terminal region of ACE2, the 20-meric EIYQAGSTPCNGVEGFNCYFP fragment had evidently evolved to compensate, by contributing its own structural organization to the complex. Note two proline residues, and a disulfide bridge (indicated with stars). This highly organized linear fragment is an almost ready-made aptamer. The design can then be completed by tying up the loose ends (indicated by arrows), to create a bicyclic, rigid assembly.

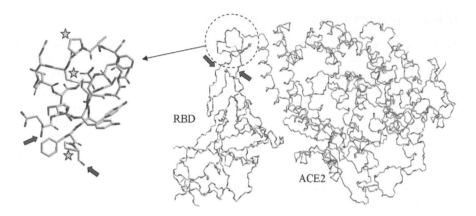

FIGURE 17.9 Manual identification of linear, highly structured epitope EIYQAGSTPC-NGVEGFNCYFP, based on PDB 6VW1.

17.6.3 Option #1: Aptamer Cyclization via Hydrocarbon Stapling

Hydrocarbon stapling is an established method for aptamer construction. Structurally semi-organized fragments (this organization requirement constitutes the general weakness of the approach), usually helices, are further structurally stabilized as macrocycles connected through side chains via the Grubbs reaction between side chains' allyl groups. Prior *in silico* design is not absolutely necessary; the spaces of connection points and side chain lengths produce a number of combinations that is manageable on the bench, through trial and error. In fact, designs involving such common structural elements can be based on prior published research. Refer, for example, to Bird et al. (2014), "Stapled HIV-1 Peptides Recapitulate Antigenic Structures and Engage Broadly Neutralizing Antibodies."

Modeling will be a preferred design method, however, when the aptamer is to be derived from some less common secondary structure. Stapling the termini of the RBD fragment shown in Figure 17.9 would again require examining combinations of side chain lengths, and this is faster and less expensively done *in silico*. The procedure would involve: (1) connecting the termini with alkene bridges, (2) partial optimization of the bridges only, (3) molecular dynamics (simulated annealing) on entire resulting structures, and (4) RMS overlay with the epitope in the original conformation, to verify the structural congruence (Figure 17.10).

17.6.4 Option #2: Grafting Epitopes on a Scaffold

In stapled peptides, discussed in the previous paragraph, the peptide's secondary structure (usually an alpha helix) is the primary source of the structural organization, with the macrocyclization playing a supporting role. Scenarios involving less organized epitopes require a different approach – the structural organization of an epitope that lacks an intrinsic structural organization is attained with the assistance of an external scaffold on which the epitope is grafted.

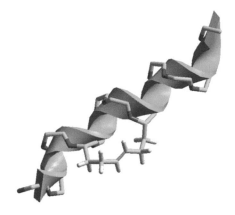

FIGURE 17.10   Hydrocarbon-stapled alpha-helical peptide.

Let us digress, to elaborate on the subject of epitope organization. Regardless of the chosen aptamer construction method, it is always useful to take a closer look at the selected epitope and explore the potential of making its structure better organized. The sequence identified by DiscoTope happens to perfectly illustrate the importance of manual structure preparation and the role of the human factor. Note in Figure 17.11 that the originally identified epitope can be expanded by just two amino acids on both termini. Then, the positively charged guanidinium group of Arg 457 fills the space surrounded by negative charges, just like in the native structure, and contributes to the structural stability of the peptide. At the C-terminus, Glu 471 expands the binding surface for the antibody to be induced, and also brings the peptide's termini closer to each other, which may prove useful, depending on the preferred cyclization strategy.

> Structural components of a protein are compatible with each other. Additional elements can structurally stabilize an epitope mimetic.

Back to the subject of epitope grafting, the work by Ofek et al. (2010) is an archetypical example of this approach. The authors employed a procedure summarized as:

1. First, the entire Protein Data Bank was searched for appropriate acceptor proteins (scaffolds) with backbone structural similarity to segments of the 2F5[antibody]-bound epitope on gp41.
2. Second, a filtering step was applied in which initial structural matches were only retained if the scaffolds could be bound by antibody without significant clashes.
3. Third, epitope side chains were transplanted at appropriate positions.
4. Fourth, additional mutations were introduced into each of the scaffolds to optimize stability, to enhance epitope exposure, and to minimize non-epitope interactions with antibody.

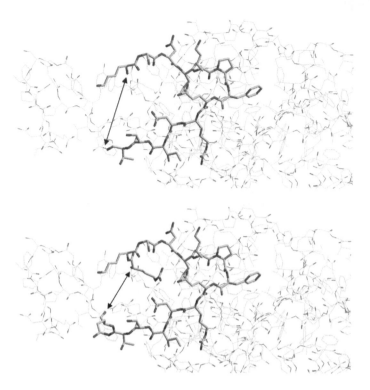

FIGURE 17.11 Identified epitope can be expanded, to take advantage of the self-compatibility within a larger fragment of the parent protein.

The approach is representative to the protein design philosophy implemented in the ROSETTA suite (David Baker's group, refer to Bhardwaj et al. (2016)) – a *de facto* industry standard as of 2020 due to its academic origins and wide dissemination.

17.6.5 OPTION #3: *DE NOVO* DESIGN

Structural organization can also be supplied to an epitope via *de novo* design of a complementary fragment. Approaches such as ROSETTA's "generate backbone and mutate" or chain growth can be employed. Since ROSETTA already has been presented in the previous section, a Monte Carlo chain growth protocol will illustrate *de novo* design of a new fragment for head-to-tail cyclization of the epitope from SARS-CoV-2 spike protein's receptor-binding domain (Figure 17.12). A Monte Carlo chain growth algorithm executes the following iterations:

1. Score is generated for the existing peptidic chain.
2. A random amino acid from a user-defined library is attached, conformational sampling is performed.
3. The new peptide is scored.

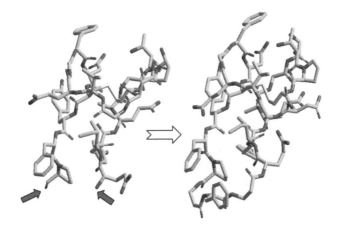

FIGURE 17.12 Head-to-tail cyclization of SARS-CoV-2 epitope EIYQAGSTPCNG-VEGFNCYFP via β-Ala-Phe-β-Ala fragment generated by a Monte Carlo chain growth procedure.

4. Fragment acceptance decision is made, based on the Boltzmann probability formulas (the Metropolis criterion) applied to the calculated scores before and after the addition of the new fragment.
5. Back to (a), until the growing chain reaches a pre-defined size, a cyclization occurs after satisfying appropriate geometric criteria, or the allowable number of iterations is exceeded (the latter to allow the algorithm to avoid dead ends).

The procedure applied* to the epitope from SARS-CoV-2 RBD identified the sequence *β-Ala-Phe-β-Ala* for a head-to-tail linker, in which phenylalanine can later be replaced with another amino acid facilitating polymer immobilization, for example. After molecular dynamics annealing, the aptamer's structure was aligned with the native RBD structure and the preservation of the epitope's geometry proved highly satisfactory (Figure 17.13).

17.6.6 Aptamer Design – Conclusion

Design of epitope-based aptamers remains, as of 2020, an emerging approach. Hydrocarbon stapling can be considered established enough within the state of the art, but being limited to already organized peptides, its area of applications is narrow. Grafting and *de novo* design remain the domain of teams interested in pushing the boundaries of science, but in the coming years the availability of new tools for protein/peptide design (AI-based approaches are presently actively developed) is expected to translate into a larger body of applied research.

---

* A proprietary implementation (Allosterix Pharmaceutical, LLC) was employed.

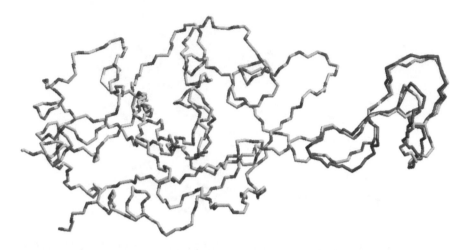

FIGURE   17.13   Aptamer   cyclo[EIYQAGSTPCNGVEGFNCYFP-{β-Ala}-Phe-{β-Ala}] aligned with the native structure of SARS-CoV-2 RBD.

TABLE 17.1
Use Case Matrix for Epitope Design

	Cell-Based Response T cell	Humoral Response B cell
Linear, Disorganized epitopes	Yes, routine. Structural organization irrelevant to epitope design.	Occasionally possible.
Linear, structured epitopes	Yes, routine. Structural organization irrelevant to epitope design.	Possible. Emerging. Structure-based approaches.
Conformational epitopes (discontinuous sequences)	No	Future (?)

17.7   IDENTIFICATION OF EPITOPES FOR T CELL RESPONSE

Soria-Guerra et al. (2015) have compiled an excellent "[…] overview of bioinformatics tools for epitope prediction: implications on vaccine development." The subject of epitope prediction was thoroughly addressed therein and will only be cursorily summarized here.

Epitope prediction for T cell response is equivalent to predicting an epitope's affinity to MHC proteins, and to TAP transporter protein. Predictions employ machine learning algorithms, trained with prior empirical knowledge. NetMHCpan (Hoof et al. 2009), for example, represents MHC molecules (primarily class I) as pseudo-sequences selected on the basis of the geometric proximity with the peptide, and pairs these pseudo-sequences with peptide sequences to create input for a neural network.

Multiple trained predictors are available as publicly available resources; the reader may also want to refer to the extensive reference list at http://tools.iedb.org/mhci/reference/.

REFERENCES

Bhardwaj, G., V.K. Mulligan, C.D. Bahl, J.M. Gilmore, P.J. Harvey, O. Cheneval, G.W. Buchko, et al. 2016. "Accurate de novo design of hyperstable constrained peptides." *Nature* 538(7625): 329–335. https://doi.org/10.1038/nature19791.

Bird, G.H., A. Irimia, G. Ofek, P.D. Kwong, I.A. Wilson, and L.D. Walensky. 2014. "Stapled HIV-1 peptides recapitulate antigenic structures and engage broadly neutralizing antibodies." *Nature Structural & Molecular Biology* 21(12): 1058–1067. https://doi.org/10.1038/nsmb.2922

Covián, C., A. Fernández-Fierro, A. Retamal-Díaz, F.E. Díaz, A.E. Vasquez, M.K. Lay, C.A. Riedel, P.A. González, S.M. Bueno, and A.M. Kalergis. 2019. "BCG-induced cross-protection and development of trained immunity: Implication for vaccine design." *Frontiers in Immunology* 10: 2806. https://doi.org/10.3389/fimmu.2019.02806.

Du, L., Y. He, Y. Zhou, S. Liu, B.J. Zheng, and S. Jiang. 2009. "The spike protein of SARS-CoV--a target for vaccine and therapeutic development." *Nature Reviews Microbiology* 7(3): 226–236. https://doi.org/10.1038/nrmicro2090.

Ellington, A.D. and J.W. Szostak. 1990. "In vitro selection of RNA molecules that bind specific ligands." *Nature* 346(6287): 818–822. https://doi.org/10.1038/346818a0.

Gilbert S.C. 2012. "T-cell-inducing vaccines – what's the future." *Immunology* 135(1): 19–26. https://doi.org/10.1111/j.1365-2567.2011.03517.x.

Haste-Andersen, P., M. Nielsen, and O. Lund. 2006. "Prediction of residues in discontinuous B-cell epitopes using protein 3D structures." *Protein Science: A Publication of the Protein Society* 15(11): 2558–2567. https://doi.org/10.1110/ps.062405906.

Hoof, I., B. Peters, J. Sidney, L.E. Pedersen, A. Sette, O. Lund, S. Buus, and M. Nielsen. 2009. "NetMHCpan, a method for MHC class I binding prediction beyond humans." *Immunogenetics* 61(1): 1–13. https://doi.org/10.1007/s00251-008-0341-z.

Kanduc, D. and Y. Shoenfeld. 2019. "Human papillomavirus epitope mimicry and autoimmunity: The molecular truth of peptide sharing." *Pathobiology: Journal of Immunopathology, Molecular and Cellular Biology* 86(5–6): 285–295. https://doi.org/10.1159/000502889.

Kringelum, J.V., C. Lundegaard, O. Lund, and M. Nielsen. 2012. "Reliable B cell epitope predictions: Impacts of method development and improved benchmarking." *PLoS Computational Biology* 8(12): e1002829. https://doi.org/10.1371/journal.pcbi.1002829.

Kwong, P.D., M.L. Doyle, D.J. Casper, C. Cicala, S.A. Leavitt, S. Majeed, T.D.Steenbeke, et al. 2002. "HIV-1 evades antibody-mediated neutralization through conformational masking of receptor-binding sites." *Nature* 420(6916): 678–682. https://doi.org/10.1038/nature01188.

MacRaild, C.A., J. Seow, S.C. Das, and R.S. Norton 2018. "Disordered epitopes as peptide vaccines." *Peptide Science (Hoboken, N.J.)* 110(3): e24067. https://doi.org/10.1002/pep2.24067.

Ofek, G., F.J. Guenaga, W.R. Schief, J. Skinner, D. Baker, R. Wyatt, and P.D. Kwong. 2010. "Elicitation of structure-specific antibodies by epitope scaffolds." *Proceedings of the*

National Academy of Sciences of the United States of America 107(42): 17880–17887. https://doi.org/10.1073/pnas.1004728107.

Platt, A., and L. Wetzler. 2013. "Innate immunity and vaccines." *Current Topics in Medicinal Chemistry* 13(20): 2597–2608. https://doi.org/10.2174/15680266113136660185

Ponomarenko, J., H.H. Bui, W. Li, N. Fusseder, P.E. Bourne, A. Sette, and B. Peters. 2008. "ElliPro: A new structure-based tool for the prediction of antibody epitopes." *BMC Bioinformatics* 9: 514. https://doi.org/10.1186/1471-2105-9-514.

Shang, J., G. Ye, K. Shi, Y. Wan, C. Luo, H. Aihara, Q. Geng, A. Auerbach, and F. Li. 2020. "Structural basis of receptor recognition by SARS-CoV-2." *Nature* 581(7807): 221–224. https://doi.org/10.1038/s41586-020-2179-yç.

Soria-Guerra, R.E., R. Nieto-Gomez, D.O. Govea-Alonso, and S. Rosales-Mendoza. 2015. "An overview of bioinformatics tools for epitope prediction: Implications on vaccine development." *Journal of Biomedical Informatics* 53: 405–414. https://doi.org/10.1016/j .jbi.2014.11.003.

Wittman, V.P., D. Woodburn, T. Nguyen, F.A. Neethling, S. Wright, and J.A. Weidanz. 2006. "Antibody targeting to a class I MHC-peptide epitope promotes tumor cell death." *Journal of Immunology (Baltimore, Md.: 1950)* 177(6): 4187–4195. https://doi.org/10.4 049/jimmunol.177.6.4187.

Molecular graphics for this chapter were prepared using PyMol and HyperChem software:

DeLano, W. L. (2009). *The PyMOL Molecular Graphics System*; DeLano Scientific: San Carlos, CA, 2002.

HyperChem(TM) Professional 8.0, Hypercube, Inc.

Index